Y0-BOC-515

ИЗДАТЕЛЬСТВО
МИР

Н. Б. БРАНДТ,
С. М. ЧУДИНОВ

ЭЛЕКТРОННАЯ СТРУКТУРА МЕТАЛЛОВ

ИЗДАТЕЛЬСТВО МОСКОВСКОГО УНИВЕРСИТЕТА

N. B. BRANDT
and S. M. CHUDINOV

electronic structure of metals

Translated from the Russian
by V. Afanasyev

MIR PUBLISHERS MOSCOW

First published 1975
Revised from the 1973 Russian edition

На английском языке

CONTENTS

INTRODUCTION

More than 70 elements in the Mendeleev Periodic System are metals. They form together with their alloys a great multitude of substances having very diverse properties. Metals differ in the magnitude of electric conductivity, in their optical, magnetic, galvano-magnetic and thermo-magnetic characteristics, and possess different capabilities of emitting electrons at heating (hot emission) and under the action of an external electric field (cold emission). Mechanical properties of metals, i.e. hardness, malleability, plasticity, etc. are also very diverse. Temperature dependences of the same characteristics of various metals also differ substantially, these differences being especially strong at low temperatures.

Notwithstanding this diversity of properties, there is a common characteristic which allows all metals to be described from a single standpoint. This is the energy spectrum of electrons in metals, or the dependence of the energy ε of electrons on the magnitude and direction of their momenta $\vec{p}$: $\varepsilon = \varepsilon(\vec{p})$. The concept of energy spectrum of electrons is the basis of the modern solid-state physics.

Great achievements of the physics of metals made in the last two decades are directly linked with an intense study of the energy spectrum. Following this way, it has been succeeded not only in explaining many specific properties of various metals, but also in discovering principally new physical effects, such as the cyclotron and magneto-phonon resonances, magnetic breakdown, or gigantic oscillations of the coefficient of sound absorption in metals in a magnetic field.

The concept of the energy spectrum of electrons has recently been applied to alloys, which are solid solutions of several metals. Studies of energy spectra of alloys have shown that one type of spectrum can be continuously reconstructed into another.

Finally, studies of the energy spectrum of metals have served as the basis for the development of a new branch of the solid-state physics, i.e. investigation of substances under combined action of strong electric and magnetic fields, strong three-dimensional or uniaxial deformations, and alloying at low and ultralow temperatures. The results achieved make it possible to solve the problem of directed variation of the characteristics of a substance and of obtaining substances possessing new properties. The possibility of changing the energy spectrum has shown that the division of substances into metals, dielectrics and semiconductors is only of conditional nature.

Construction of the energy spectrum of a metal in the most general form assumes that all possible energy states of electrons should be considered. In a particular case of an individual energy zone the dependence $\varepsilon_i = \varepsilon_i(\vec{p})$ is called the law of dispersion (here i is the number of the zone).

A convenient method for the description of energy spectra is introduction of Fermi surface, a clear geometrical concept which characterizes the properties of conduction electrons and determines all possible values of the momenta (both in magnitude and direction) that electrons may have in a metal.

A Fermi surface is a combination of constant-energy surfaces $\varepsilon_i(\vec{p}) =$ const in various bands and usually has a very complicated form.

At the beginning of this century, it has been established that the most specific property of metals, i.e. their high electric conductivity, is linked with the presence of "free" electrons in them. For instance, in the experiments of Tolman and Stewart [1], a coil rotating on its axis was connected to a galvanometer. At a sharp braking of the coil, a current pulse was formed, whose direction corresponded to the motion of negatively charged particles having the charge-to-mass ratio e/m close to the ratio e/m_0 of free electrons.

In the experiments carried out by E. Riecke [2], a strong current was passed for a long time through three cylinders made of different metals which were tightly pressed to each other. Upon disconnecting the cylinders, no transfer of matter, apart from that caused by diffusion, was revealed. Thus, it has been confirmed that electric current is not related to the transfer of ions forming the crystal lattice of a substance.

All the effects related to cold and hot emission of electrons are also an evidence of the existence of free electrons in metals.

Metallic conductors have been found to have no voltage threshold beginning from which an electric current would be formed in the conductor up to the very low temperatures. With improvements in the sensitivity of measuring instruments, lower and lower currents could be measured in metals. It has been concluded that transmission of a charge in a metal can occur under the action of an arbitrarily weak electric field and is not connected with the preliminary ionization of atoms of the metal. In other words, atoms in a metal are always partly ionized and an electric current is the motion of collective electrons.

The characteristic lustre of a polished surface of a metal, i.e. its capability of reflecting electromagnetic waves, also is an indication to the existence of free electrons in metals.

The first attempt to explain electric conduction was made by K. Drude in 1900 [3]. Some years later, Drude's ideas were developed by H. Lorentz [4]. According to the Drude-Lorentz theory, a metal was regarded as a potential box filled with a gas of free electrons obeying Boltzmann's statistics.

From the modern point of view, this model is absolutely incorrect, though it helped in explaining the phenomena of electric and thermal conduction and their interrelation, expressed by the Wiedemann-Franz law. This circumstance is not astonishing, because any gas of charged particles possesses similar properties.

The Drude-Lorentz theory failed to explain the phenomena in which the nature and specific properties of current carriers are essential. For instance, it follows from that theory that an ideal gas of electrons in a metal must possess a strong paramagnetism. This prediction of the theory, however, contradicts the experimental facts that most metals are weakly diamagnetic. Experiments have also shown that the molar specific heat of an electron gas is substantially lower than that of an ideal monatomic gas, which is approximately equal to 3 cal/mol degree. It may be added that the immense totality of diverse physical properties of metals discovered in the course of 70 years cannot be explained by the Drude-Lorentz theory.

Further development of the electron theory of metals is connected with the works of Sommerfeld carried out in 1926-30 [5]. He used the same model of metal as in the Drude-Lorentz theory, but took into account the quantum nature of electrons and employed the Fermi-Dirac statistics to describe the system of electrons. Thus, Sommerfeld managed to explain the absence of strong paramagnetism and high thermal capacity in an electron gas. The Sommerfeld theory, however, also failed in solving the principal problem of the solid-state physics, i.e. to explain the nature of the differences between metals, semiconductors, and dielectrics.

An essential drawback of the theories described is that the crystal lattice was excluded from consideration. The first works that took into account the interaction of electrons with the lattice relate to 30-s and to the names of Bloch [6, 7] and Brillouin [8, 9]. They have played a substantial part in the formation of modern views on the energy spectrum of electrons. Established in those works were a number of fundamental concepts related to the motion of electrons in the lattice (for instance, the concepts of quasi-momentum, Brillouin zones, etc.) that have retained their significance up to the present. But those works, which have become classical, regarded a certain abstract metal and did not explain why the properties of real metals were very diverse.

To the beginning of 50-s, a great number of various experimental data was accumulated which could not be explained by the theoretical concepts existing at that time. It was found, for instance, that for very pure metals the dependence of electric resistance on the orientation of monocrystalline specimens in a magnetic field at low temperatures was of an extremely complicated form. Studies of oscillational effects indicated to the existence in metals of various groups of current carriers strongly differing in their properties. It was still unclear why valence electrons, which seemed to be under the same conditions, formed groups with different physical properties.

These and many other facts have been given adequate explanation in the theory of electron energy spectrum of metals which was mainly developed during the last two decades. A great contribution to the development of this theory has been made by Soviet physicists I. M. Lifshits, A. A. Abrikosov, M. Ya. Azbel, V. L. Gurevich, M. I. Kaganov, A. M. Kosevich, V. G. Peschansky [10-18]. The development of various methods and interpretation of the theory are also connected with the names of J. M. Ziman, Ch. Kittel, W. A. Harrison, A. B. Pippard, N. F. Mott, H. L. Jones, D. Pines, etc. [20-35].

The Initial Model of Metal. According to the modern views, a metal may be regarded as a combination of a system of positively charged oscillating ions which form a quasi-periodical space structure (crystal lattice) and a system of relatively free collective valence electrons filling the lattice. The differences between two metals may be related to the different valences $\varkappa$ of atoms, the peculiarities of their electronic structure, and the symmetry of the crystal lattice.

The theoretical description of a metal in the frames of such a model results in a quantum-mechanical problem of a system of $(N + \varkappa N)$ interacting bodies (N being the number of atoms in the lattice) whose solution is still impossible at present. But, if we are not to solve the problem of exact calculation of the energy spectrum of electrons in metals (note that even the strict solution of the problem would not provide an exact quantitative picture because of the absence of adequate data on the form of the periodic potential of ions in the lattice), the general qualitative representation of the nature of the spectrum can be found on the basis of a simpler model.

Simplification of the initial model, which will be made in the book, is based on a vast experimental material accumulated through the studies of the electron properties of metals carried out in recent years. For that reason, the simpler model that can be obtained in such a way reflects sufficiently fully the properties of real metals.

CHAPTER ONE

THE CRYSTAL LATTICE

1-1. METAL FORMATION THROUGH CONDENSATION OF NEUTRAL ATOMS. THE METALLIC BOND

The nature of the forces that form ionic bonds in the lattice of a metal can be explained by considering the formation of a metal from the gaseous phase.

At temperatures below that required for thermal ionization, vapours of a metal, as any other gases, consist of electrically neutral atoms and possess no electric conductivity. In other words, there are no free electrons in the vapours of a metal, and the vapour is a dielectric.

On the other hand, atoms in a solid (or liquid) metal are ionized at any temperature. It can therefore be assumed that in the process when neutral atoms are brought closer together, metallic conductivity forms at a certain specific distance between them, irrespective of the state of aggregation of the substance at that moment. This assumption has been confirmed experimentally by Kikoin [36] by measuring the electric conductivity of mercury vapours under pressure.

Mercury vapours were compressed isothermally at a temperature above the critical point. An increase of pressure in that case only caused atoms to come closer together, but could not cause the vapour to condense to liquid or solid. On the other hand, the temperature selected was insufficient for thermal ionization of atoms, since at the initial low pressure the gas possessed no conductivity.

The appearance of metallic conductivity in mercury vapours at compression was only observed when the mean distances between atoms were reduced practically to the values characteristic of the interionic distances in liquid mercury. Ionization of atoms was caused by that the outermost electron shells of neighbouring atoms began to overlap. Namely this process results in collectivization of valence electrons and the formation of a system consisting of positively charged atoms of the structures and of collective electrons.

Thus, the formation of metallic conductivity at condensation of neutral atoms depends on what is the mean distance between the atoms at which the forces of their mutual attraction and repulsion come to an equilibrium.

The forces of attraction between sufficiently distant neutral atoms are related to the existence of fluctuating dipole moments. In the electric field of such a dipole moment, polarization of the surrounding atoms occurs, i.e. the atoms become dipoles. The interaction of dipoles results in that attraction forces are formed.

At distances r greater than the size of a dipole, the strength E of an electric field decreases by the relationship $E \sim 1/r^3$. The dipole moments $\vec{p}$ of neighbouring atoms induced by this field are proportional to E. The energy U of interaction of two polarized atoms, which is equal to the energy of the dipole in the electric field $-(\vec{p}\vec{E})$, is in turn proportional to E^2 and decreases with distance as $1/r^6$.

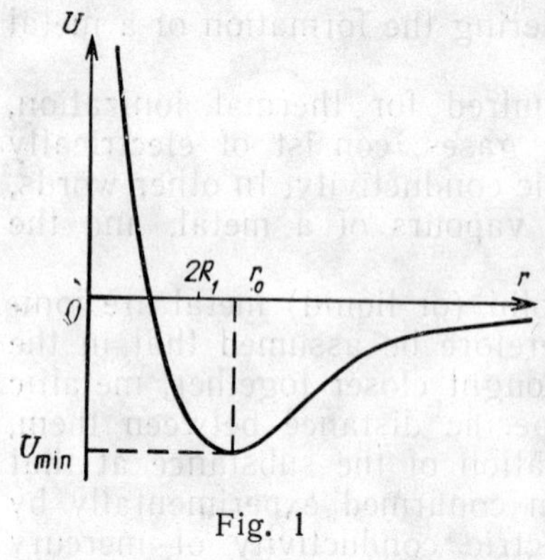

Fig. 1

The polarization effect produces forces of attraction between atoms, $F = -\frac{\partial U}{\partial r} \sim \frac{1}{r^7}$, which are termed the Van der Waals forces. These forces cause neutral atoms to approach and condense to solid, if their thermal motion does not impede the process.

With atoms being brought sufficiently close together, their electron shells begin to deform, so that repulsive forces are formed, which increase sharply with a decrease of r. These repulsive forces are the consequence of the Pauli exclusion principle and are termed the exchange forces.

With the same distance between atoms, exchange forces increase as the outermost electron shell is being filled and attain the maximum for the atoms of the noble gas elements (He, Ne, Ar, etc.). For these substances, an equilibrium between the Van der Waals forces and exchange forces is attained at very small deformations of the electron shells of neighbouring atoms, at which overlapping between the shells is practically absent and the atoms remain non-ionized. For that reason, crystals of noble gases are perfect dielectrics. The dependence of the energy of interaction U between atoms on the distance r between them in the lattice of such substances is shown in Fig. 1.

An equilibrium between attractive and repulsive forces corresponds to the minimum of energy U at $r = r_0$. The formation of a solid dielectric is equivalent to that the equilibrium distance r_0

between the atoms exceeds the doubled radius, $2R_1$, of the outermost electron shell. Note that the equilibrium can be shifted in the direction of lower values of r if the Van der Waals forces are supplemented with forces of external pressure on the substance, as was demonstrated in Kikoin's experiments mentioned earlier.

Let us consider a case when the outermost electron shells of atoms are unfilled. The energy of interaction between atoms at their approaching then will not reach the minimum at $r \sim 2R_1$. Further approaching will result in overlapping of the electron shells and collectivization of valence electrons. Beginning from that moment, the nature of interatomic bond is changed, since the free electrons increase the forces of attraction. The electrons located between positively charged ions attract the latter with a force greater than the force of repulsion between these ions, since the distance between each ion and electrons is smaller than that between ions. Any change of the lattice parameters will then cause a variation of electron density between ions, which in turn causes the formation of forces tending to restore the equilibrium of the lattice. This type of bond is called metallic bond. The forces of repulsion between ions are now related to the deformation of the inner filled electron shell of an ion having an average radius $R_2 < R_1$.

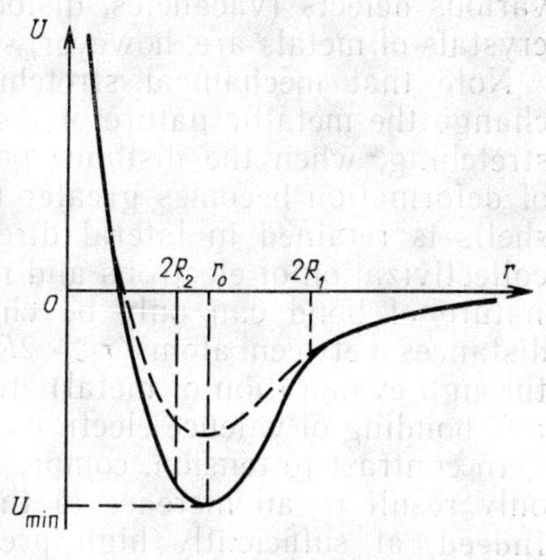

Fig. 2

The formation of a metallic bond changes the relationship for the energy of interaction between atoms at $r \sim 2R_1$, the equilibrium distance r_0 now being between the values $2R_2$ and $2R_1$ (Fig. 2).

A metallic bond has no saturation, i.e. a third, a fourth atom, etc. may be added to two atoms. A metal is similar to a gigantic molecule consisting of a system of positively charged ions; stability of this system is ensured by the existence of collective valence electrons.

If the metal is stretched in some direction, the distances between the ions in that direction will increase. The probability for electrons to be located between the pulled apart ions is also increased, and forces are formed that tend to return the ions to their initial positions. On the contrary, if the metal is compressed in some direction, electrons are forced out from interionic spaces in the direction, which results in that repulsive forces will prevail.

The stability of a metallic bond against strong displacements of ions in the lattice relative to each other may explain the capability of metals of withstanding high plastic deformations, their good malleability, etc.

As distinct from metals, substances having the chemical type of bond are extremely brittle and their mechanical properties are anisotropic. Hardness and brittleness of some technical metals and alloys may be explained by that their lattices are strongly distorted by the presence of impurities and a large number of various defects (vacancies, dislocations, etc.). Pure perfect monocrystals of metals are, however, soft and plastic.

Note that mechanical stretching of a piece of metal cannot change the metallic nature of the bond between atoms. With such stretching, when the distance between atoms, r, in the direction of deformation becomes greater than $2R_1$, overlapping of electron shells is retained in lateral directions. This overlapping ensures collectivization of electrons and retention of the metallic bond. The nature of bond can only be changed by increasing the mutual distances between atoms $r > 2R_1$ in all directions (for instance, through evaporation of metal). In that case, deionization of atoms and bonding of valence electrons occur.

In contrast to tension, compression, either axial or uniform, can only result in an increase of the degree of ionization of atoms. Indeed, at sufficiently high pressures the inner electron shells become overlapped, which will result in collectivization of electrons at these shells. Similarly, a strong compression of dielectric crystals should cause the formation of a metallic bond between atoms and transition to the metallic state. In the limit of ultra-high pressures, all substances at temperatures below the critical point should transform to the state of a "perfect" metal whose lattice consists of atomic nuclei and is filled with collective electrons.

It is of interest to note that in the formation of a metallic bond under pressure the energy of attraction increases substantially, which in some cases results in the formation of a local minimum of the energy of interaction of atoms as a function of the distance between them. The metallic phase thus formed may turn to be stable after relieving of the pressure. Such a situation may, for instance, be expected at compression of atomic, or molecular, hydrogen. Theoretical estimations show that at a pressure of several million atmospheres and sufficiently low temperatures, atomic hydrogen should transform into metal, which will retain its metastable state at a subsequent decrease of the pressure (the pressure corresponding to the transformation of molecular hydrogen into the metallic phase is, by coarse estimates, 3 to 4 mln. atm.).

1-2. THE METALLIC BOND AND CRYSTAL LATTICE

The forces of metallic bond are to a large extent isotropic in their nature. It then follows that ionic structures in the lattice of a metal must be packed so as solid spheres are under the action of an external pressure, i.e. must occupy the least volume.

The packing of spheres is the denser, the greater volume V_{sph} they occupy in the whole volume V of the lattice. The density of packing can be characterized by the ratio V_{sph}/V and also by the number of nearest neighbouring atoms (spheres) surrounding an

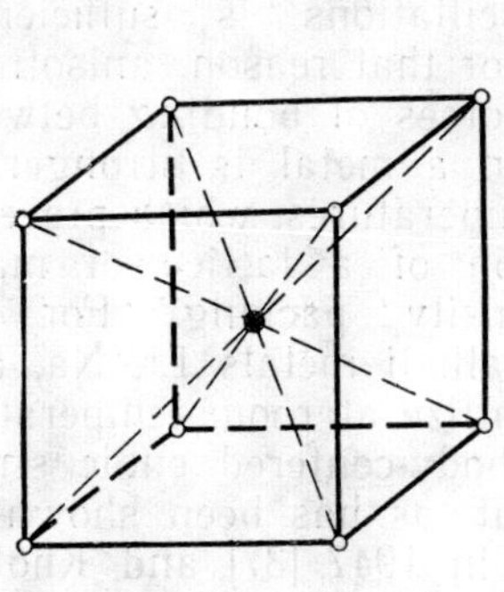

Fig. 3

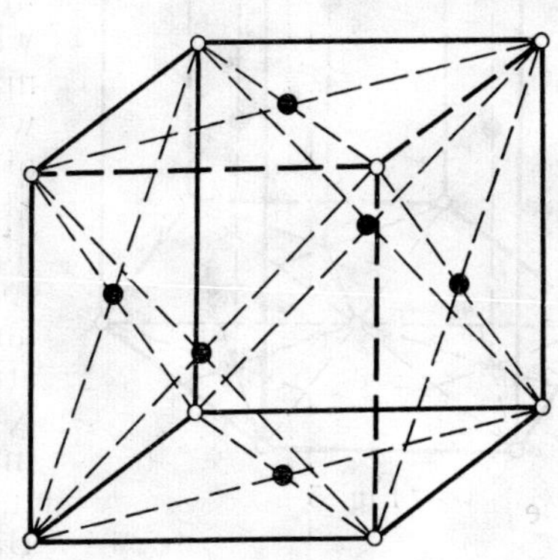

Fig. 4

atom (sphere) in the lattice. The number of nearest neighbours is termed the coordination number z.

Consider the density of packing of atoms in simplest-type lattices. For a simple cubic lattice in which atoms are located at the corners of cubes, the coordination number is evidently equal to $z = 6$, and the V_{sph}/V ratio is $\frac{\pi}{6} \simeq 0.523$.

In a body-centered cubic lattice (b.c.c.), atoms are located at the corners of cubes and in the centre of each cube at the intersection of its body diagonals (Fig. 3). In a b.c.c. lattice, $z = 8$ and $V_{sph}/V = \frac{\sqrt{3}\,\pi}{8} \simeq 0.681$.

The highest density of packing corresponds to $z = 12$ and $\frac{V_{sph}}{V} = \frac{\sqrt{2}\,\pi}{6} \simeq 0.740$. This packing is attained in face-centered cubic (f.c.c.) and close-packed hexagonal (c.p.h.) lattices. In a face-centered cubic lattice, atoms are located at the corners of cubes and in the centres of faces of each cube (Fig. 4); the close-packed hexagonal lattice consists of six trihedral prisms whose height c relates to the side a of the equilateral triangle in the base of the prism as $\frac{c}{a} = \sqrt{\frac{8}{3}} \simeq 1.633$. Atoms are located at the corners of the prisms, and also in the centres of the three prisms which are symmetrical relative to the hexagonal axis (Fig. 5).

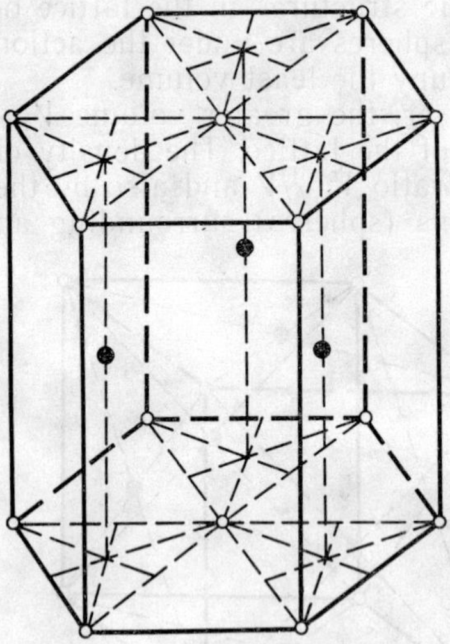

Fig. 5

The face-centered cubic and close-packed hexagonal lattices evidently suit in the best way the nature of metallic bond. But the maximum-density packing of atoms in metals is only attainable when the disturbing action of thermal oscillations is sufficiently weak. For that reason, anisotropy of the forces of bonding between atoms in a metal is stronger at high temperatures, which prevents formation of a lattice of maximum-density packing. For instance, alkali metals Li, Na, and K crystallize at room temperature into a body-centered cubic structure. But, as has been shown by Barrett in 1947 [37] and Khotkevich in 1952 [38], these metals undergo polymorphic phase transformations when being cooled with the b.c.c. lattice being recombined into a f.c.c. lattice.

1-3. THE NATURE OF OSCILLATIONS OF ATOMS IN A CRYSTAL LATTICE

The crystal structure of a metal is not strictly periodical in each given instant of time because of oscillating motion of ions. Oscillations of ions disturb the motion of electrons in the lattice. In order to describe the system of a metal, it is required, strictly speaking, to take into account the motions of all its component particles, i.e. electrons and nuclei. The picture may be substantially simplified, however, if we consider that atomic nuclei are relatively heavy and move with substantially lower speeds than electrons. Electrons adapt so rapidly to the motion of nuclei that their state in each moment is practically determined by position of the nuclei at that moment.

Such a description of a system is termed adiabatic approximation. First, it makes it possible to consider each nucleus and respective electrons at the inner shells (ion-core) as an integral whole, and second, to describe the state of valence electrons in any instant of time as if their ion-cores were fixed in correspondence with their instantaneous position at that moment.

Adiabatic approximation has been introduced by Born and Oppenheimer; its theoretical substantiation may be found in textbooks on quantum theory of solids (for instance, in [39], [40]).

Let us consider oscillatons of ion-cores in a metal lattice, first without taking into account the motion of valence electrons. Under this assumption, oscillations of ion-cores in metals do not differ in their nature from oscillations of atoms in the lattice of a dielectric. In this connection, ion-cores (ions) will be further called atoms for simplicity (where it will not cause misunderstanding).

At a temperature T other than zero, atoms in a lattice participate in two kinds of oscillation which are of different nature. First, they perform thermal oscillations. The energy of these oscillations is determined by temperature and becomes zero at $T = 0°$ K. Second, the atoms are in the motion of purely quantum origin which is termed zero oscillations; these exist even at $T = 0°$ K.

Zero oscillations are the consequence of the relationship between Heisenberg's uncertainties for the coordinate x and momentum p of a particle: $\Delta x \Delta p \gtrsim \hbar$ ($h = 2\pi\hbar = 6.62 \cdot 10^{-27}$ erg·sec is Planck constant). This relationship implies that a quantum particle cannot be at rest ($p = 0$, $\Delta p = 0$) in a definite point of space ($\Delta x = 0$). Localization of the particles ($\Delta x \to 0$) results in increased uncertainty of their momentum Δp and kinetic energy $\frac{(\Delta p)^2}{2M}$ (where M is the mass of particle) and is unfavourable from the standpoint of energy losses.

The energy ε_0 of zero oscillations actually corresponds to the minimum value of the total energy which a particle can possess. The amplitude Δx_0 of zero oscillations is $\frac{\hbar}{\sqrt{2\varepsilon_0 M}}$. For solids, it is usually much smaller than the interatomic distances a in the lattice. If the amplitude of zero oscillations is commensurable with the mean distance between atoms, no crystalline structure can be formed in such a system at normal pressure even at $T = 0°$ K and the substance will remain liquid.

The sole known example of such a substance is liquid helium. At normal pressure, the isotopes He^3 and He^4 remain liquid at any low temperature. Helium can be transformed to the solid state only by applying an excess external pressure (for instance, He^4 solidifies at $T = 4°$ K and a pressure of approximately 100 atm). Solid helium isotopes turn to have very specific properties.

If zero oscillations of atoms are to be characterized by a dimensionless parameter $\gamma = \frac{\Delta x_0}{a} = \frac{\hbar}{a\sqrt{2\varepsilon_0 M}}$, then for Ne, He^4 and He^3 in solid state the values of γ are respectively 0.6, 2.7, and 3.1. These numbers indicate that quantum effects are quite strong for Ne and critical for He^4 and He^3. Generally speaking, at $\gamma > 1$ a solid may have a periodic structure, as a crystal, but its atoms are not necessarily bonded with definite nodes of the lattice, since there exists a finite probability of displacement of an atom from one node to another. Such solids have therefore some features of liquids.

Atoms in a liquid are at definite mean distances from each other (so that the liquid retains its volume), but their positions are not fixed in space (the liquid cannot retain the shape). When moving in a liquid, atoms possess a sufficiently high kinetic energy and "jump" through the potential pits formed by the nearest neighbours and thus do not form a lattice. But liquids are observed to tend to solid-state order and certain anisotropy of the forces of interaction. If a single crystal is melted down carefully, without introducing any additional perturbations, then a single crystal of the same orientation is formed upon cooling of the melt.

For a transformation from the liquid to solid state to occur, it is required that the atoms, which have been freely moving relative to each other, acquired some fixed positions of equilibrium. Such ordering of atoms occurs during cooling of a liquid down to temperatures T close to the melting point T_{melt} and determines the periodic (as time-averaged) structure of the lattice as a whole. The amplitude A of thermal oscillations of atoms is proportional to $\sqrt{T}$, and their frequency ν is the natural frequency of the oscillations caused by the quasi-elastic force of bonding between atoms. The frequency ν is determined by the mass M of an atom and the coefficient β of quasi-elastic force $f = -\beta x$:

$$\nu \simeq \frac{1}{2\pi}\sqrt{\frac{\beta}{M}} \sim 10^{13} \text{ Hz}$$

At sufficiently low temperatures, the nature of oscillations of atoms in a lattice acquires a new feature. While at $T \lesssim T_{melt}$ the atoms oscillate about fixed equilibrium positions with random phases (not coordinated), then, as the temperature is being reduced, a correlation is gradually established between the motions of individual atoms, which result in that they begin to oscillate coordinately. This means that at low temperatures a solid can be considered as a continuum of points, i.e. as a continuous medium. In this range of temperatures, the positions of equilibrium, about which the atoms oscillate with a frequency $\nu \simeq \frac{1}{2\pi}\sqrt{\frac{\beta}{M}}$, are themselves in an oscillating motion with lower frequencies. Near the absolute zero, the nature of motion of atoms changes once more. The low-frequency oscillations of equilibrium positions attenuate at $T \to 0°$ K. Zero oscillations are again performed about fixed positions of equilibrium.

Thus, thermal oscillations of atoms in a lattice are qualitatively different in two temperature regions determined by the ratio between the energy of thermal oscillations of an atom $\sim kT$ (where $k = 1.38 \cdot 10^{-16}$ erg/degree is Boltzmann constant) and the energy of bonding between atoms in the solid: in the high-temperature region where $kT \sim U_{bon}$, $T < T_{melt}$, and in the low-temperature region where $kT \ll U_{bon}$.

In the first region and for time intervals $\tau < 10^{-13}$ sec, the lattice is not a periodic structure. A 'momentary photograph' of the lattice would show that the atoms are located chaotically, being displaced at random from their equilibrium positions. But for time intervals $\tau \gg 10^{-13}$ sec, when averaging of displacements of atoms from their equilibrium positions occurs, the periodicity of the lattice becomes perfect and reflects the periodicity of equilibrium positions of individual atoms.

Physically, the absence of periodicity of a crystal lattice in this temperature region is of importance only for very rapid processes, for instance, for a moving particle which covers the interatomic distance in an interval much less than 10^{-13} sec. For sufficiently slow physical processes (with the characteristic time $\tau \gg 10^{-13}$ sec), the lattice may be regarded as a strictly periodic combination of fixed atoms located in their equilibrium positions.

In the low-temperature region ($kT \ll U_{bon}$), the energy of bonding between atoms is so great compared with the energy of oscillations that an atom, when being shifted from its equilibrium position in some direction, entrains neighbouring atoms in that direction. Namely this circumstance causes a coordinated motion of neighbouring atoms, similar to the propagation of sound waves in solids. Within this temperature region it is possible to consider a solid as an integral whole with a specific pattern of the frequencies of natural oscillations.

The temperature conditionally separating these two regions, which differ in the nature of thermal oscillations of atoms, is termed Debye temperature, Θ_D. Each of the temperature regions described above is strictly determined by only one of the following inequilities, either $T \ll \Theta_D$ or $\Theta_D < T < T_{melt}$. This means that Θ_D actually defines a certain temperature interval within which the nature of thermal motion of atoms is changed qualitatively.

Though Θ_D cannot be determined exactly in the general case, it is a very convenient physical parameter and enters the expressions for a number of integral characteristics of solids, such as heat capacity, heat conductivity, etc. Debye temperature of a given substance depends on the energy of bonding U_{bon} of atoms, and therefore, on melting point T_{melt}. Quantitative relationship between Θ_D and T_{melt} has been established by Lindeman's empirical formula [41]:

$$\Theta_D \simeq BT_{melt}^{1/2}\tilde{A}^{-5/6}D_0^{1/3} \tag{1.1}$$

where $\tilde{A}$ is the mean atomic weight, D_0 is the density of substance, and B is a constant. B has been found empirically to be close to 120 (with A expressed in grammes and D_0 in grammes per cubic centimetre).

Let us show, by way of two examples, that knowledge of the physical parameter Θ_D is of high importance.

1. Debye temperature determines the temperature at which single crystals must be annealed to remove stresses in the lattice. Annealing at temperatures close to T_{melt} worsens the structure of a crystal, instead of improving it, because of the formation of vacancies, inclusions, etc. On the other hand, annealing at $T < \Theta_D$ is ineffective, since the energy of bonding at such temperatures is so high compared with kT that thermal oscillations cannot remove local stresses and distortions of the lattice. The most effective temperature for annealing is $T \gtrsim \Theta_D$. For instance, for copper ($\Theta_D = 315°$ K) annealing is quite effective at a temperature as low as 493° K (220° C). When cured at this temperature for some days, copper becomes very soft (for copper, $T_{melt} = 1083°$ C).

2. Debye temperature determines the probability that some or other phase of a solid will remain metastable.

Consider a substance in which phase transformations from one crystalline modification to another occur at successive variations of temperature, $T_1 < T_2 < T_3$, etc. Within the temperature regions $T < T_1$, $T_1 < T < T_2$, $T_2 < T < T_3$, etc., phases 1, 2, 3, etc. correspond to an equilibrium state of the lattice. Examples of substances having different phases are β- and α-tin, α-, β-, and γ-iron, etc.

At temperatures above Debye temperature the energy of thermal oscillations is sufficient to bring a lattice to an equilibrium. Then it is very easy to retain, for instance, phase 2 in metastable state at a drop of temperature, if the transition from phase 2 to phase 1 occurs at a temperature T_1 below Debye temperature Θ_{D2} for phase 2. With an inverse relationship ($\Theta_{D2} < T_1$) phase 2 can be kept in metastable state only by very rapid cooling. In this process, the probability of retaining phase 2 is the smaller, the lower Θ_{D2} is compared with T_1.

1-4. THE NATURAL-FREQUENCY SPECTRUM OF A ONE-DIMENSIONAL LATTICE

A simple and physically clear model of thermal oscillations should depict both the processes of heat propagation and the state of thermal equilibrium. Propagation of heat is related to oscillations in the form of running waves. The energy flux in a wave determines the heat flux. On the other hand, thermal equilibrium is equivalent to a state in which heat fluxes are absent. Such a state corresponds to a combination of standing waves, since each standing wave is the sum of two running waves of the same amplitude propagation in opposite directions.

In order to describe thermal oscillations of a lattice, let us first elucidate what natural frequencies can exist in a system consisting of a large number $N \gg 1$ of particles interconnected by quasi-elastic forces.

As is known, oscillations of a system of N bodies can be decomposed into normal oscillations each of which corresponds to one degree of freedom of the system. The problem of constructing the spectrum of normal oscillations in the lattice of a solid was first examined by Born and Kármán in 1912 [42].

Note that in a lattice consisting of a finite number of atoms the positions of atoms in the depth of lattice are not equivalent to the positions of those at its boundaries. In order to describe the oscillations, it is required to know the boundary conditions, which may be different depending on the problem being solved. For instance, the atoms at the boundary may be considered either completely bound or, on the contrary, free; this, in turn, determines the nature of the standing waves being formed. On the other hand, a continuous mode of running waves can be maintained in a lattice by bringing the boundary atoms into forced oscillations, for instance, by passing a heat flow across the boundary.

When considering the natural oscillations of a lattice, the positions of all its atoms are conveniently assumed to be equivalent, irrespective of their positions in the crystal. It is clear by intuition that boundary conditions cannot have a strong effect on the behaviour of a large number of particles, at least with sufficiently short-ranged forces of bonding between them. Born and Kármán have shown mathematically that for large systems ($N \gg 1$) this assumption gives only a negligible error. They also proposed a method by which it is possible to eliminate non-equivalency of positions of atoms inside a crystal and at its boundaries.

The method consists in introducing cyclic boundary conditions, i.e. extending the crystal lattice periodically in all directions, the period in the given direction being taken equal to the size of the crystal in that direction. Cyclic boundary conditions make it possible to neglect the effects related to the boundaries of a crystal and greatly simplify the solution of the problem of finding the normal modes of a system. By introducing these conditions, it is possible to use the property of translational invariance of a perfect crystalline lattice of infinite size.

With cyclic boundary conditions, a crystal turns to be surrounded from all sides by an infinite number of its own copies. The idea of introducing cyclic boundary conditions is based on a physically evident assumption that in a crystal of a sufficiently large size the nature of oscillations of an atom about the given node should not differ in any respect from the nature of oscillations about any other node.

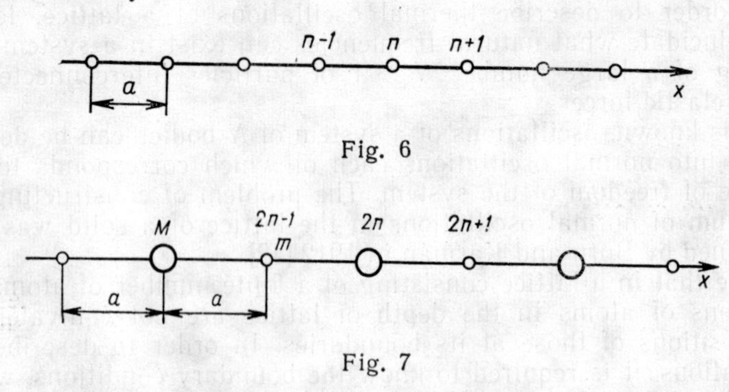

Fig. 6

Fig. 7

The principal qualitative peculiarities of vibrations of a lattice may be elucidated by considering two simplest one-dimensional models.

The first model is a one-dimensional chain composed of N atoms of the same kind. In an equilibrium position, all atoms of such a chain are located at the same distance a from each other (Fig. 6).

The second model is a one-dimensional chain consisting of atoms of two different kinds (for instance, differing in their masses, m and M, with $m < M$). An equilibrium position of this chain is similar to that of the first model (see Fig. 7).

One-dimensional models imply that each atom can be displaced only along the axis of the chain. The atoms in the chains will be thought of to be interconnected by quasi-elastic forces with a coefficient β. For simplicity, the interaction of each atom with only two nearest neighbours will be considered. If more distant neighbours were taken into consideration this would bring in nothing principally different in the nature of oscillations.

For a one-dimensional chain of atoms, cyclic boundary conditions can be introduced quite naturally in the following way: if the chain is turned into a ring by connecting its ends, the non-equivalence of the positions of atoms in the middle of the chain and at its ends can be removed (Fig. 8). In such a ring, a displacement of an atom having the number n from its equilibrium position should evidently coincide with the displacement of an atom numbered $n+N$, since by going round the ring completely, we come to the same atom. This automatically results in that the chain is extended periodically with a period $L=Na$. (For convenience, L denotes the length of the chain plus one interval equal to a; with cyclic extension, the period is defined by this quantity, rather than by L alone). On the other hand, shaping the chain into a closed ring completely eliminates its boundaries.

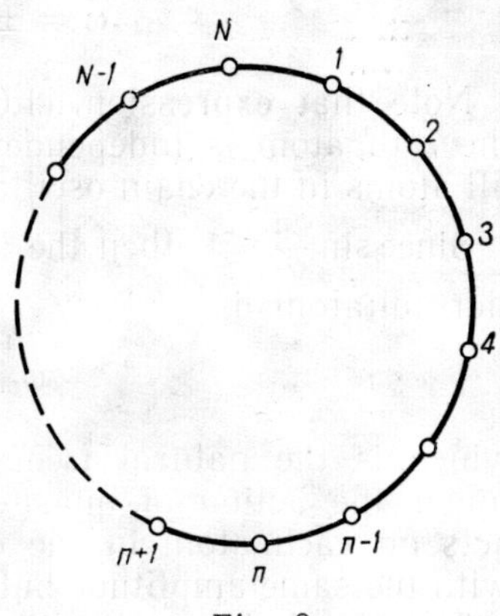

Fig. 8

1. Oscillations of a one-dimensional chain consisting of atoms of the same kind. Let us follow the motion of an atom with an arbitrary number n in the chain shown in Fig. 6. Let ξ be the displacement of each atom from its equilibrium position (Fig. 9). To be definite, displacements to the right will be considered positive.

In an equilibrium position, the force f exerted on each atom by its neighbours is equal to zero. With arbitrary displacements the

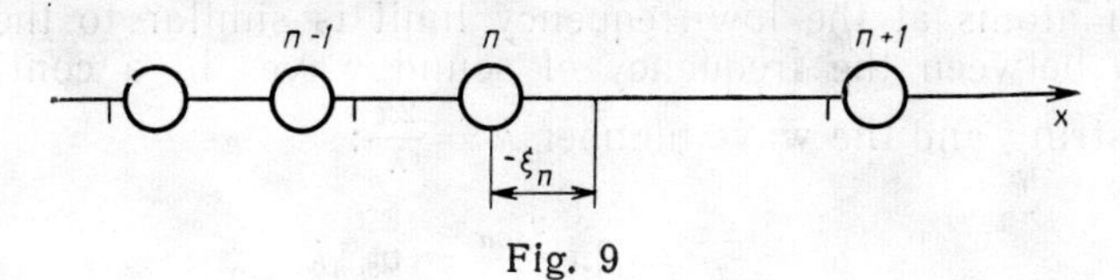

Fig. 9

force f_n acting on the atom with the number n is proportional to the variation of equilibrium distances between the atoms:

$$f_n = \beta(\xi_{n+1} - \xi_n) - \beta(\xi_n - \xi_{n-1}) \tag{1.2}$$

The equation of motion of the n-th atom is of the form:

$$M\ddot{\xi}_n = \beta(\xi_{n+1} + \xi_{n-1} - 2\xi_n) \tag{1.3}$$

where M is the mass of an atom.

Solution of Eq. (1.3) can be found in the form

$$\xi_n = \zeta e^{i(\omega t + nka)} \tag{1.4}$$

where ω is the cyclic frequency, ζ is the amplitude of oscillation, and k is a constant. Substituting this solution into (1.3) gives:

$$-\omega^2 M = \beta(e^{ika} + e^{-ika} - 2) \tag{1.5}$$

whence

$$\omega = \pm 2\sqrt{\frac{\beta}{M}} \sin\frac{ka}{2} \tag{1.6}$$

Note that expression (1.6) for the frequency of oscillation of the n-th atom is independent of the number n. This means that all atoms in the chain oscillate with the same frequency.

Since $\sin\frac{ka}{2} \leqslant 1$, then the upper limit of the frequency of oscillations of atoms is

$$\omega_{max} = 2\sqrt{\frac{\beta}{M}} \tag{1.7}$$

which is the natural frequency of oscillation of a single atom under the action of quasi-elastic force $f = -4\beta\xi$. Such a force acts on each atom in the chain if neighbouring atoms oscillate with the same amplitude but in opposite phase.

The physical meaning of the constant k may be shown by considering low-frequency oscillations $\omega \ll \omega_{max}$. It is then obvious that $\sin\frac{ka}{2} \ll 1$, and the expression for ω becomes:

$$\omega \simeq \pm\left(a\sqrt{\frac{\beta}{M}}\right)k \tag{1.8}$$

The linear relationship between ω and k for oscillations of a chain of atoms at the low-frequency limit is similar to the relationship between the frequency of sound waves in a continuous elastic string and the wave number $k = \frac{2\pi}{\lambda}$:

$$\omega = \frac{2\pi}{T} = 2\pi\frac{v_{son}}{\lambda} = v_{son}k \tag{1.9}$$

where λ is the wavelength and v_{son} is the velocity of sound propagation in the string.

Indeed, for long-wave oscillations, the discreteness of the structure of the chain consisting of individual atoms is immaterial. With an approximation $\lambda \gg a$, the chain of atoms may be considered as an elastic homogeneous thread. By comparing expressions (1.8) and (1.9) it may be seen that $a\sqrt{\frac{\beta}{M}}$ has the sense of the velocity of propagation of elastic (sound) waves in the chain at $\omega \to 0$, and the constant k, the sense of the wave number $2\pi/\lambda$. Low-frequency oscillations in a linear chain can be represented in the form of a running sound wave $\sim e^{i(\omega t + kx)}$, where the discrete coordinate na along the chain corresponding to the equilibrium position of the n-th atom is replaced at $\lambda \gg a$ with a continuous coordinate x [1]). Then the neighbouring atoms should

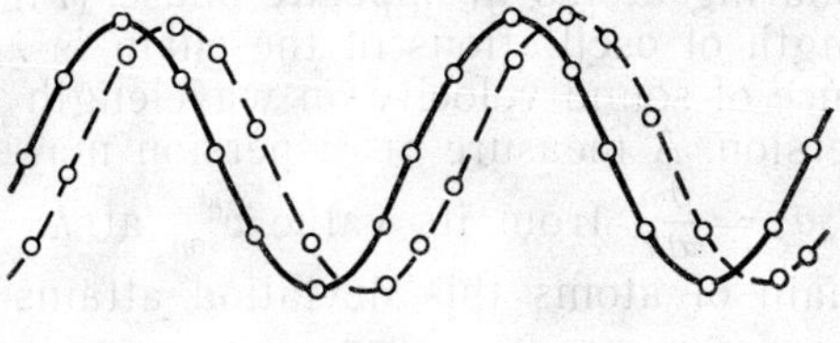

Fig. 10

oscillate with the same amplitude and a negligible phase shift $\Delta\varphi = 2\pi \frac{a}{\lambda} \ll 1$ relative to each other.

This wave running along the chain of atoms can be visualized in the form of transverse oscillations of the atoms (Fig. 10). The positions of atoms of the chain for two fixed instants of time are shown in that figure. Oscillations of one-dimensional chains will be further illustrated, for convenience, by considering transverse oscillations of atoms of a definite polarization.

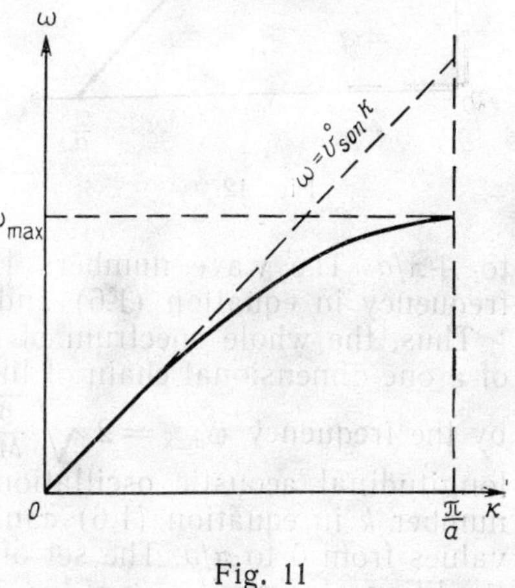

Fig. 11

The relationship between ω and k (1.6) is called the dispersion equation. In the high-frequency region $\omega \sim \omega_{max}$ it differs substantially from expression (1.8) which is valid for sound waves in a chain (Fig. 11). A departure of the k/ω relationship for oscillations of a chain from the linear law may be thought of as a reduction of the sound velocity with the wave number increasing to the value π/a. We then can speak about the group velocity of a sound wave, $v_g = \frac{\partial\omega}{\partial k}$, which is the velocity of transfer of vibrational energy. It follows from (1.6) that $v_g = \pm \left(a\sqrt{\frac{\beta}{M}}\right)\cos\frac{ka}{2} = v^0_{son}\cos\frac{ka}{2}$, where v^0_{son} is the velocity of sound waves, equal to $a\sqrt{\beta/M}$,

[1]) Negative values of wave numbers k correspond to running waves in the reverse direction.

at $\omega \to 0$ (Fig. 12). With $k = \pi/a$, v_g becomes zero, and ω attains its limit value $\omega = \omega_{max}$. At $k = \pi/a$, a running wave turns into a standing one. This picture corresponds to oscillations of neighbouring atoms in opposite phase (Fig. 13). The minimum wavelength of oscillations of the chain is $\lambda_{min} = 2a$.

The dependence of sound velocity on wavelength (or frequency) is termed dispersion. A measure of dispersion may be a deviation of the velocity $v_g = \frac{\partial \omega}{\partial k}$ from its value v^0_{son} at $k = 0$. In a one-dimensional chain of atoms this deviation attains the maximum at $k = \pi/a$ at which $v_g = \frac{\partial \omega}{\partial k}$ becomes zero.

All physically different values of the wave number are within the interval from $-\pi/a$ to $+\pi/a$. The wave numbers $|k| > \pi/a$ give no new values of frequency in equation (1.6) and have no physical sense.

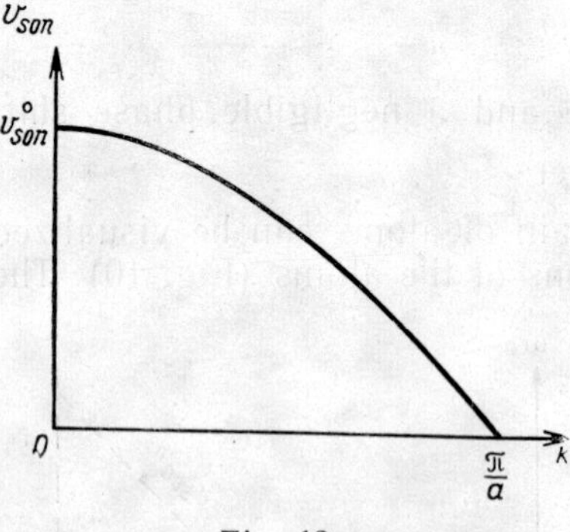

Fig. 12

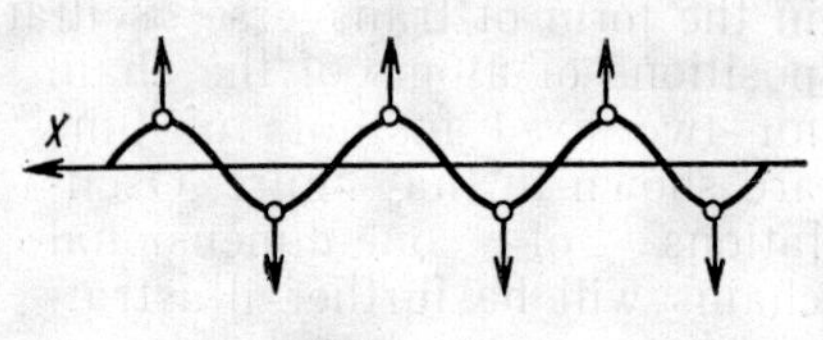

Fig. 13

Thus, the whole spectrum of frequencies of natural oscillations of a one-dimensional chain of like atoms is limited from the above by the frequency $\omega_{max} = 2\sqrt{\frac{\beta}{M}}$ and consists of a single band of longitudinal acoustic oscillations. In a discrete chain, the wave number k in equation (1.6) cannot acquire a continuous series of values from 0 to π/a. The set of wave numbers k must satisfy the condition of cyclicity $\xi_n = \xi_{n+N}$ which reflects the discrete nature of the chain and from which it follows [see (1.4)] that $e^{inka} = e^{i(n+N)ka}$ or $e^{iNka} = 1$. The last equality makes it possible to write a discrete set of wave numbers

$$k_q = \frac{2\pi}{Na} q = \frac{2\pi}{L} q \tag{1.10}$$

where $q = \pm 1, \pm 2, \ldots, \pm \frac{N}{2}$.

The maximum value of $|q|$ is determined by the maximum value of $|k| = \pi/a$. Note that oscillations corresponding to $k = \pm\pi/a$ are only possible in a chain consisting of an even number of atoms $N = 2N'$, where N' is an integer. If N is sufficiently large

($N \gg 1$), this condition puts no principal limitations on the model of a lattice. Let us assume, for simplicity, that this condition is satisfied. Then a single normal mode will correspond to each value of k_q in the set (1.10). The whole spectrum will evidently be composed of a combination of N normal oscillations, one half of them being represented by waves running in the negative direction ($q > 0$), and the other half, in the positive direction ($q < 0$) of the chain. The sum of two running waves with $k = +\frac{2\pi}{L} q$ and $k = -\frac{2\pi}{L} q$, taken with the same amplitude, is equivalent to a standing wave.

2. Oscillations of a one-dimensional chain consisting of atoms of two kinds. Let us assume, as shown in Fig. 7, that atoms with masses M and m (where, for instance, $M > m$) are located respectively in even and odd nodes of a chain. By analogy with (1.3), equations of motion of two neighbouring atoms with the numbers $2n$ and $2n+1$ can be written as

$$\begin{aligned} M\ddot{\xi}_{2n} &= \beta(\xi_{2n+1} + \xi_{2n-1} - 2\xi_{2n}) \\ m\ddot{\xi}_{2n+1} &= \beta(\xi_{2n+2} + \xi_{2n} - 2\xi_{2n+1}) \end{aligned} \tag{1.11}$$

where ξ_{2n} and ξ_{2n+1} are the corresponding displacements of the $2n$-th and $(2n+1)$-th atoms from the equilibrium, and β has the same sense as before. The solution of equation (1.11) will be sought in the form similar to (1.4), taking into account that the atoms have now different masses and will therefore oscillate with different amplitudes ζ and η:

$$\begin{aligned} \xi_{2n} &= \zeta e^{i(\omega t + 2nka)} \\ \xi_{2n+1} &= \eta e^{i[\omega t + (2n+1)ka]} \end{aligned} \tag{1.12}$$

Substitution of (1.12) into (1.11) gives a set of uniform linear equations in ζ and η:

$$\begin{aligned} -\omega^2 M\zeta &= \beta\eta(e^{ika} + e^{-ika}) - 2\beta\zeta \\ -\omega^2 m\eta &= \beta\zeta(e^{ika} + e^{-ika}) - 2\beta\eta \end{aligned} \tag{1.13}$$

The set of equations (1.13) has a nontrivial solution only when the determinant of its coefficient equals zero:

$$\begin{vmatrix} 2\beta - \omega^2 M, & -2\beta\cos ka \\ -2\beta\cos ka, & 2\beta - \omega^2 m \end{vmatrix} = 0 \tag{1.14}$$

Evaluating the determinant, we get a biquadratic equation for the frequency of oscillations ω, whose solution gives two different bands of the natural-frequency spectrum, ω_+ and ω_-

$$\omega_\pm^2 = \beta\left(\frac{1}{m} + \frac{1}{M}\right) \pm \beta\sqrt{\left(\frac{1}{m} + \frac{1}{M}\right)^2 - \frac{4\sin^2 ka}{Mm}} \tag{1.15}$$

(Note that ω_+ and ω_- are also independent of n and are the natural frequencies of oscillations of any atom in the chain.)

Let us consider the behaviour of the solutions at small values of the argument $ka \ll 1$. Under this assumption, the frequencies ω_+ and ω_- are:

$$\begin{aligned} \omega_+ &\simeq \pm \sqrt{2\beta\left(\frac{1}{m}+\frac{1}{M}\right)} \\ \omega_- &\simeq \pm \left(a\sqrt{2\beta\,\frac{1}{M+m}}\right)k \end{aligned} \tag{1.16}$$

By comparing (1.16) and (1.8), it can be concluded that the relationship between ω_- and k describes a band of longitudinal acoustic oscillations. The velocity of sound for the long-wave limit is given by the expression

$$v^0_{son} = a\sqrt{2\beta\,\frac{1}{m+M}} \tag{1.17}$$

which at $m = M$ determines the velocity of sound $v^0_{son} = a\sqrt{\beta/M}$ in a monatomic chain.

Thus, in addition to the acoustic band $\omega_-(k)$, an additional band $\omega_+(k)$ has appeared in the spectrum of oscillations. Its physical meaning can be explained by comparing the ratios of amplitudes of oscillations ζ/η in both bands at small values of k ($ka \ll 1$). These ratios can be found by substituting the corresponding frequency, either ω_+ or ω_-, into any of equations (1.13). For instance, substituting ω_- from (1.16) into (1.13), we find that $(\zeta/\eta)_- = +1$. This expression implies that oscillations of neighbouring atoms in the chain occur in phase and with the same amplitude. Oscillations of this kind are characteristic of acoustic waves.

Substituting ω_+ from (1.16) into (1.13), we get for the amplitude ratio

$$\left(\frac{\eta}{\zeta}\right)_+ = -\frac{M}{m} \tag{1.18}$$

whence it follows that the expression $B = M\zeta + m\eta$ is equal to zero. But $\frac{B}{M+m}$ is, in essence, the amplitude of displacement of the mass centre of an elementary cell in a chain consisting of atoms of two kinds. Thus, the branch $\omega_+(k)$ at small values of k corresponds to such oscillations of atoms in the chain at which the mass centre of each cell remains fixed. It may be shown that this property of oscillations in an $\omega_+(k)$ branch is retained at any values of k.

Oscillations of this kind were first found in NaCl and KBr crystals whose lattices consist of positively and negatively charged ions. Oscillations of unlikely charged ions satisfying the condition (1.18) (in a particular case at $m = M$, these are anti-phase oscillations of neighbouring atoms with the same amplitude) can

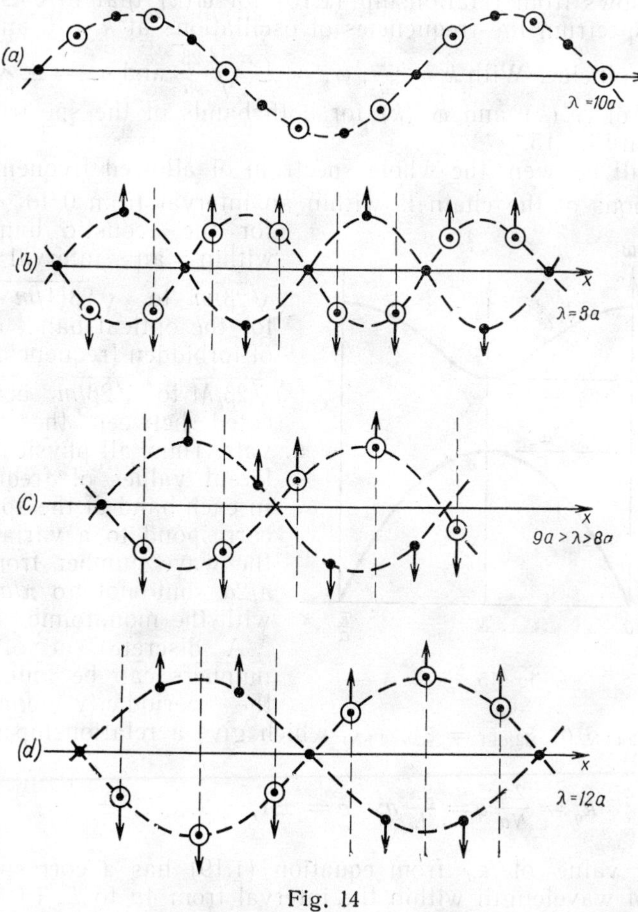

Fig. 14

be excited by light waves, because of which this type of oscillations is termed optical, and the corresponding band of $\omega_+(k)$ spectrum is called the optical band.

Figure 14 shows schematically lateral acoustic (a) and optical (b, c and d) oscillations of a linear chain consisting of atoms of two kinds. Vertical dotted lines in b, c and d separate neighbouring elementary cells from each other consisting, in combination, of two atoms whose mass centre remains fixed. As can be seen, a cell includes two halves of the nearest atoms shown in bright circles and an atom of another kind located between them (black circle). Selection of such an elementary cell makes it possible to satisfy the condition (1.18).

It follows from relationship (1.15) for $\omega(k)$ that in each band of the spectrum the frequencies of oscillations at $k = 0$ and $k = \pi/a$ coincide. With $k = \frac{\pi}{2a}$, $\omega_+ = \pm\sqrt{\frac{2\beta}{m}}$ and $\omega_- = \pm\sqrt{\frac{2\beta}{M}}$. Graphs of $\omega_+(k)$ and $\omega_-(k)$ for both bands of the spectrum are shown in Fig. 15.

As will be seen, the whole spectrum of allowed frequencies of oscillations of the chain is within an interval from 0 to $\sqrt{2\beta/M}$ for the acoustic band and within an interval from $\sqrt{2\beta/m}$ to $\sqrt{2\beta(1/m + 1/M)}$ for the optical band, a band of forbidden frequencies from $\sqrt{2\beta/M}$ to $\sqrt{2\beta/m}$ being located between these intervals. Thus, all physically different values of frequencies in each band of the spectrum correspond to a variation of the wave number from 0 to $\pi/2a$, but not to π/a, as is with the monatomic chain.

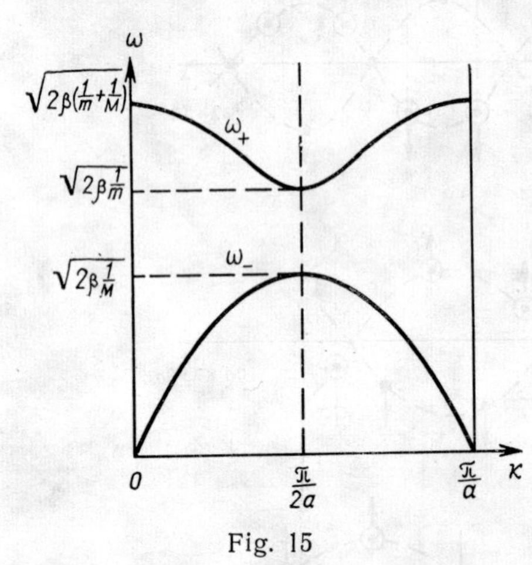

Fig. 15

A discrete set of wave numbers can be found from the periodicity conditions $\xi_{2n} = \xi_{2n+N}$ or $\xi_{2n+1} = \xi_{2n+1+N}$, which give a relationship similar to (1.10):

$$k_q = \frac{2\pi}{Na} q = \frac{2\pi}{L} q; \quad q = \pm 1, \pm 2, \ldots, \pm \frac{N}{4} \qquad (1.19)$$

Each value of k_q from equation (1.19) has a corresponding value of wavelength within the interval from $4a$ to L. In a chain consisting of atoms of two kinds, two types of oscillation with the same wavelength $\lambda(k)$ are possible: acoustical and optical. The total number of normal modes of the chain in both bands of spectrum, corresponding to all possible values of k within the interval from $-\pi/2a$ to $+\pi/2a$, is again equal to the number of atoms, N, in the chain. The interval of frequencies of optical oscillations of a one-dimensional chain is located higher than that of acoustical oscillations. The frequency range of the optical band becomes very narrow at $M \gg m$. Optical frequencies in that case are close to the limiting value of allowed frequency

$\omega_{max} = \sqrt{2\beta(1/m + 1/M)}$, whereas wave numbers (or wavelengths) of these oscillations can vary in a wide range from 0 to $\pi/2a$.

Thus, the limiting case of $M \gg m$ shows clearly that an optical oscillation cannot be represented by a combination of two opposite running waves and therefore differs in its nature from an acoustical oscillation.

Consider the group velocity of oscillations $v_g = \frac{\partial\omega}{\partial k}$ in various bands of the spectrum. As follows from (1.15), at $k = 0$ $v_{g-} = \frac{\partial\omega_-}{\partial k}$ coincides with the velocity of sound (1.17) and $v_{g+} = \frac{\partial\omega_+}{\partial k} = 0$. At $k = \pi/2a$, group velocities of acoustical $\frac{\partial\omega_-}{\partial k}$ and optical $\frac{\partial\omega_+}{\partial k}$ oscillations are equal to zero.

At values of k close to $\pi/2a$ (i.e. at $ka = \frac{\pi}{2} - \delta$ where $\delta \ll 1$), amplitude ratios in the corresponding bands of the spectrum are

$$(\eta/\zeta)_- = \frac{\delta}{1 - \frac{m}{M}} > 0$$

$$(\zeta/\eta)_+ = -\frac{\delta}{\frac{M}{m} - 1} < 0 \tag{1.20}$$

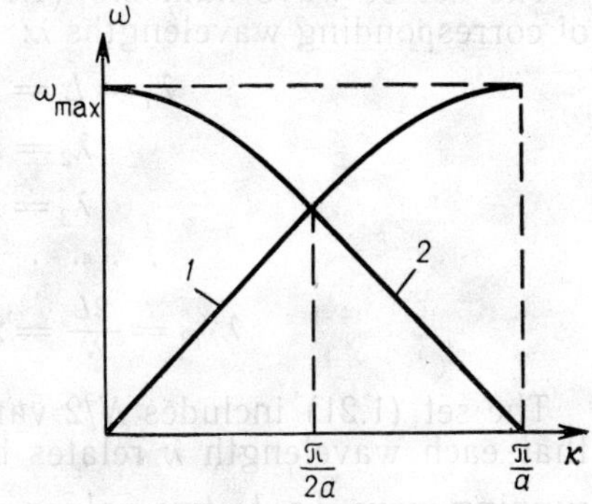

Fig. 16

Expressions (1.20) imply that as the limit value $k = \pi/2a$ is being approached (i.e. with $\delta \to 0$), a reduction of the amplitude of oscillations of lighter atoms (of mass m) occurs in the acoustical band and of heavier atoms (of mass M), in the optical band, neighbouring atoms oscillating in phase in the acoustical band (as at small values of k), and in anti-phase in the optical band. At $k = \pi/2a$, oscillations in both bands, ω_- and ω_+, are in the form of a standing wave with nodes at the fixed lighter atoms in the former case and at the fixed heavier atoms, in the latter.

With a limit transition from a chain consisting of atoms of two kinds ($m \neq M$) to a monatomic chain ($m \to M$), the optical band in the interval $\pi/2a \leqslant k \leqslant \pi/a$ becomes an extension of the acoustical band in the interval $0 \leqslant k \leqslant \pi/2a$, and the band of forbidden frequencies becomes zero. Then the tangent $\frac{\partial\omega}{\partial k}$ in point $k = \frac{\pi}{2a}$ attains its finite value.

Thus, upon passing to the limit $m \to M$, the period of a chain is halved and the spectrum of acoustical and optical oscillations degenerates into two identical acoustical bands *1* and *2* in Fig. 16 (for branch *2*, the group velocity $v_g = \frac{\partial \omega}{\partial k} < 0$, that is, corresponds to waves running in the opposite direction).

1-5. PHONONS. THERMAL OSCILLATIONS IN A ONE-DIMENSIONAL LATTICE

We will again consider the set (1.10) of wave numbers of normal oscillations of a monatomic chain. Any value of k from (1.10) is a multiple of the minimum wave number $k_{min} = \frac{2\pi}{L}$. The spectrum of wave numbers is equidistant with a step k_{min}.

The set of wave numbers (1.10) may be correlated with a set of corresponding wavelengths λ:

$$\begin{gathered}
\lambda_1 = L = Na = \lambda_{max} \\
\lambda_2 = L/2 \\
\lambda_3 = L/3 \\
\dots\dots\dots\dots \\
\lambda_{N/2} = \frac{2L}{N} = 2a = \lambda_{min}
\end{gathered} \tag{1.21}$$

The set (1.21) includes $N/2$ various wavelengths. Let us remind that each wavelength λ relates both to a direct and a reciprocal running wave, i.e. to two values of wave number $k = \pm \frac{2\pi}{\lambda}$. Thus, set (1.21) characterizes N different normal modes or N different states.

Note that with the boundary conditions equivalent to standing waves in the chain (atoms at the boundaries are either free or completely bound) we obtain a set of $N - 1$ wavelengths characterizing $N - 1$ different states:

$$\begin{gathered}
\lambda_1 = 2(N-1)a = \lambda_{max} \\
\lambda_2 = (N-1)a \\
\dots\dots\dots\dots \\
\lambda_{N-1} = 2a
\end{gathered} \tag{1.22}$$

Standing waves equivalent to the set (1.22) are illustrated in Fig. 17*a* for a case of nodes and in Fig. 17*b* for a case of loops at the boundaries.

With $N \gg 1$, the sets (1.21) and (1.22) practically describe the same number of states. For simplicity, we shall further consider only the positive k in (1.10) and the corresponding set of λ (1.21). For each value of k from (1.10), the frequency ω of oscillations can be found by means of (1.6). A set of such frequencies is not equidistant. It can be approximately taken to be equidistant with a step $\omega_{min} = v^0_{son} k_{min}$ in the region $\omega \ll \omega_{max}$. With an increase of frequency, the interval between neighbouring frequencies redu-

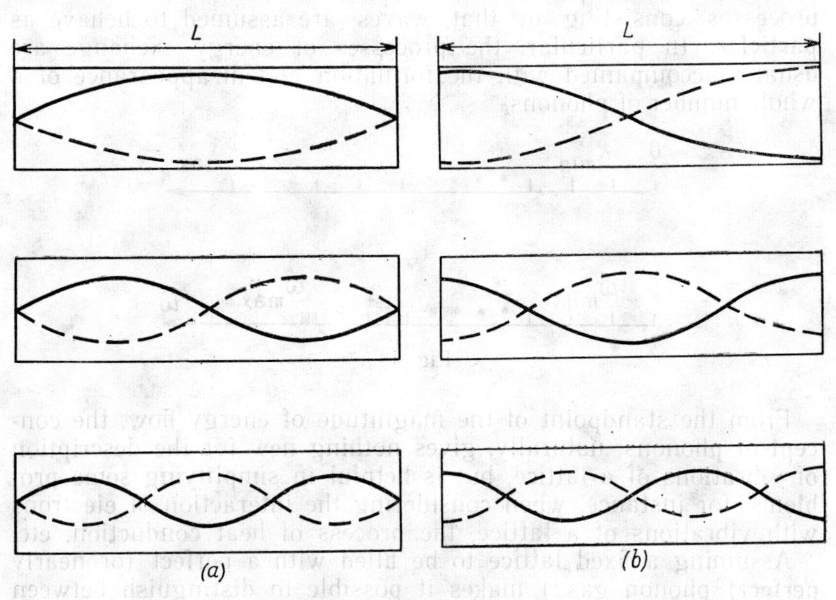

Fig. 17

ces and tends to zero, i.e. $\omega \to \omega_{max}$. The frequency spectrum is condensed to the limiting value ω_{max}.

Replacing the non-equidistant frequency spectrum by an equidistant one with a step $v^0_{son} k_{min}$ gives a false limit frequency $v^0_{son} k_{max} = v^0_{son} \frac{\pi}{a}$, which is $\pi/2$ times greater than the actual value of ω_{max} according to (1.6) $\left(\omega_{max} = \frac{2}{a} v^0_{son}\right)$. The sets of frequencies ω and wave numbers k are compared in Fig. 18, where k_{min} and ω_{min} are shown for clarity by sections of equal length.

Each normal mode of the lattice is equivalent to a linear monochromatic oscillator. The energy ε of an elastic wave for frequency ω cannot acquire arbitrary values, but must be equal to a whole number of quanta $\hbar\omega$. Each individual quantum of energy of the normal mode of the lattice may be conveniently correlated with a quasi-particle of energy $\varepsilon = \hbar\omega$ and momentum $p =$

$=\hbar k$. The velocity of quasi-particle is determined by the group velocity of propagation of oscillation, $v_g = \frac{\partial \omega}{\partial k} = \frac{\partial \varepsilon}{\partial p}$.

This quantum of energy of normal modes has been termed the phonon by Frenkel, by analogy with the name of the quantum of an electromagnetic field, the photon.

The concept of phonons reflects the quantum aspect of wave processes consisting in that waves are assumed to behave as particles. In particular, the processes of energy exchange are usually accompanied with the formation and disappearance of a whole number of phonons.

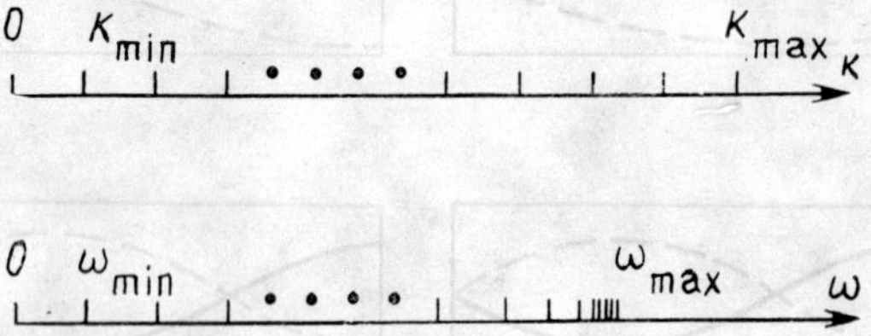

Fig. 18

From the standpoint of the magnitude of energy flow, the concept of phonons, naturally, gives nothing new for the description of vibrations of a lattice, but is helpful in simplifying some problems, for instance, when considering the interaction of electrons with vibrations of a lattice, the process of heat conduction, etc.

Assuming a fixed lattice to be filled with a perfect (or nearly perfect) phonon gas [1]) makes it possible to distinguish between the effects related to periodicity of a lattice and those caused by oscillations of atoms. In a crystal of finite dimensions, the normal equilibrium oscillations are standing waves in a system of coordinates related to the mass centre. Such normal oscillations cannot transfer momentum. But if the waves in a body are excited from the outside, a momentum transfer relative to the mass centre will take place.

Let us now consider thermal excitation of phonons. The energy of a linear oscillator of frequency ω can acquire discrete values $\varepsilon_n = \left(n + \frac{1}{2}\right)\hbar\omega$, where $n = 0, 1, 2$, etc. Quantization of the energy results in quantization of the amplitude x_0 of oscillations of the oscillator. According to the correspondence principle

$$\varepsilon_n = \frac{1}{2} M\omega^2 x_0$$

[1]) Under the harmonic approximation, phonons do not interact and form a perfect gas. Interaction of phonons related to the anharmonicity of oscillations of a lattice will be considered later.

whence

$$x_0 = \sqrt{\frac{2h}{M\omega}\left(n + \frac{1}{2}\right)}$$

where M is the mass of oscillator. Uncertainty of the value x_0 coincides with the amplitude of zero oscillations $\Delta x_0 = \sqrt{\frac{\hbar}{M\omega}}$ corresponding to a state having the quantum number $n = 0$.

According to the Gibbs distribution, the probability that an oscillator with frequency ω and temperature T will be in an energy state ε_n is equal to $De^{-\varepsilon_n/kT}$, where the constant D is independent of ε_n. The mean energy of the oscillator at temperature T may be represented as

$$\bar{\varepsilon} = \frac{\sum_{n=0}^{\infty} \varepsilon_n D e^{-\varepsilon_n/kT}}{\sum_{n=0}^{\infty} D e^{-\varepsilon_n/kT}} \tag{1.23}$$

whence

$$\varepsilon = \frac{\hbar\omega}{2} + \frac{\sum_{n=0}^{\infty} n\hbar\omega e^{-\frac{n\hbar\omega}{kT}}}{\sum_{n=0}^{\infty} e^{-\frac{n\hbar\omega}{kT}}} \tag{1.24}$$

Denoting $\frac{\hbar\omega}{kT} = w$, it is easy to sum up

$$\bar{\varepsilon} = \frac{\hbar\omega}{2} + \frac{\hbar\omega\,(e^{-w} + 2e^{-2w} + \ldots)}{1 + e^{-w} + e^{-2w} + \ldots} =$$

$$= \frac{\hbar\omega}{2} + (-\hbar\omega)\frac{d}{dw}[\ln(1 + e^{-w} + e^{-2w} + \ldots)] =$$

$$= \frac{\hbar\omega}{2} - \hbar\omega\frac{d}{dw}\left[\ln\frac{1}{1 - e^{w}}\right]$$

and finally get

$$\bar{\varepsilon} = \frac{\hbar\omega}{2} + \frac{\hbar\omega}{e^{\hbar\omega/kT} - 1} \tag{1.25}$$

The first term in the right-hand side of (1.25) is the energy of zero oscillations. In order to describe the thermal excitation of the oscillator, its energy will be counted from the level of zero-oscillation energy, which essentially means that thermal oscillations will be considered separately from zero oscillations. The mean energy $\bar{\varepsilon}$ of the oscillator with the natural frequency ω at temperature T should now be written as

$$\bar{\varepsilon} = \frac{\hbar\omega}{e^{\hbar\omega/kT} - 1} \tag{1.26}$$

Consider an oscillator with a frequency corresponding to one of the normal oscillations ω_i of the chain. An energy quantum of this frequency is equal to $\hbar\omega_i$. Formula (1.26) can then be interpreted as the energy quantum $\hbar\omega_i$ multiplied by the mean number n_i of these quanta at temperature T:

$$\bar{\varepsilon} = \bar{n}_i \hbar\omega_i \tag{1.27}$$

where

$$\bar{n}_i = \frac{1}{e^{\hbar\omega_i/kT} - 1}$$

When $kT < \hbar\omega_i$, the probability of excitation of a quantum with energy $\hbar\omega_i$ is low and the mean number of such quanta $\bar{n}_i \simeq \simeq e^{-\hbar\omega_i/kT}$ is exponentially small; $\bar{n}_i$ becomes equal to unity at a definite temperature $T' = \frac{\hbar\omega_i}{k \ln 2} \simeq 1.44 \frac{\hbar\omega_i}{k}$. At a further definite temperature T'', two quanta $\hbar\omega_i$ will be excited, at T''', three quanta, etc. Excitation of each new quantum causes a step-like increase of the amplitude of oscillations of the oscillator.

With passing over to the concept of phonons, expression (1.27) should be treated as the mean number of phonons of frequency ω_i excited at temperature T. A phonon with a frequency ω_i actually becomes excited at a temperature $T' \simeq 1.44 \frac{\hbar\omega_i}{k}$. With temperature being increased further, the number of phonons with frequency ω_i increases and at $kT \gg \hbar\omega_i$ attains the mean value:

$$\bar{n}_i \simeq \frac{1}{1 + \frac{\hbar\omega_i}{kT} - 1} = \frac{kT}{\hbar\omega_i} \tag{1.28}$$

The mean total energy of all the phonons of this frequency is:

$$\bar{\varepsilon}_i = \bar{n}_i \hbar\omega_i \simeq \frac{kT}{\hbar\omega_i} \hbar\omega_i = kT \tag{1.29}$$

At the temperature of absolute zero, no phonons exist. With an increase of temperature, two processes occur simultaneously: (1) high-frequency phonons are excited more and more and (2) the number of excited low-frequency phonons increases. The first process is finished at a definite temperature T^* when the phonon with the maximum possible spectrum frequency becomes excited, after which only the number of phonons of each frequency continues to increase.

The temperature T^* is the limiting temperature below which a body can be considered approximately as a continuum. In its meaning it coincides with Debye temperature Θ_D introduced in Sec. 1-3. Thus, for a one-dimensional chain of atoms, Θ_D can be defined as

$$\Theta_D = T^* \simeq \frac{\hbar\omega_{max}}{k} = \frac{2\hbar v^0_{son}}{ka} \tag{1.30}$$

1-6. SPECIFIC HEAT OF A LATTICE. EINSTEIN'S MODEL

In order to calculate the energy of thermal oscillations of a lattice at a temperature T, we have to find the total energy of all the phonons excited in the lattice at that temperature. For that, we evidently should sum up the expression (1.26) over all frequencies from the whole set characterizing natural oscillations of the lattice. For a one-dimensional monatomic lattice, for instance, summation is done only along the acoustic band up to $\omega = \omega_{max}$, whereas for a one-dimensional chain consisting of atoms of two kinds, summation will include the frequencies of both the acoustical and optical bands.

It will be emphasized that the equilibrium value of thermal energy should be calculated by summing up the energies of oscillations corresponding both to direct and reverse running waves. This corresponds to the summation for the complete set of wave numbers (wave vectors) k characterizing the spectrum of equilibrium thermal oscillations.

Thus, thermal energy ε_{lat} of a lattice can be written as

$$\varepsilon_{lat} = \sum_k \frac{\hbar\omega(k)}{e^{\frac{\hbar\omega(k)}{kT}} - 1} \tag{1.31}$$

where summation is to be made over the whole spectrum of allowed wave vectors. Expression (1.31) is a general one and holds true for any type of system, characteristics of each particular case being determined only by the peculiarities of the oscillation spectrum.

Calculation of ε_{lat} by formula (1.31) in the most general case may be done for the region of high temperatures at which all frequencies of an oscillation spectrum up to the maximum frequency ω_{max} are excited. Indeed, let the temperature T be such that $kT > \hbar\omega_{max}$. Then the total energy $\bar{\varepsilon}$ of phonons of any frequency ω_i will be equal to kT and the sum (1.31) reduces to NkT, where N is the number of normal oscillations, or of degrees of freedom of the system.

For one-dimensional lattices, as has been indicated earlier, N coinsides with the number of atoms in the chain. The number of degrees of freedom of a three-dimensional lattice consisting of N_a atoms is equal to $3N_a$. Specific heat at constant volume, C_v, for a three-dimensional lattice in the range of high temperatures is

$$C_v = \frac{\partial \varepsilon_{lat}}{\partial T} = Nk = 3N_a k \tag{1.32}$$

When calculated per one gram-molecule of substance, $C_v = 3N_a k = 3R \simeq 6$ cal/ mol degree (where $R = 1.93$ cal/degree·mole is Mendeleev's constant). This result, known as the Dulong-Petit law, is in good agreement with the observed values of molar specific heat of many solids, including metals, at temperatures above Debye temperature.

The simplest model of an oscillation spectrum, which made it possible to calculate ε_{lat} by formula (1.31) at any temperature, was first suggested by Einstein in 1911 [44]. In that model, various frequencies of oscillations of a lattice were replaced by a single characteristic frequency ω_E.

The thermal energy of a lattice with approximation by Einstein's model is

$$\varepsilon_{lat} = \frac{\hbar\omega_E}{e^{\hbar\omega_E/kT} - 1} N \tag{1.33}$$

where N is the number of degrees of freedom.

At $T > \hbar\omega_E/k$, Einstein's model also gives the Dulong-Petit law. This result is self-evident, since at high temperatures the total energy of each normal mode is equal to kT and the thermal energy of the lattice is no more dependent on the particular kind of oscillation spectrum. Summation over the types of phonons at high temperatures becomes equivalent to multiplication of kT by the number of degrees of freedom of the system.

The characteristic frequency ω_E in Einstein's model should be close in its sense to the limit frequency ω_{max} in a phonon spectrum. Consequently, the order of magnitude of ω_E is determined by Debye temperature, $\omega_E \sim k\Theta_D/\hbar$.

At low temperatures, $T \leqslant \Theta_D$, Einstein's model gives an exponential dependence of ε_{lat} and C_c on temperature:

$$\varepsilon_{lat} \simeq N\hbar\omega_E e^{-\hbar\omega_E/kT}$$
$$C_v = kN\left(\frac{\hbar\omega_E}{kT}\right)^2 e^{-\hbar\omega_E/kT} \tag{1.34}$$

which does not agree with the dependence $C_v \sim T^3$ that has been found experimentally for this region and is known as Debye's law. The disagreement can be related to that phonons of different frequencies are excited in the temperature region $T < \Theta_D$. Under such conditions, Einstein's model, which uses a single characteristic frequency ω_E for describing the whole spectrum of oscillations, is too a coarse approximation.

In the region of sufficiently low temperatures, however, where only the phonons corresponding to the lowest frequency $\omega_{\min}$ remain excited in the oscillation spectrum, oscillations of the lattice can be described by Einstein's model, but with a different characteristic frequency ω_E equal to $\omega_{\min} = \frac{2\pi}{L} v^0_{son}$. The temperature dependence of specific heat C_v will then again become exponential and be determined by expression (1.34).

Let us define the temperature region within which the dependence of specific heat on temperature should be exponential. Its boundary is determined by the inequality

$$T \lesssim \frac{\hbar\omega_{\min}}{k} = \frac{2\pi\hbar v^0_{son}}{kL}$$

At a sonic velocity $v^0_{son} = 5 \cdot 10^5$ cm/sec (such is the order of magnitude of v^0_{son} in most solids), $T \lesssim \left(\frac{2.4 \cdot 10^{-5}}{L\,[\text{cm}]}\right)$°K, from which it can be seen that at really attainable temperatures $T \sim (0.2\text{-}0.02)$°K, Einstein's model is only applicable to bodies having microscopically small linear dimensions $L \sim (1\text{-}10)$ microns. For bodies having dimensions of approximately 1 mm and greater, Debye's law remains practically valid up to the lowest attainable temperatures ~ 0.001° K.

1-7. SPECIFIC HEAT OF A LATTICE AT TEMPERATURES BELOW Θ_D

The thermal energy of a lattice (1.31) can be approximately estimated for an intermediate temperature region $\frac{\hbar\omega_{\min}}{k} < T \ll \Theta_D$ within which only a portion of the spectrum of thermal oscillations is excited. Such estimation will be made for cases of one, two, and three dimensions. From the definition of a one-dimensional chain (1.30), and also from formula (1.27), it follows that at $T \ll \Theta_D$ only acoustic phonons will be excited in the oscillation spectrum.

The optical portion of the spectrum is at its high-frequency end and is excited only at $T \gtrsim \Theta_D$. Therefore, to estimate the thermal energy at intermediate temperatures, it suffices to consider the simplest lattices consisting of like atoms, in which there are no optical oscillations.

1. A one-dimensional lattice. In a one-dimensional lattice at a temperature $T \ll \Theta_D$, of all the possible normal modes, only those are excited that have the frequencies ω for which $\hbar\omega < kT$. The excitation boundary corresponds to the frequency $\omega' \sim kT/\hbar$. Low-frequency oscillations with $\omega \ll \omega'$ consist of a large number of energy quanta (correspond to a large number of phonons), while high-frequency ones, with $\omega \sim \omega'$, of a small number of quanta.

The probability of excitation of phonons with frequencies $\omega > \omega'$ is exponentially small. Since the boundary frequency ω' is substantially smaller than $\omega_{max} \simeq \frac{k\Theta_D}{\hbar}$, then at $\omega \leqslant \omega'$, non-equidistancy of the frequency spectrum can be ignored. This corresponds to the initial linear portion of the dispersion relationship (1.6). The wave number of an oscillation with a frequency ω' is approximately equal to $k' \simeq \omega'/v^0_{son}$.

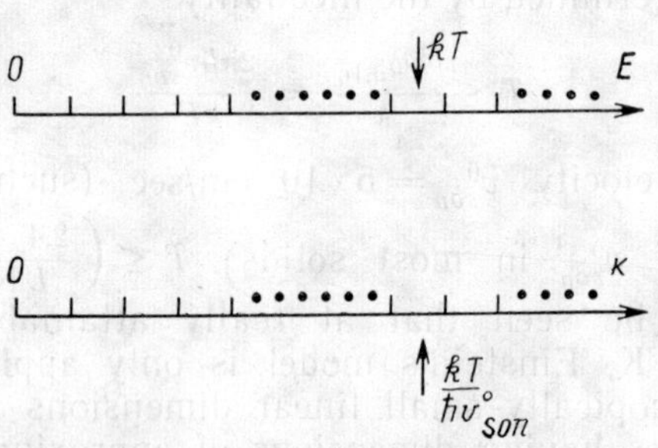

Fig. 19

Figure 19 shows the energy states and wave numbers of the excited portion of an oscillation spectrum. This portion is bounded by the values $\sim kT$ for energy and $\sim kT/\hbar v^0_{son}$ for wave numbers. The condition $\hbar\omega < kT$ holds true for any of excited frequencies ω, and therefore, the total energy of phonons of each frequency ω is $\sim kT$ according to expression (1.29).

As can be seen from Fig. 19, the number of different frequencies of phonons, which in the one-dimensional case considered coincides with the number of normal modes, is approximately equal to the ratio of boundary energy kT to $\hbar\omega_{min}$ (or the ratio of boundary wave number $kT/\hbar v^0_{son}$ to k_{min}). Multiplication of this number by the total energy of phonons of each frequency gives the total thermal energy ε^I_{lat} of oscillations of the lattice:

$$\varepsilon^{\mathrm{I}}_{lat} \approx \frac{kT}{\hbar\omega_{min}} kT \sim T^2 \tag{1.35}$$

where the superscript I at ε_{lat} indicates that the expression holds only true for a one-dimensional lattice. Hence it follows that specific heat of a one-dimensional lattice should be proportional to T:

$$C^{\mathrm{I}}_v = \frac{\partial \varepsilon^{\mathrm{I}}_{lat}}{\partial T} \sim T \tag{1.36}$$

2. A two-dimensional lattice. Oscillations of a two-dimensional lattice may be considered on the basis of the analysis of oscillations of a one-dimensional chain of atoms, but at passing to two dimensions there appears a new property of oscillations related to the possibility of two different polarizations.

In the limiting case of long waves, the lattice is essentially an elastic continuum (continuous medium) whose internal structure is so fine that has no noticeable effect on the properties of oscillations. In such a medium, both longitudinal and transverse acoustic waves can propagate. With longitudinal polarization, the vector of displacement of each atom is parallel with the direction of wave propagation. Oscillations of this type are similar to acoustic waves in a one-dimensional chain.

With transverse polarized waves, the vector of displacement is perpendicular to the direction of wave propagation. The velocity $v_{son\parallel}$ of a longitudinally polarized wave usually exceeds the velocity $v_{son\perp}$ of a wave of mode polarization. This circumstance is linked with that longitudinal oscillations are compression waves for which the elastic forces are greater than those for transverse shear waves.

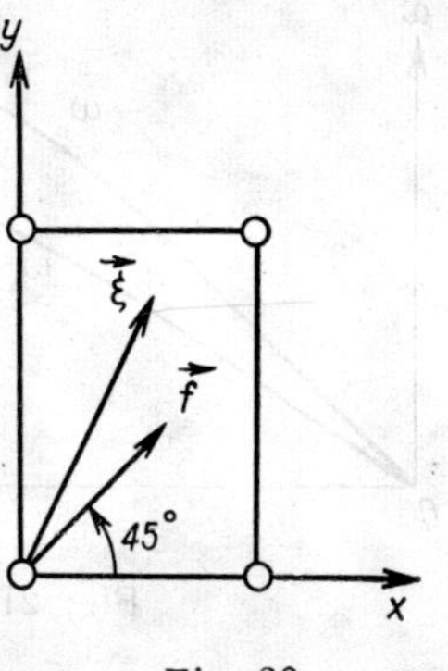

Fig. 20

Such a simple pattern of oscillations is only valid for a continuous isotropic medium. In a discrete anisotropic medium, the velocity of wave propagation depends on the magnitude and direction of wave vector. Decomposition of oscillations into transverse and longitudinal ones becomes conditional, since the displacement vector in such a medium is not strictly parallel or perpendicular to the wave vector $\vec{k}$. This can be shown on a simple example of a plane rectangular lattice in which the coefficients of quasi-elastic forces β_x and β_y in the directions of the x and y axis are not equal to one another. An elementary cell of such a lattice is shown in Fig. 20.

Let $\beta_x > \beta_y$ and the force displacing the atom 1 in the cell be directed at an angle of 45 degrees to the x axis, so that its components along the x and y axis are equal to one another. Since $\beta_x > \beta_y$, atom 1 will be displaced along the y axis more than along the x axis. The displacement vector $\vec{\xi}$ of the atom will not coincide in direction with the force vector $\vec{f}$ (see Fig. 20), the result being that a transverse wave will be formed simultaneously at excitation of a longitudinal wave. Exceptions are oscillations along the directions of symmetry in the lattice.

The solution of the general problem of oscillations of atoms in a discrete anisotropic two-dimensional lattice gives an algebraic equation of the order 2γ in terms of ω^2, where γ is the number of atoms in a unit cell of the lattice. This equation determines dispersion relationships $\omega_i = \omega_i(\vec{k})$ for each of 2γ bands of the oscillation spectrum.

For instance, if a cell contains a single atom ($\gamma = 1$), the spectrum will consist of two acoustic bands. For the selected directions of the wave vector $\vec{k}$ along the axes of symmetry of the lattice, one of the bands corresponds to longitudinally polarized, and the other, to transversely polarized oscillations. The corresponding dispersion relationships for such directions of $\vec{k}$ will be denoted as $\omega_{\parallel}=\omega_{\parallel}(\vec{k})$ and $\omega_{\perp} = \omega_{\perp}(\vec{k})$.

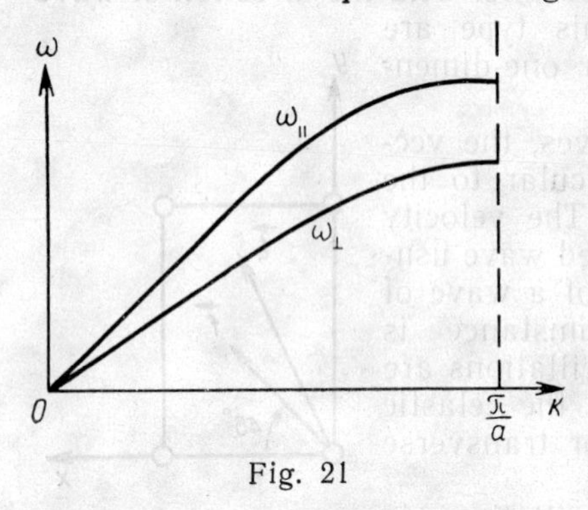

Fig. 21

For an arbitrary direction in the lattice, there are also two dispersion relationships, one of which passes over into $\omega_{\parallel}(\vec{k})$ and the other, into $\omega_{\perp}(\vec{k})$ as vector $\vec{k}$ is rotated toward the direction of the axis of symmetry. Then it may be assumed that, for an arbitrary direction, one of the bands also corresponds to longitudinally polarized, and the other to transversely polarized oscillations, as has been found for the selected directions of $\vec{k}$. But with an arbitrary direction in the lattice, the displacement vector for each type of oscillation is no more strictly parallel or strictly perpendicular to the direction of wave propagation. One of the waves turns to be "more or less" polarized transversely and the other, "more or less" polarized longitudinally.

For each acoustic band of the oscillational spectrum, the relationship $\omega_i(|\vec{k}|)$ is linear in the vicinity of point $|\vec{k}| = 0$ and its slope is equal to the corresponding sound velocity in that direction. As $|\vec{k}|$ is being increased, dispersion will be sooner or later formed depending on direction in the lattice. For each direction chosen, dispersion curves of "longitudinally" and "transversely" polarized acoustic waves can be constructed (Fig. 21). But these dispersion curves can no more characterize the spectrum of oscillations, since their shape depends on direction in the lattice. The spectrum of oscillations can be described completely, if relationships $\omega_i(\vec{k})$ are given for all directions. These relationships can be conveniently represented graphically by curves of constant frequency $\omega_i(\vec{k}) =$ $= \text{const}$.

For an isotropic medium and without account of dispersion, the curves of constant frequency for the two bands of the spectrum have the form of concentric circles whose radii $|\vec{k}| = \sqrt{k_x^2 + k_y^2}$ for each value of ω are determined by the sound velocity v_{son}^0:

$$\begin{aligned} k_x^2 + k_y^2 &= \frac{\omega^2}{(v_{son\parallel}^0)^2} \text{ with longitudinal polarization} \\ k_x^2 + k_y^2 &= \frac{\omega^2}{(v_{son\perp}^0)^2} \text{ with transverse polarization} \end{aligned} \tag{1.37}$$

Since $v_{son\parallel}^0 > v_{son\perp}^0$ then for the same frequency of both transversely and longitudinally polarized waves, the circle of constant

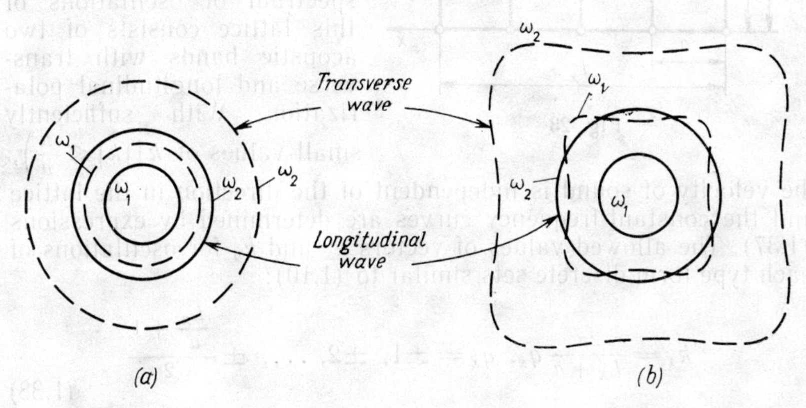

Fig. 22

frequency in the first case will be farther from the centre than in the second, as shown in Fig. 22*a*, where each type of polarization is represented by two circles corresponding to two values of frequency, ω_1 and ω_2 ($\omega_1 < \omega_2$).

For oscillations of a discrete anisotropic lattice, the shape of constant-frequency curves can differ greatly from that of a circle. Two curves of constant frequency are also shown in Fig. 22*b* for each type of polarization corresponding to two values of frequency, ω_1 and ω_2 ($\omega_1 < \omega_2$).

Since sound velocity depends both on the type of polarization of oscillations and the direction of their propagation, it is impossible to introduce a unified Debye temperature by means of a relationship of a type of (1.30). The physical parameter Θ_D, which has the meaning of Debye temperature in some or other physical process, is introduced as a certain averaged value and is dependent on the method it has been found. (For instance, the parameters Θ_D introduced to describe the temperature dependence of specific heat or electric conductivity, may differ noticeably from one another.)

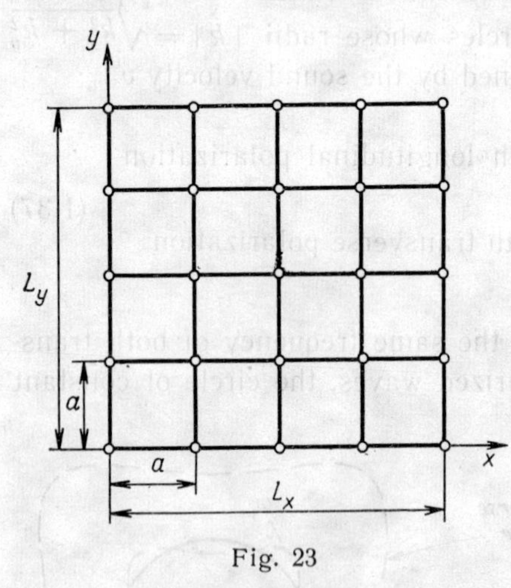

Fig. 23

In order to estimate the energy of thermal oscillations of a two-dimensional system, let us consider the simplest type of lattice, i.e. a monatomic square lattice with the dimensions L_x and L_y along the x and y axis respectively (Fig. 23). The spectrum of oscillations of this lattice consists of two acoustic bands with transverse and longitudinal polarization. With sufficiently small values of $\vec{k}\left(|\vec{k}| \ll \frac{\pi}{a}\right)$, the velocity of sound is independent of the direction in the lattice and the constant-frequency curves are determined by expressions (1.37). The allowed values of vectors k_x and k_y for oscillations of each type form discrete sets similar to (1.10):

$$k_x = \frac{2\pi}{L_x + a} q_x, \quad q_x = \pm 1, \ \pm 2, \ \ldots, \ \pm \frac{\frac{L_x}{a} + 1}{2}$$
$$k_y = \frac{2\pi}{L_y + a} q_y, \quad q_y = \pm 1, \ \pm 2, \ \ldots, \ \pm \frac{\frac{L_y}{a} + 1}{2} \tag{1.38}$$

In a two-dimensional square lattice, only such oscillations can propagate for which the projections of wave vectors $\vec{k}$ onto the x and y axis coincide with vectors k_x and k_y from sets (1.38). The allowed values of wave vectors are shown in Fig. 24.

In the isotropic approximation, an acoustic spectrum can be characterized by two different Debye temperatures, for longitudinal and transverse oscillations:

$$\Theta_{D\parallel} \approx \frac{2\hbar v^0_{son\parallel}}{ka}$$
$$\Theta_{D\perp} \approx \frac{2\hbar v^0_{son\perp}}{ka} \tag{1.39}$$

Similar to the one-dimensional case at $T \ll \min\{\Theta_{D\parallel}, \Theta_{D\perp}\}$ the frequency spectrum to the excitation boundary $\omega' \sim kT/\hbar$ is nearly equidistant. Note that in a two-dimensional case the number of normal oscillations in a lattice, excited at a definite temperature T, is determined not by the number of different frequencies ω satisfying the condition $\omega < \frac{kT}{\hbar}$, but by the number of different wave vectors whose projections onto the x and y axis are limited by the values $kT/\hbar v^0_{son\parallel}$ and $kT/\hbar v^0_{son\perp}$ for waves of longitudinal and transverse polarization respectively. The same frequency ω then has a greater number of corresponding normal oscillations whose wave vectors satisfy relationships (1.37).

The total number of normal oscillations of transverse polarization excited in a square lattice at temperature T, which is equal to the total number of different allowed wave vectors, can be found approximately as the ratio of the area of a square $(2kT/\hbar v^0_{son\parallel})^2$ to an elementary area $(k_x)_{\min}(k_y)_{\min}$ occupied by a single allowed

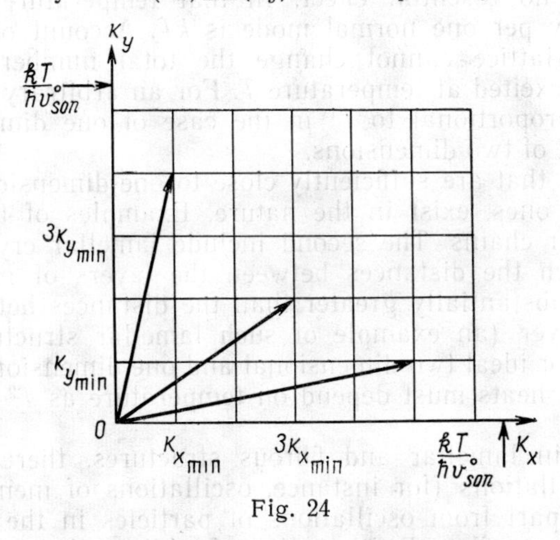

Fig. 24

state (see Fig. 24). Similarly, the total number of normal oscillations of transverse polarization is determined by the ratio $(2kT/\hbar v^0_{son\perp})^2 : (k_x)_{\min}(k_y)_{\min}$. The number of normal oscillations of both types of polarization can therefore be approximately found from the expression:

$$\frac{1}{(k_x)_{\min}(k_y)_{\min}}\left(\frac{2kT}{\hbar}\right)^2\left[\left(\frac{1}{v^0_{son\parallel}}\right)^2+\left(\frac{1}{v^0_{son\perp}}\right)^2\right]\sim T^2 \quad (1.40)$$

As in the one-dimensional case, for all frequencies ω of excited phonons, the inequality $\hbar\omega < kT$ holds true, and therefore, according to (1.29), the total energy of all phonons of the given frequency, both transversely and laterally polarized, is equal to kT.

The thermal energy ε_{lat}^{II} of the lattice is found by multiplying kT by the number of different oscillations (1.40) and is proportional to T^3. It then follows that the specific heat C_v^{II} of a two-dimensional lattice is in quadratic dependence on temperature:

$$C_v^{II} = \frac{\partial \varepsilon_{lat}^{II}}{\partial T} \sim T^2 \tag{1.41}$$

It may be concluded from the consideration of simplest one- and two-dimensional structures that the temperature dependence for the specific heat of a lattice is determined by the number of dimensions of the lattice. This conclusion is true for lattices of any arbitrary type. Indeed, at intermediate temperatures, when only the low-frequency portion of a spectrum is excited, dispersion of oscillations has no essential effect. In that temperature region the mean energy per one normal mode is kT. Account of the anisotropy of a lattice cannot change the total number of normal oscillations excited at temperature T. For an arbitrary lattice, this number is proportional to T^1 in the case of one dimension, and to T^2, in that of two dimensions.

Structures that are sufficiently close to one-dimensional or two-dimensional ones, exist in the nature. Examples of the first are long polymer chains. The second include lamellar crystalline bodies in which the distances between the layers of molecules or atoms are substantially greater than the distances between particles in a layer (an example of such lamellar structure is flaky graphite). For ideal two-dimensional and one-dimensional systems, their specific heats must depend on temperature as T^2 and T^1 respectively.

Actually, in lamellar and fibrous structures, there may arise bending oscillations (for instance, oscillations of membrane type in layers) apart from oscillations of particles in the plane of a layer or along a fibre. Participation of additional degrees of freedom of a system in the spectrum of oscillations results in an increase of the exponents α at T in the temperature law for specific heat $C \sim T^{\alpha}$. The experimental value of α for real fibrous structures turns to be greater than unity, and for lamellar structures, $2 < \alpha < 3$.

3. A three-dimensional lattice. The spectrum of oscillations in a three-dimensional lattice consists of 3γ bands, each of which can be described by the dispersion equation $\omega_i = \omega(\vec{k})$, where $i = 1, 2, \ldots, 3\gamma$. In lattices having a single atom in a unit cell ($\gamma = 1$), there are three acoustic bands of the spectrum, for which the dispersion laws, generally speaking, differ in all directions of vector $\vec{k}$.

In the direction of the axes of symmetry of the lattice, for each frequency ω there are one longitudinally polarized and two transversely polarized waves propagating with the velocities $v_{son\,\parallel}$, $v_{son\,\perp 1}$, and $v_{son\,\perp 2}$ respectively. The directions of polarization of two last waves are mutually perpendicular. The velocities of propagation of lateral oscillations along an axis of symmetry of a fourth or higher order do not depend on which of two mutually orthogonal directions has the displacement vector $\vec{\xi}$ in a plane perpendicular to $\vec{k}$. Two laterally polarized bands then become degenerated: $v_{son\,\perp 1} = v_{son\,\perp 2}$. Along such directions of the wave vector $\vec{k}$, the constant-frequency surfaces for the transversely polarized bands of the spectrum touch one another.

For an arbitrary direction in a three-dimensional anisotropic lattice, the displacement vector is not strictly parallel or strictly perpendicular to the wave vector. As in a two-dimensional lattice, polarization of this band of the spectrum in an arbitrary direction can be conditionally determined by the polarization of that band into which it passes to a continuous rotation of $\vec{k}$ in the direction of the axis of symmetry.

In an isotropic three-dimensional medium without dispersion, the constant-frequency surfaces for the three acoustic bands of the spectrum are concentric spheres. For laterally polarized oscillations, the spheres are twice degenerated relative to two mutually perpendicular polarizations $\left(v^{0}_{son\,\perp 1} = v^{0}_{son\,\perp 2}\right)$

$$
\begin{aligned}
k_x^2 + k_y^2 + k_z^2 &= \frac{\omega^2}{\left(v^{0}_{son\,\parallel}\right)^2} \\
k_x^2 + k_y^2 + k_z^2 &= \frac{\omega^2}{\left(v^{0}_{son\,\perp 1}\right)^2} = \frac{\omega^2}{\left(v^{2}_{son\,\perp 2}\right)^2}
\end{aligned}
\tag{1.42}
$$

Sections through constant-frequency surfaces by various planes passing through the centre of symmetry are shown in Fig. 25 for the lattice of aluminium. The points of degeneration of two laterally polarized branches of the spectrum are located on the directions of the type $\langle 001 \rangle$. The figure shows constant-frequency surfaces typical of face-centered cubic lattices.

The thermal energy ε^{III}_{lat} of a three-dimensional lattice in the region of intermediate temperatures can be estimated, similar to a two-dimensional case, without accounting for the anisotropy and discreteness of the lattice. Under such an approximation, the branches of transversely polarized oscillations are degenerated and the lattice is characterized by two parameters: $\Theta_{D\,\parallel}$ and $\Theta_{D\,\perp}$, which are Debye temperatures corresponding to the longitudinal and transverse oscillations.

The region of intermediate temperatures is limited from above by the inequality $T \ll \min\{\Theta_{D\parallel}, \Theta_{D\perp}\}$. The procedure of calculating thermal energy is similar to that considered in Secs. **1-1** and **1-2**.

It can be shown that the number of normal oscillations excited in the lattice at temperature T is proportional to T^3 and the ther-

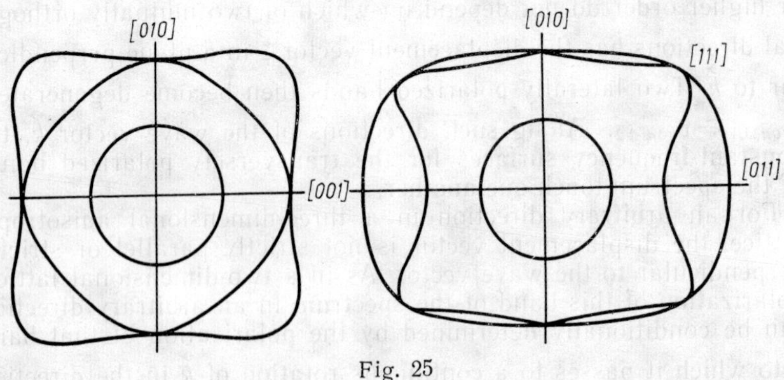

Fig. 25

mal energy of the lattice, $\varepsilon_{lat}^{III} \sim T^4$. Hence, Debye's law for the specific heat capacity of a three-dimensional lattice is:

$$C_v^{III} = \frac{\partial \varepsilon_{lat}^{III}}{\partial T} \sim T^3 \qquad (1.43)$$

1-8. THE SPECTRAL DENSITY OF PHONONS

Heat capacity of a lattice at any temperature can be calculated from the general expression (1.31) for the thermal energy of lattices. Let us remind that summation must here be made over the allowed values of wave vectors in each band of the oscillation spectrum.

As has been shown in Secs. **1-5** and **1-7**, the sets of allowed wave vectors along the Cartesian axes for lattices of the simplest type in cases of one, two and three dimensions are equidistant, and the wave vectors themselves are distributed in the $\vec{k}$-space with a constant density. In particular, for a three-dimensional case, the volume of $\vec{k}$-space per one state is $\frac{2\pi}{L_x} \cdot \frac{2\pi}{L_y} \cdot \frac{2\pi}{L_z} = \frac{8\pi^3}{V}$, where V is the total volume of the crystal. It then follows that the number of allowed states in a unit volume of $\vec{k}$-space (the density of allowed states) is $1\Big/\frac{8\pi^3}{V}$. This quantity is determined only by the number of dimensions of the lattice and is independent of the type of lattice and shape of crystal.

The set of allowed wave vectors is practically quasi-continuous, since the number of atoms N_a in a lattice is enormous. Thus, we can use integration instead of summation with respect to $\vec{k}$ within each band of the spectrum in formula (1.31), so that ε_{lat} will be written in the following form

$$\varepsilon_{lat} = \sum_{\vec{k}} \frac{\hbar\omega(\vec{k})}{e^{\hbar\omega(\vec{k})/kT}-1} \to \sum_{i=1}^{3\gamma} \left(\frac{V}{8\pi^3} \int_{\vec{k}} \frac{\hbar\omega_i(\vec{k})}{e^{\hbar\omega_i(\vec{k})/kT}-1} d\vec{k} \right)$$

Calculation of ε_{lat} is possible if the law of dispersion $\omega_i = \omega_i(\vec{k})$ is known for each band of the spectrum. Dispersion law is usually calculated theoretically for a simplest model of lattice.

It is advisable to pass over from the integration with respect to $\vec{k}$ to that with respect to frequency ω in formula (1.44) by introducing a new unknown function $D(\omega)$ by means of the relationship

$$\frac{V}{8\pi^3} \sum_{i=1}^{3\gamma} d\vec{k}_i = D(\omega)\, d\omega \tag{1.44}$$

$D(\omega)$ is termed the distribution function, or the spectral density of phonons. The quantity $D(\omega)\,d\omega$ determines the total number of oscillations per a frequency interval from ω to $\omega + d\omega$ and $\int_0^\infty D(\omega)\, d\omega$ is equal to the total number of normal oscillations or the number of degrees of freedom of the lattice. The thermal energy of the lattice can then be written as follows:

$$\varepsilon_{lat} = \int_0^\infty \frac{\hbar\omega}{e^{\hbar\omega/kT}-1} D(\omega)\, d\omega \tag{1.45}$$

Various experimental methods have been developed for determining the spectral density of phonons (for instance, by scattering of slow neutrons or X-rays in a lattice).

Thus, $D(\omega)$ being known, ε_{lat} can be calculated for the given temperature by formula (1.45) by numerical integration. An analytical expression for $D(\omega)$ can only be obtained for the simplest models of phonon spectrum, i.e. Einstein's and Debye's models.

In Einstein's model, which has been introduced in Sec. **1-6**, the whole spectrum of oscillations of a lattice is approximately described by oscillations with a single characteristic frequency ω_E. The spectral density of phonons $D^E(\omega)$ in this model is obviously proportional to the δ-function of argument $\omega - \omega_E$:

$$D^E(\omega) = B\delta(\omega - \omega_E) \tag{1.46}$$

The proportionality factor B is determined by the number of degrees of freedom of the lattice. For instance, for a crystal with two atoms in a cell

$$\int_0^\infty D^E(\omega)\,d\omega = B = 6N_a \tag{1.47}$$

In the Debye model of phonon spectrum, a real crystal is approximated by an isotropic dispersionless medium in which the spectrum of oscillations is limited by the magnitude of the wave vector within the region $|\vec{k}| \leqslant k_D$. Debye's wave number k_D is a characteristic parameter of the model and is found from the condition that the total number of allowed states within a sphere of radius k_D must be equal to the number of normal oscillations in one of the bands of the spectrum. With the assumptions underlying Debye's model, we can write

$$\frac{V}{8\pi^3} \cdot \frac{4}{3}\pi k_D^3 = 3N_a \tag{1.48}$$

or

$$k_D = 2\pi \left(\frac{9n_a}{4\pi}\right)^{1/3} \tag{1.49}$$

where n_a is the concentration of atoms in the lattice. The spectral density $D^D(\omega)$ in Debye's model is determined, according to (1.44), by the following function:

$$D^D(\omega) = \begin{cases} \dfrac{3\omega^2 V}{2\pi^2 v_0^3} & \text{at} \quad \omega \leqslant \omega_D \\ 0 & \text{at} \quad \omega > \omega_D \end{cases} \tag{1.50}$$

where $\omega_D = v_0 k_D$ and

$$\frac{3}{v_0^3} = \frac{1}{(v^0_{son\parallel})^3} + \frac{2}{(v^0_{son\perp})^3}$$

A diagram of Debye's function $D^D(\omega)$ is shown in Fig. 26. Its shape differs substantially from that of the spectral density of phonons in real solids. For comparison, Fig. 27 shows the function $D(\omega)$ for aluminium obtained by Walker [46] in 1956.

Debye's theory of heat capacity based on the modular spectral function (1.50) enjoys large success, possibly owing to that heat capacity depends only weakly on the specific kind of phonon distribution function, since this function enters only the integrand.

The exact form of the spectral density and especially the position of peaks of $D(\omega)$ is of great importance in determining many other physical characteristics of solids (for instance, when studying superconductivity of metals, interpreting magneto-acoustic effects, experimenting in tunnel spectroscopy, etc.). In this connection, it is of interest to study what qualitative peculiarities of the phonon spectrum of a real crystal are not reflected in the modular function $D^D(\omega)$ (1.50). For that purpose, let us analyse the assumptions used in Debye's model.

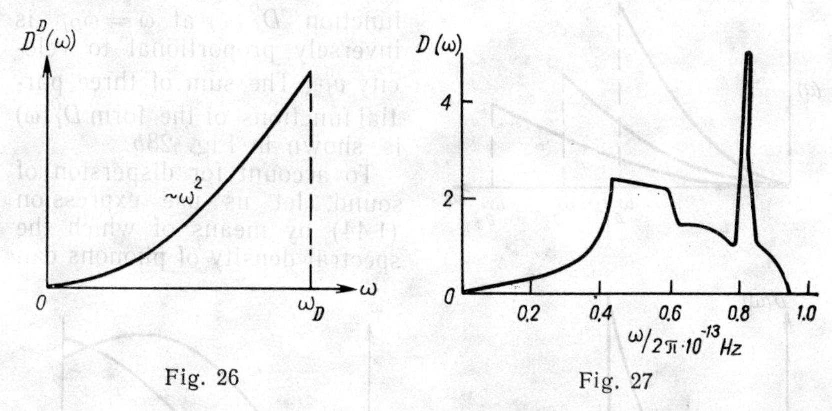

Fig. 26

Fig. 27

To begin with, we have to take into account that sound velocities in a real crystal are different for each band of the spectrum of acoustic oscillations. Therefore, the spectral densities of phonons in various bands of the spectrum must differ from each other.

The total spectral density $D^D(\omega)$ of acoustic oscillations of a crystal with a single atom in a unit cell should be represented as the sum of three partial functions of the form

$$D_j^D(\omega) = \frac{\omega^2 V}{2\pi^2 \left(v_j^0\right)^3}, \qquad j = 1,\ 2,\ 3$$

which include the velocities of propagation of oscillations, $v^0_{son\,\|}$, $v^0_{son\,\perp 1}$, and $v^0_{son\perp\,2}$ corresponding to each band of the spectrum.

The boundary frequencies ω_{Dj} which limit each of the partial functions $D_j^D(\omega)$ are found from the condition

$$\int\limits_0^{\omega_{D_j}} D_j^D(\omega)\, d\omega = N_a$$

Thus, the areas under each of the curves $D_j^D(\omega)$ are equal, and therefore, a greater value of sound velocity v_j^0 has a corresponding greater boundary frequency ω_{Dj} (see Fig. 28*a*)

$$\omega_{Dj} = v_j^0 \left(\frac{6\pi^2 N_a}{V} \right)^{1/3}$$

Without account of anisotropy and dispersion of the medium at $\omega < \omega_{\max j}$, the relationships $\omega = v_j^0 |\vec{k}|$ and $\omega_{Dj} = v_j^0 k_{\max}$ hold true. Under such an assumption, the height of a peak of the function $D_j^D(\omega)$ at $\omega = \omega_{Dj}$ is inversely proportional to velocity v_j^0. The sum of three partial functions of the form $D_j^D(\omega)$ is shown in Fig. 28*b*.

To account for dispersion of sound, let us use expression (1.44) by means of which the spectral density of phonons can

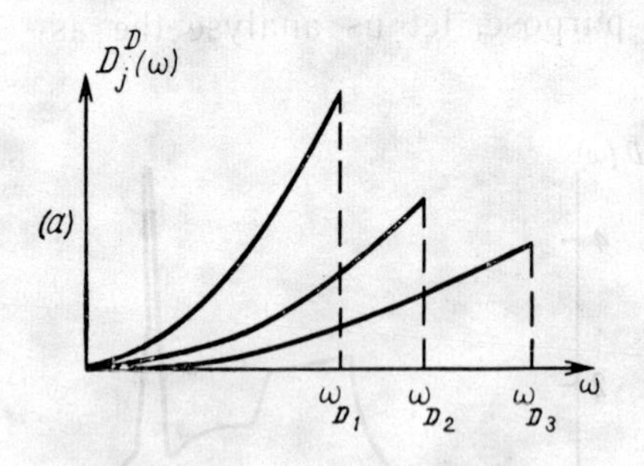

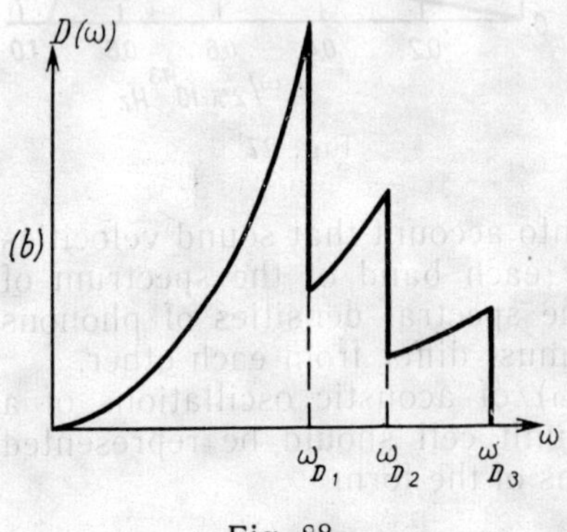

Fig. 28

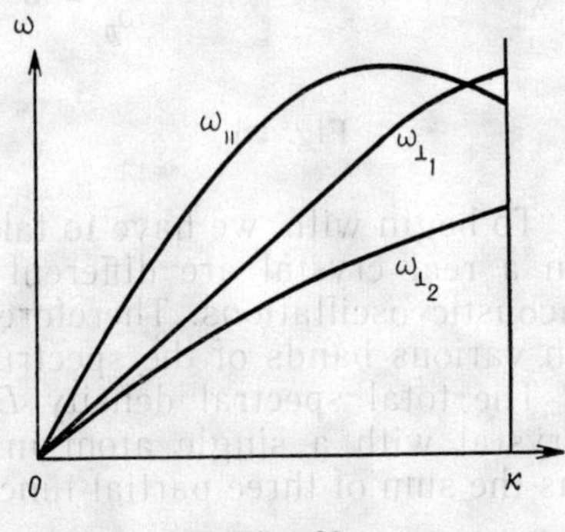

Fig. 29

be introduced. As can be seen from this expression, with a fixed direction of wave vector $\vec{k}$, $D_j(\omega)$ is proportional to the derivative $\frac{\partial |\vec{k}|}{\partial \omega}$ or inversely proportional to the group velocity of propagation of oscillations. Thus, a reduction of sound velocity must result in a departure of the function $D_j(\omega)$ from the parabolic form $D_j^D(\omega)$.

Dispersion curves for ω as a function of $|\vec{k}|$ corresponding to a fixed direction of k for each band of the acoustic spectrum in a three-dimensional lattice have a more complex shape than the simplest dispersion curve for a monatomic chain (see Fig. 29).

In particular, the group velocity of oscillations in some or other band may not turn to zero in one case, whereas in another case

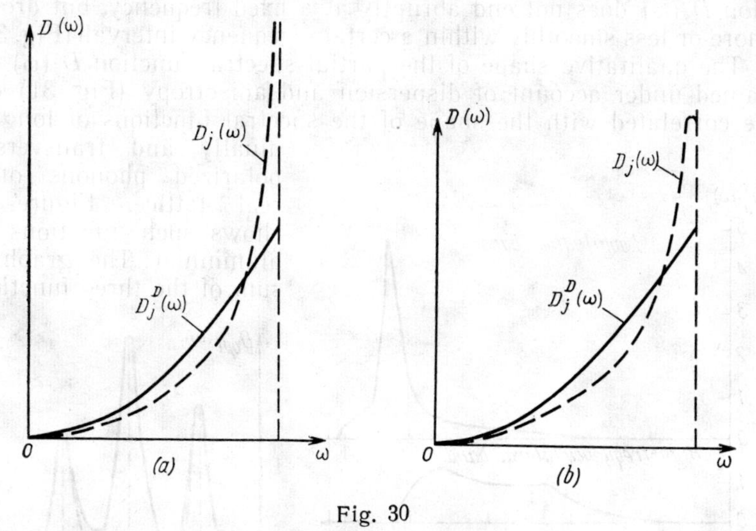

Fig. 30

$\frac{\partial \omega}{\partial |\vec{k}|}$ turns to zero at the boundary value of wave vector or its smaller values. With $\frac{\partial \omega}{\partial |\vec{k}|}$ turning to zero at a definite frequency $\omega = \omega'$, this causes a sharp increase of $D(\omega)$ near this frequency and an infinite singularity at $\omega = \omega'$. In the simplest case, ω' coincides with the boundary frequency $\omega_{\max j}$.

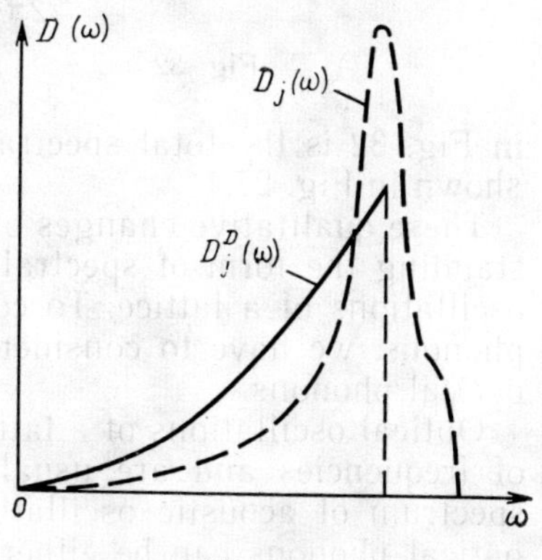

Fig. 31

The shape of the function $D_j(\omega)$ under such conditions is shown qualitatively by the dotted line in Fig. 30*a*, where the solid line is the function $D_j^D(\omega)$ without account of dispersion. Let us recall that the areas under both curves must be equal. In a real case, the infinite singularity of $D_j(\omega)$ in point $\omega = \omega'$ transforms into a more or less pronounced peak (Fig. 30*b*).

Finally, it must be taken into account that the surfaces of constant frequency in *k*-space in an anisotropic lattice can differ strongly from the spherical shape and, as a consequence, the boundary wave vector will have different values for different directions in the crystal. This circumstance results in that the function $D_j(\omega)$ does not end abruptly at a fixed frequency, but droops more or less smoothly within a certain frequency interval (Fig. 31).

The qualitative shape of the partial spectral function $D_j(\omega)$ obtained under account of dispersion and anisotropy (Fig. 31) can be correlated with the shape of the spectral functions of longitudinally and transversely polarized phonons of a real lattice. Figure 32 shows such functions for aluminium. The graphical sum of the three functions in Fig. 32 is the total spectral density of phonons for aluminium, shown in Fig. 27.

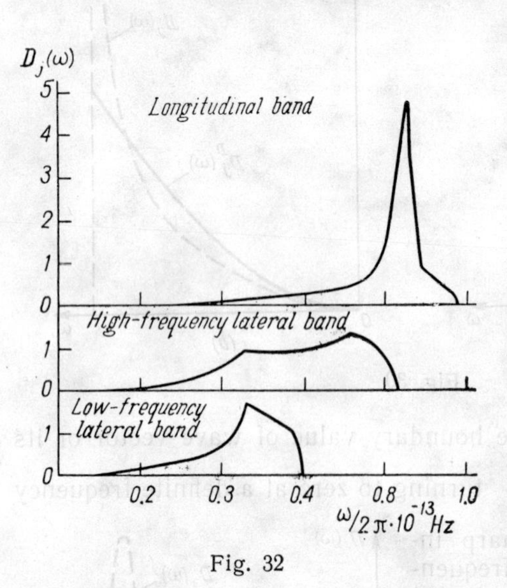

Fig. 32

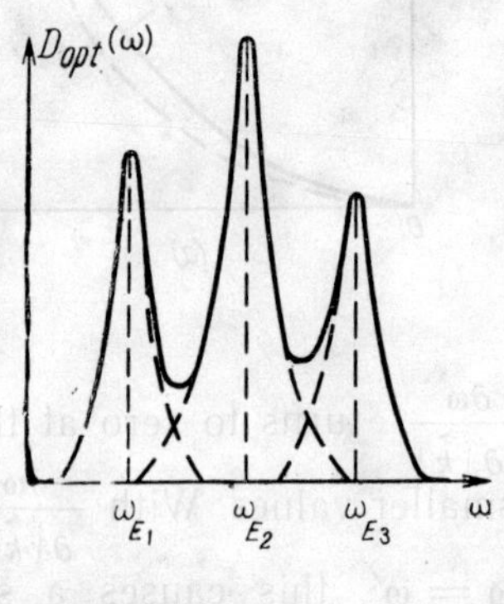

Fig. 33

These qualitative changes of Debye's function can help in understanding the form of spectral density of the acoustical portion of oscillations of a lattice. To construct the total spectral density of phonons, we have to consider the composition of the spectrum of optical phonons.

Optical oscillations of a lattice occupy a relatively narrow range of frequencies and are usually separated from the frequency of spectrum of acoustic oscillations. In a three-dimensional crystal, optical phonons can be either longitudinally or transversely polarized. As has been indicated, the spectral density of an optical band of oscillations can be approximated in Einstein's model with a spectral function of a certain characteristic frequency ω_E. A partial spectral density of optical phonons of one kind of polarization in a real crystal can be represented by a curve with a sharp maximum at the characteristic frequency ω_{Ej}, the area below the curve being equal to N_a. The total spectral density of optical phonons in the general case (Fig. 33) is the sum of three curves (dotted lines) with the maxima at characteristic frequencies ω_{E1}, ω_{E2}, and ω_{E3}.

Acoustical and optical phonons play different parts in a solid. At low or intermediate temperatures only acoustical oscillations are mainly excited in the lattice. They are responsible for thermal energy of the body. Optical phonons are excited only at temperatures $T \gtrsim \Theta_D$. It is usual to say that optical phonons "freeze out" at a decrease of temperature.

1-9. INTERACTION OF PHONONS

Monochromatic phonons of different frequencies that were considered in Sec. **1-5** do not interact with each other and form a perfect gas of quasi-particles. But this concept of phonons is only valid for harmonic oscillations of atoms in a lattice. Actually, atoms are not harmonic oscillators, the degree of departure from harmonicity (anharmonism) determining the effectiveness of interaction of phonons.

Anharmonism of oscillations of atoms can also be related to the expansion of bodies at heating. The coefficient of linear expansion may be a measure of anharmonism of oscillations. Let us study this phenomenon in more detail.

The potential energy U of a harmonic oscillator is a parabolic function of displacement ξ of an atom from its equilibrium position at $r = r_0$ $(\xi = r - r_0)$:

$$U(\xi) = \int_0^{\xi} \beta z \, dz = \frac{\beta \xi^2}{2} \tag{1.51}$$

where β is the coefficient of quasi-elastic restoring force. Relationship (1.51) is a symmetrical function of ξ (Fig. 34).

At $T = 0°$ K the oscillator occupies the lowest energy level in the parabolic potential pit, which corresponds to the energy of zero oscillations. With an increase of temperature, the oscillator passes onto the higher and higher energy levels. On each level, the distance between the bands of the parabola determines the doubled amplitude of oscillations (in Fig. 34, W_T denotes the total energy of oscillator at temperature T). But this increase of the amplitude of oscillations cannot cause an atom to be displaced from its equilibrium position $r = r_0$ (which determines its time-average position), and therefore, cannot change the mean distances r_0 between atoms. This means that under harmonic approximation no thermal expansion of bodies exists.

A real curve of potential energy is assymmetrical. It can be approximated by a parabola only near the bottom of the potential pit (Fig. 35). With an increase of interatomic distance the energy of interaction varies more slowly than would be expected from the parabolic law, while with a decrease of that distance it changes more rapidly than by that law. The nature of deviation of potential energy from the parabolic relationship can be described by a correction proportional to ξ^3:

$$U(\xi)=\frac{1}{2}\beta\xi^2-\frac{1}{2}g\xi^3 \tag{1.52}$$

The coefficient g characterizes the degree of anharmonicity of oscillations of atoms.

As can be seen from Fig. 35, an increase of temperature results now not only in an increase of the amplitude of oscillations, but

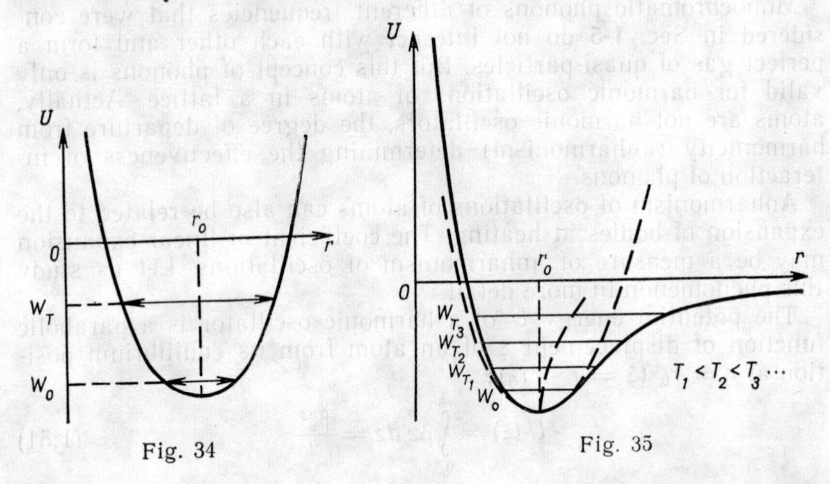

Fig. 34

Fig. 35

also in a displacement of the equilibrium position in the direction of higher values of r_0. This causes thermal expansion of bodies. Let the coefficient of linear expansion α in the known relationship $l=l_0(1+\alpha T)$ (where l and l_0 are linear dimensions of a body at a temperature T and at $T=0°$ K respectively) be correlated with the anharmonicity coefficient g. The mean displacement $\bar{\xi}$ from the equilibrium position $r=r_0$ at temperature T can be calculated by means of Gibbs' distribution function for physical parameters of oscillator, $f=De^{-\frac{U(\xi)}{kT}}$, where D is a constant:

$$\bar{\xi}=\frac{D\int\limits_{-\infty}^{+\infty}\xi e^{-\frac{U(\xi)}{kT}}d\xi}{D\int\limits_{-\infty}^{+\infty}e^{-\frac{U(\xi)}{kT}}d\xi}\approx\frac{\sqrt{\pi}}{4}\cdot\frac{\left(\frac{2kT}{\beta}\right)^{1/2}\frac{g}{kT}}{\left(\frac{2\pi kT}{\beta}\right)^{1/2}}=g\frac{kT}{\beta^2} \tag{1.53}$$

[the approximate expression (1.52) for $U(\xi)$ being used here]. The relative elongation of the body at temperature T is:

$$\frac{l-l_0}{l_0}=\frac{\xi N_a}{r_0 N_a}=\frac{gkT}{r_0\beta^2}=\alpha T \tag{1.54}$$

Hence

$$\alpha=\frac{k}{r_0\beta^2}\,g \tag{1.55}$$

The coefficient of linear expansion of bodies is directly proportional to the anharmonicity coefficient and can be used to find the latter.

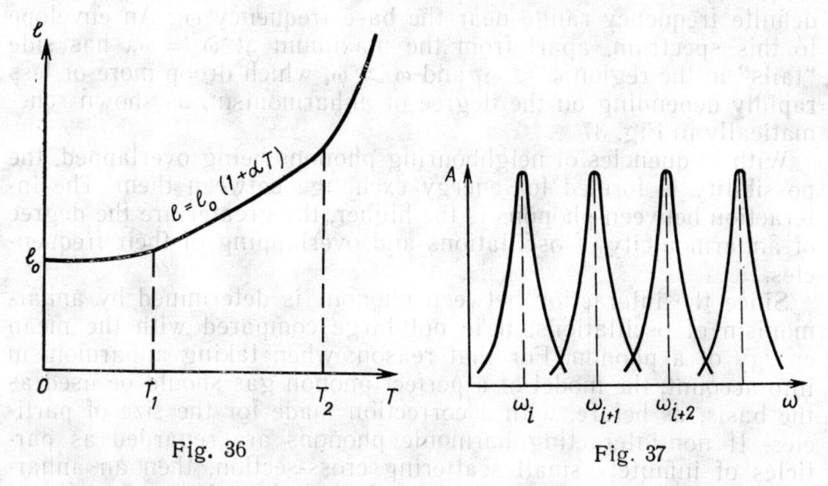

Fig. 36

Fig. 37

A quantum-mechanical analogue of expression (1.53) can be obtained by the correspondence principle by replacing the mean energy of a classic oscillator with frequency ω, which is equal to kT, by the mean energy of a quantum oscillator (1.26) $\bar{\varepsilon}=$ $=\frac{\hbar\omega}{e^{\frac{\hbar\omega}{kT}}-1}$. Thus

$$\alpha T=\frac{\bar{\xi}}{r_0}\rightarrow g\,\frac{\bar{\varepsilon}}{r_0\beta^2}=\frac{g}{r_0\beta^2}\cdot\frac{\hbar\omega}{e^{\frac{\hbar\omega}{kT}}-1} \tag{1.56}$$

We can expect from expression (1.56) that the coefficient of linear expansion must decrease abruptly as soon as the temperature drops below the excitation temperature $T\approx\hbar\omega/k$ of the given oscillator. At $T\rightarrow 0°$ K, α will tend to zero. This pattern of variation of the coefficient of linear expansion has actually been observed in experiments.

A linear pattern of variation of the length of a rod with temperature should only take place within a certain range of medium temperatures (Fig. 36). At low temperatures, α tends to zero and thermal expansion is practically absent. On the other hand, higher anharmonic terms proportional to ξ^5, ξ^7, etc. become more essential in the dependence of U on ξ at temperatures near the melting point T_{melt}. Account of these terms gives a deviation from the linear law of thermal expansion.

Let us now consider how the interaction of phonons is related to anharmonism. Appearance of anharmonism results in that the frequency spectrum of each phonon becomes continuous within a definite frequency range near the base frequency ω_i. An envelope to this spectrum, apart from the maximum at $\omega = \omega_i$, has side "tails" in the region $\omega < \omega_i$ and $\omega > \omega_i$ which droop more or less rapidly depending on the degree of anharmonism, as shown schematically in Fig. 37.

With frequencies of neighbouring phonons being overlapped, the possibility is formed for energy exchange between them. The interaction between phonons is the higher, the greater are the degree of anharmonicity of oscillations and overlapping of their frequencies.

Since the interaction between phonons is determined by anharmonism of oscillations, it is not large compared with the mean energy of a phonon. For that reason, when taking anharmonism into account, the model of a perfect phonon gas should be used as the basis, as before, with a correction made for the size of particles. If non-interacting harmonic phonons are regarded as particles of infinitely small scattering cross-section, then an anharmonic phonon must be assigned a finite value of the effective radius of scattering r_{eff}. It is usually assumed that this value is proportional to the first power of g. By analogy with scattering of particles in a weakly non-perfect gas, the constant of phonon-phonon interaction, $\mu_{ph\text{-}ph}$, is taken to be

$$\mu_{ph\text{-}ph} = \pi r_{eff}^2 \cdot n_{ph} \sim \pi g^2 n_{ph}$$

where n_{ph} is the density of phonons. The free-path length of phonons l_{ph} is inversely proportional to $\mu_{ph\text{-}ph}$. With a decrease of temperature, l_{ph} increases sharply owing to a strong reduction of n_{ph} (the "freezing-out" effect).

CHAPTER TWO

THE SYSTEM OF VALENCE ELECTRONS

2-1. THE BEHAVIOUR OF THE ELECTRON SYSTEM IN A METAL

As has been mentioned in the Introduction, the most characteristic properties of metals are connected with the presence of free electrons in them. But the term "free electrons in metal" is incorrect and only implies that valence electrons are not bonded with individual atoms and can travel over the whole volume of the metal. The behaviour of an individual electron in the metal can then differ substantially from that of a free isolated electron in vacuum.

The adiabatic approximation makes it possible to consider the system of valence electrons in a crystalline lattice by neglecting the motion of ion-cores. The concept of a perfect (or weakly imperfect) phonon gas, in turn, makes it possible to assume that the crystalline lattice is strictly periodical. Thus the system of valence electrons can be regarded as a system of $\varkappa N$ quasi-free charged particles in an ideal crystalline lattice. This electron system, in addition, is confined within a potential box formed by the surface of the crystal.

Let us see what are the peculiarities of the behaviour of the system of electrons in a metal.

The first peculiarity is connected with that an electron is an elementary particle with a spin equal to $^1/_2$, because of which the system of electrons obeys Pauli's exclusion principle and is described by Fermi-Dirac statistics. In a system of fermions, each energy level can only be occupied by two particles with opposite spins. Thus, the system of quasi-free electrons filling the lattice can be either a Fermi gas or a Fermi liquid. The liquid and gaseous models are treated in more detail in the next section.

The second peculiarity of behaviour of the electron system in a metal is related to the wave nature of electrons. It consists in that, with any small interaction of an electron with each ion of the lattice separately, its interaction with the whole lattice at a certain value of the momentum may become critical for the dynamics of its motion, because of interference of the electron waves reflected from individual ions. At these values of momentum, an electron wave cannot propagate in the lattice, the results being the appearance of bands of allowed and forbidden energy levels and formation of a band energy spectrum of electrons. Construction and study of band spectra is the main topic of most chapters in this book.

Finally, in order to describe the system of electrons in a metal, we have to consider the interaction of electrons with each other and with external fields.

The first problem will be solved within the limits of the one-electron approximation. Transition from the description of the electron system in a metal to that of an individual electron is made by introducing the concept of self-consistent field (the Hartree approximation) in which the given electron is present. The substantiation of this procedure is beyond the scope of the book. We have only to note that each electron in a metal is in a field of periodic potencial $V(\vec{r})$ whose period coincides with that of the crystal lattice.

The interaction of conduction electrons with the fields external relative to the metal (electric, magnetic, etc.) will be considered under the quasi-classical approach, which is applicable when the characteristic dimensions of the path of an electron exceed substantially its de Broglie wavelength λ_B.

Let us discuss it in more detail. When speaking of the motion of electrons in external fields, it is natural to assume these fields such that do not alter substantially the structure of the crystal proper. The latter means that the forces set up by the fields are small compared with the interatomic or intercrystalline forces. Such fields should vary only slightly at distances of the order of the lattice constant a, i.e. should be sufficiently smooth and uniform. With such assumptions, the path of an electron usually exceeds substantially the dimensions of the atom.

On the other hand, as will be shown later, interaction with external fields is mainly due to electrons having a wavelength $\lambda_B \sim a$.

Thus, the quasi-classical approach holds true for external fields, which are of interest for us.

2-2. THE CONCEPT OF FERMI SURFACE

According to Fermi-Dirac statistics, the distribution of electrons over energy levels ε_n in a metal is described by the function of the form

$$f_n = \frac{1}{e^{\frac{\varepsilon_n - \varepsilon_F}{kT}} + 1}, \quad n = 1, 2, \ldots \tag{2.1}$$

This distribution means that in a system consisting of N electrons, $N/2$ of the lowest energy levels are occupied at the temperature of absolute zero and the absence of external excitation, with two electrons being present on each level. The energy boundary of filling is the energy level $\varepsilon = \varepsilon_F$ corresponding to the Fermi boundary energy; this level enters as a parameter into the distribution function (2.1). At $T = 0°$ K, all the energy levels below ε_F are filled and those higher than ε_F, empty.

The Fermi energy ε_F in a system consisting of N electrons coincides with the energy level having the number $n = N/2$. Therefore, an increase of the number of electrons N in the system, all other conditions being equal, results in an increase of ε_F. The distribution of electrons over levels (2.1) is only related to the observance of the Pauli principle and is independent of the interaction between electrons and the dimensions of the potential box within which the system is confined. The latter factors determine the magnitude of the energy levels ε_n proper.

According to the Pauli principle, an electron can only interact with an external perturbation (electric, magnetic, or thermal) when the energy state into which it will pass upon this interaction is free. Otherwise the external field will be unable to change the energy of the electron.

Let the change of energy of an electron in an external field be $\Delta\varepsilon$. Since all the levels below Fermi energy are occupied at $T = 0°$ K, only those electrons can pass to the free level at interaction with the external field that are located at a depth not greater than $\Delta\varepsilon$ from ε_F. At $\Delta\varepsilon \ll \varepsilon_F$, the proportion of such electrons is of the order of magnitude of $\Delta\varepsilon/\varepsilon_F$.

In real metals, the magnitude of Fermi energy ε_F can reach 5-10 eV. Let us estimate the variation $\Delta\varepsilon$ of the energy of electrons at their interaction with external fields.

An increment of the energy of electron, $\Delta\varepsilon$, in an electric field $\vec{E}$ along the free-path length l is $|e|El$, which is practically equal numerically to 10^{-4}-10^{-6} eV.

With the classical motion of a charged particle in a magnetic field $\vec{H}$ its energy does not change, since the Lorentz force $F_H = \frac{e}{c}|\vec{v}\vec{H}|$ acting on the charge is always directed perpendicular to the velocity $\vec{v}$ of motion. The energy variation $\Delta\varepsilon$ of an electron in a magnetic field can be explained by the quantum nature of its motion and is equal to $\hbar\omega$, where $\omega = \frac{|e|H}{m_0 c}$ (m_0 being the mass of free electron) is termed the cyclotron, or Larmor, frequency.

Experimentally attainable magnetic fields are of the order of magnitude of approximately 10^6 oersteds. The energy variation in such fields for most metals does not exceed 10^{-3} ε_F. At interaction with the oscillations of the lattice at a temperature T the energy of an electron varies by a magnitude of the order of kT ($kT \sim$ ~ 0.03 eV at room temperature). Therefore, the ratio kT/ε_F does not exceed 0.05 even at the maximum temperatures at which a metal is still solid ($T \sim 2000°$ K). In that connection the system of electrons in a metal is degenerated.

These estimations show that from the total number of N electrons in a system, only a small portion of particles which are located within a narrow energy layer near the Fermi level participate in interaction with external fields. In other words, when considering the interaction of electrons with external perturbations, we can neglect the majority of electrons which are in states sufficiently distant from Fermi energy. This assumption is incorrect only for such external actions as interaction of electrons with emissions or fast particles when the energy of interaction may substantially exceed ε_F. Not only all valence electrons, but also the electrons of atomic structures can participate in these processes.

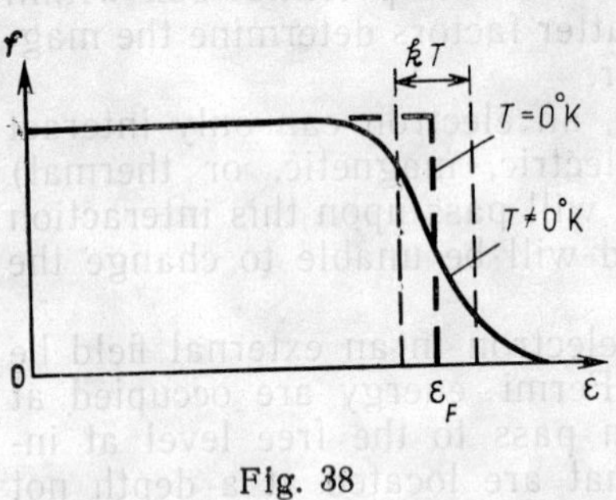

Fig. 38

With a temperature other than absolute zero, the boundary of Fermi distribution becomes "blurred" by a magnitude of an order of kT. The function of Fermi-Dirac distribution for a finite temperature is shown in Fig. 38, where the dotted lines denote the Fermi step at $T = 0°$ K.

In a system consisting of N electrons ($N \sim 10^{23}$ cm^{-3}), the distances between energy levels are of the order of $\varepsilon_F/N \approx$ $\approx 10^{-22}$ eV, i.e. negligible compared with ε_F. For that reason the energy spectrum of electrons in a metal is usually said to be quasi-continuous.

Let us now consider the peculiarities of the models of a perfect Fermi gas and Fermi liquid for the electron system in a metal. As has been indicated, the difference between these models is connected with positions of the energy levels ε_n. In a model of non-interacting particles (the model of a perfect Fermi gas), the total energy of the system is composed of the energies of individual electrons, while the energy levels ε_n are independent of the number of particles and coincide with the energy levels of an individual electron which is confined within the potential box formed by the faces of the crystal.

When a number of particles with energies close to ε_F are added to or removed from the system, the energy states of other particles of a perfect Fermi gas do not change. In other words, the loss of a particle remains "unnoticed" by the gas system from the standpoint of its energy spectrum.

In a case of strongly interacting particles (a Fermi liquid) [1]), the addition or removal of a number of particles changes not only the total energy of the whole system by the energy of those particles, but also the energy states of the remaining particles. Thus, at a variation of the number of particles a liquid system undergoes a complete energy recombination. Let it be emphasized once more that the distribution of electrons over energy levels is independent of the model, and therefore, both for a perfect gas or a perfect liquid, only a small portion of particles of a system, of the order of $\Delta\varepsilon/\varepsilon_F$ can respond to external perturbations. For the system of electrons in a metal considered as an integral whole, only the model of Fermi liquid is applicable, since the energy of Coulomb interaction between electrons is of the order of ε_F.

Landau's theory of Fermi liquids is based on that the energy of strongly bonded particles at a small deviation from the equilibrium (main) state of a system can be represented as a sum of energies of certain interacting fictitious particles termed quasi-particles.

The concept of quasi-particles introduced to describe the system of electrons in a metal is quite similar to that of quasi-particles, phonons, which is used to describe the oscillations of a system of strongly bonded atoms in a lattice.

Let us recall that transition to phonons is based on expanding the energy of interaction of atoms in a lattice into degrees of displacement of atoms from equilibrium with an accuracy to second-order terms. Since first-order terms are absent in this expansion (the energy of the main state corresponds to the minimum of energy of interaction), the quadratic form obtained is the energy of a system of connected oscillators. A linear transformation of coordinates which makes it possible to diagonalize the quadratic form leads to a system of non-interacting oscillators (phonons), the energy of the system being the sum of the energies of individual oscillators.

Landau's idea consists in that a homogeneous system, composed of a large number of particles, has excited states of the same type as an oscillating lattice. In other words, the properties of any system can be described by means of the model of quasi-particles.

Statistics of quasi-particles is not related uniquely to the statistics of the particles of a system. For instance, phonons can be described by Bose-Einstein statistics irrespective of the spin of the atoms constituting the lattice. Quasi-particles introduced to describe the system of electrons in a metal have a half-integral

[1]) The theory of Fermi liquid was developed by L. D. Landau.

spin and obey Fermi-Dirac statistics. Thus, according to Landau's theory, quasi-particles in a metal behave like a perfect Fermi gas.

The physical properties of a metal (i.e. the nature of its behaviour relative to external actions) is determined by the quasi-particles located near the Fermi level. With each exciting action in such a system, a quasi-particle appears above the Fermi level and simultaneously a free energy level (hole) is formed below that level.

The behaviour of elementary excitations may in general differ substantially from that of free electrons. The main peculiarity of elementary excitations is that their lifetime τ (which is determined by the probability of scattering $\omega \sim 1/\tau$) decreases rapidly at moving further from the Fermi level. If p_F is the momentum of a quasi-particle at the Fermi level and p is an arbitrary value of momentum, then $\tau \sim \frac{1}{(p - p_F)^2}$. It then follows that quasi-particles having a sufficiently long lifetime are located at energy levels near $\varepsilon = \varepsilon_F$.

For simplicity, quasi-particles in a metal will be further called simply "electrons", this term taking into account all the peculiarities connected with the quasi-particular description of the electron system.

By considering the elementary excitations only in direct vicinity of Fermi energy we can largely simplify the description of the system of electrons in a metal. Instead of determining the dispersion law $\varepsilon = \varepsilon(\vec{p})$ in the general case for the whole momentum space, it is now sufficient to determine the relationship between the energy and momentum at a constant value of energy $\varepsilon = \text{const} = \varepsilon_F$.

The equation $\varepsilon(\vec{p}) = \text{const} = \varepsilon_F$ in momentum space determines the surface of constant energy which is termed the Fermi surface. This may be visualized as a rigid structure on which (or in direct vicinity of which) a gas of elementary excitations is located. The properties of the electrons with the energy $\varepsilon = \varepsilon_F$ are fully determined by the kind (shape) of Fermi surface and the nature of its variation at small variations of Fermi energy.

Let us remind that in order to use the Fermi surface for description of the electron system in a metal, it is only needed that the Pauli principle be obeyed and the density of electrons be high (of an order of $1/a^3$), which ensures that the condition $\varepsilon_F \gg \Delta\varepsilon$ is fulfilled at an interaction with external fields.

Since in the absence of external fields the total energy of a non-relativistic electron is $\varepsilon = \frac{p^2}{2m_0}$, the Fermi surface for a system of free electrons is a sphere of radius $p_F = \sqrt{2m_0\varepsilon_F}$. Such a mo del of the system of free electrons is convenient for illustration of various properties of degenerated Fermi systems and will be employed more than once in this book.

2-3. THE EFFECT OF CRYSTAL LATTICE ON THE MOTION OF AN ELECTRON

Let us now consider the interaction of electrons with a crystal lattice. Note, first of all, that the potential of an individual ion of a metal lattice is more short-acting than the Coulomb potential $\varphi = -\frac{\varkappa |e|}{r}$ of an equivalent isolated ion (here $\varkappa |e|$ is the charge

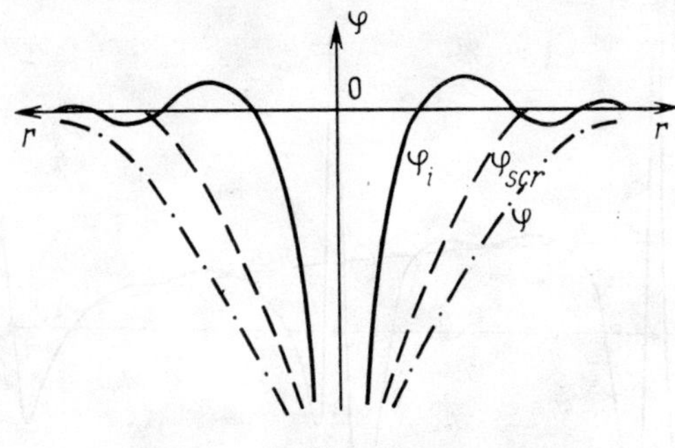

Fig. 39

of the ion). Screening of the potential of individual ions is observed owing to the presence of collective valence electrons filling the lattice. The formula to describe the screened potential of an ion, proposed initially by Debye, is as follows:

$$\varphi_{scr} = -\frac{\varkappa |e|}{r} e^{-r/r_D} \tag{2.2}$$

where r_D is a certain effective radius of interaction between an electron and an ion.

By modern views, screening with electrons having an energy close to the Fermi energy results in that the potential φ_i of an ion is not only short-acting, but also oscillating. Curves of these three potentials, φ, φ_{scr}, and φ_i, are shown in Fig. 39.

Owing to the screening effect, the potential energy of an electron can be taken constant over a substantial portion of the lattice space, and the motion of the electron in that part of space regarded as that of a free particle. At distances from the centres of ions exceeding the effective radius of screening, the wave function ψ of an electron is close to the wave function ψ^0 of a free electron, which is a plane wave function of the form $\psi^0 = \text{const}\ e^{i\vec{kr}}$ (where

$\vec{k}$ is the wave vector of a free electron). This namely is the main principle underlying the quasi-free electron approximation that makes it possible to describe many properties of conduction electrons in metals.

The state of an electron in direct vicinity of the centre of an ion is similar to its bound state, i.e. its wave function oscillates strongly and is close in its shape to the atomic wave function ψ_α.

Thus, the wave function ψ of an electron in the lattice can be regarded as a combination of a plane wave and quickly oscillating wave functions of bound states inside ion-cores.

The real part of the wave function of an electron in the lattice is shown diagrammatically in Fig. 40, where the dotted line repre-

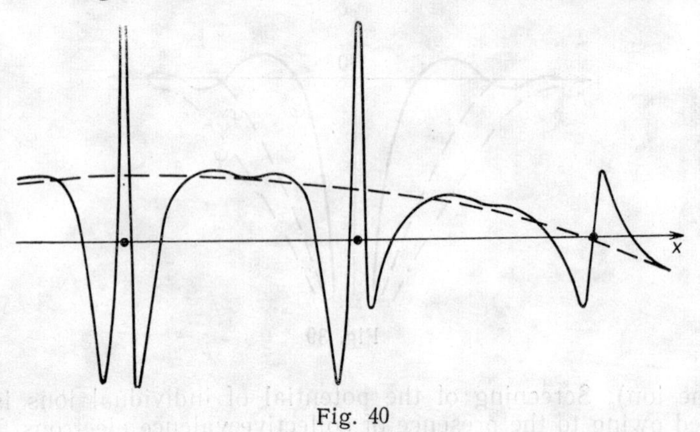

Fig. 40

sents the plane wave and solid circles, the centres of ions. A correct wave function ψ of an electron must reflect the fact that valence electrons form no bound states near individual ions. In accordance with the Pauli principle, the function ψ must then be orthogonal with the wave functions of bound states ψ_α. This condition is satisfied by the following combination of a plane wave and functions ψ_α:

$$\psi = \text{const}\left[e^{i\vec{k}\vec{r}} - \sum_{\alpha} \psi_\alpha(\vec{r}) \int \psi_\alpha^*(\vec{r}')\, e^{i\vec{k}\vec{r}'}\, d\vec{r}'\right] \tag{2.3}$$

A function of the form (2.3) is termed an orthogonalized plane wave (OPW). We have actually described the idea behind the OPW method which was proposed in its initial form by Herring in 1939 and has become one of the most widely used methods to describe electrons in metals. Apart from being visual physically, this method is also convenient in that an orthogonalized plane wave is in itself a sufficiently close approximation of the wave function of an electron. In other words, expansion of a wave function into OPW functions gives a rapidly converging series.

The OPW method is the basis of another method, that of pseudo-potential, which makes it possible to describe and analyse the pattern of motion of valence electrons in a metal.

Let us consider in more detail the physical aspect of pseudo-potential. A strong increase of the kinetic energy of a conduction electron inside an ion compensates almost completely the negative potential energy of the electron in the electric field of an atomic nucleus.

The total energy of the electron inside the ion turns to be equivalent to the total energy of an almost free electron in a field of a weak potential which cannot form a bound state. For that reason, instead of describing the motion of an electron with a complex wave function shown in Fig. 40 through deep potential pits of nuclei (Fig. 39), we can sufficiently accurately describe the motion of the electron through the whole volume of the crystal by a wave function which is close to a plane wave propagating in the field of a weak effective potential (pseudo-potential).

Pseudo-potential also has the periodicity of the crystalline lattice. Let us emphasize that the concept of pseudo-potential has been introduced only as a means to describe the motion of an electron from the standpoint of physical consequences. The motion of a real electron in a lattice is complicated and therefore cannot be reduced to a weakly excited motion of an almost free electron, since excitation inside atomic structures are quite strong. But, because of conservation of energy and momentum at intersections of ion-cores, the electron passes beyond the screening radius practically in the same state as at the entry to the field of ion.

If we disregard the processes occurring inside an ion, then the motion of an electron in a metal lattice will be equivalent to the motion of a weakly excited quasi-free particle with what is called the pseudo-wave function (which is close to a plane wave weakly modulated with the period of the lattice).

The true potential $V(\vec{r})$ in which an electron moves in the crystal lattice is a periodic (with the period of the lattice) repetition of screened potentials of individual ions having the form φ_i (Fig. 39). The motion of an electron in a field of $V(\vec{r})$ is described by Schrödinger's equation:

$$\left[-\frac{\hbar^2}{2m_0}\nabla^2 + V(\vec{r})\right]\psi = \varepsilon_n \psi \tag{2.4}$$

whose solution is the wave function ψ of an electron shown in Fig. 40. (Here ε_n are eigenvalues of the energy of electron, dependent on quantum number n.)

Substitution of an expansion of the wave function ψ into a series of OPW functions into Eq. (2.4) after certain transformation gives Schrödinger's equation with a new potential $V_{eff}(\vec{r})$ for a wave function $\tilde{\psi}$, which is a combination of only plane waves of the type $e^{i(\vec{k}+\vec{g})\vec{r}}$ (where $\vec{g}$ is a vector multiple of $\frac{2\pi}{|\vec{a}|}$, $\vec{a}$ being the period of the lattice in a definite direction).

This circumstance was first noticed by Phillips and Kleinman [50], who not only found the form of pseudo-potential $V_{eff}(\vec{r})$, but also showed its matrix elements calculated for the wave functions of a free electron are small compared with the corresponding energy levels of a free electron.

Thus, solution of the equation of pseudo-potential can be carried out by means of the perturbation theory using the wave functions of a free electron as a zero approximation.

We have described the method of pseudo-potential purely schematically, without touching its drawbacks. The actual situation is far from being so simple. The introduction of pseudo-potential is, on the one hand, not a single-valued operation, the final solution being substantially dependent on the selected form of pseudo-potential. On the other hand, the pseudo-potential itself is a non-local operator [it depends not only on the coordinate $\vec{r}$ as a real periodic potential $V(\vec{r})$ of the lattice, but is a functional of atomic wave functions ψ_α]:

$$V_{eff}(\vec{r})\,\tilde{\psi}(\vec{r}) = V(\vec{r})\,\tilde{\psi}(\vec{r}) + V_R(\vec{r})\,\tilde{\psi}(\vec{r})$$

where

$$V_R(\vec{r})\,\tilde{\psi}(\vec{r}) = \sum_\alpha (\varepsilon_n - \varepsilon_\alpha)\,\psi_\alpha(\vec{r}) \int \psi^*_\alpha(\vec{r}')\,e^{i\vec{k}\vec{r}'}\,d\vec{r}'$$

ε_α being the eigenvalue of the energy of atomic electrons.

A number of more or less substantiated forms of pseudo-potential, convenient for calculations, have been proposed (pseudo-potentials of A. Animalu [51], Heine-Abarenkov [52], etc.). Various aspects of the method of pseudo-potential were discussed by one of its authors, W. Harrison [53].

The introduction of pseudo-potential is a substantial mathematical simplification based on a certain approximate picture of the motion of electrons in the lattice. The smallness of pseudo-potential may explain why the electron in a metal can in many cases be regarded as a quasi-free particle. It is essential to emphasize here that the consequences obtained from the theory of pseudo-potential agree well with the known experimental data. The pseudo-potential of some or other metal cannot be calculated purely analytically or, as one says, from the first principles, but this is not necessary.

The theory of pseudo-potential is a powerful tool for determining some physical parameters of the energy spectrum of electrons (for instance, energy levels in various points of a crystal) in terms of other parameters which have been measured experimentally (for instance, zone overlapping, energy gaps, etc.), the shape and parameters of the pseudo-potential being then matched with the experimental data available.

In this way there were calculated the main levels of the energy spectrum of electrons for many metals and alloys, for instance, the metals of groups I, II, III, and IV of the Periodic System, semi-metals Bi, Sb, As, alloys Bi-Sb, Pb-Sn, etc.

Let us consider the peculiarities of the wave function of electrons in a metal. The periodic nature of the potential $V(\vec{r})$, which follows from the periodic structure of the crystal, gives the condition of translational invariance for the wave function of an electron

$$|\psi(\vec{r})|^2 = |\psi(\vec{r}+\vec{a})|^2 \tag{2.5}$$

where $\vec{a} = n_1\vec{a}_1 + n_2\vec{a}_2 + n_3\vec{a}_3$; $\vec{a}_1$, $\vec{a}_2$, $\vec{a}_3$ = vectors of the main translations in the lattice; n_1, n_2, n_3 = whole numbers.

Condition (2.5) corresponds to that the density of probability of finding an electron is a periodic function with the period of the lattice. In other words, it implies that the behaviour of an electron in the given cell of a crystal does not differ in any respect from its behaviour in any other cell.

As earlier, the influence of the boundaries of the crystal can be eliminated by introducing cyclic boundary conditions, which result in a periodic repetition of the crystal in all directions.

It follows from expression (2.5) that with the argument of the wave function ψ shifted by a vector $\vec{a}$, the wave function itself is multiplied by a phase multiplier $C(\vec{a})$:

$$\psi(\vec{r}+\vec{a}) = C(\vec{a})\,\psi(\vec{r}) \tag{2.6}$$

where $C(\vec{a})$ satisfies the condition

$$|C(\vec{a})|^2 = 1 \tag{2.7}$$

We make a shift by n_1 periods in the direction, say, of vector $\vec{a}_1$. Let $n_1 = n_1' + n_1''$, where n_1' and n_1'' are integers. The transition from point $\vec{r}$ to point $\vec{r} + n_1\vec{a}_1$ can be made either directly or consequently through an intermediate point $\vec{r} + n_1'\vec{a}_1$. In such cases the wave function $\psi(\vec{r}+n_1\vec{a}_1)$ can be given either as

$$\psi(\vec{r}+n_1\vec{a}_1)=C(n_1\vec{a}_1)\,\psi(\vec{r})$$

or as

$$\psi(\vec{r}+n_1\vec{a}_1)=C(n_1''\vec{a}_1)\,\psi(\vec{r}+n_1'\vec{a}_1)=C(n_1''\vec{a}_1)\,C(n_1'\vec{a}_1)\,\psi(\vec{r})$$

Hence

$$C(n_1\vec{a}_1)=C(n_1''\vec{a}_1)\,C(n_1'\vec{a}_1)$$

Therefore, the logarithm of the phase multiplier $C(n_1\vec{a}_1)$ is an additive function:

$$\ln C(n_1\vec{a}_1)=\ln C(n_1'\vec{a}_1)+\ln C(n_1''\vec{a}_1)$$

Comparing the last expression with the sum $n_1\vec{a}_1=n_1'\vec{a}_1+n_1''\vec{a}_1$, we can conclude that $\ln C(n_1\vec{a}_1)$ is proportional to $|n_1\vec{a}_1|$. With condition (2.7) taken into account, the multiplier $C(n_1\vec{a}_1)$ must be written in the form

$$C(n_1\vec{a}_1)=e^{in_1\vec{k}\vec{a}_1} \tag{2.8}$$

where $\vec{k}$ is a real vector. Thus, with the argument being shifted by a vector $\vec{a}$, the wave function is transformed as follows:

$$\psi(\vec{r}+\vec{a})=e^{i\vec{k}\vec{a}}\,\psi(\vec{r}) \tag{2.9}$$

Let us consider the function $u(\vec{r})$, equal to $e^{-i\vec{k}\vec{r}}\,\psi(\vec{r})$, and shift the argument by a vector $\vec{a}$:

$$u(\vec{r}+\vec{a})=e^{-i\vec{k}(\vec{k}+\vec{a})}\,\psi(\vec{r}+\vec{a})=e^{-i\vec{k}\vec{r}}\,\psi(\vec{r})=u(\vec{r})$$

As can be seen, the function $u(\vec{r})=e^{-i\vec{k}\vec{r}}\,\psi(\vec{r})$ is a periodic function, its period being equal to that of the lattice.

Thus, the wave function of an electron in a lattice in the most general case can be given in the form

$$\psi(\vec{r})=u(\vec{r})\,e^{i\vec{k}\vec{r}} \tag{2.10}$$

where the function $u(\vec{r})$ satisfies the condition of periodicity:

$$u(\vec{r}+\vec{a})=u(\vec{r}) \tag{2.11}$$

and vector $\vec{k}$ is a real vector.

The fact that the wave function of an electron in a crystal can be written in the form (2.10) is a general consequence of the translational symmetry of a lattice and constitutes the content of Bloch's theorem.

Expression (2.10) implies that the wave function has the form of a plane wave modulated with the lattice period. Obviously, the same form must have the pseudo-wave function $\tilde{\psi}$ which satisfies Schrödinger's equation with a pseudo-potential $V_{eff}(r)$. This follows from the periodicity of the pseudo-potential, its period being that of the lattice.

Since the pseudo-potential is small, it then follows additionally that the periodic function $u(\vec{r})$ must only slightly differ from a constant and must only weakly modulate the plane wave $e^{i\vec{k}r}$[1]. This makes it possible, when analysing the motion of an electron in a lattice, to use the model of almost free electrons as the first approximation. For this purpose it is sufficient to consider the motion of an electron in the field of an arbitrarily low effective periodic potential $V_{eff}(\vec{r})$.

Construction of Fermi surfaces for electrons in metals can be successfully made by using Harrison's method (described later in the book), which is based on the assumption that it is possible to pass over continuously from the behaviour of an electron in a lattice to that of a free electron, if the pseudo-potential is made to tend to zero. With a transition $V_{eff}(\vec{r}) \to 0$, vector $\vec{k}$, which enters the wave function (2.10) as a parameter, must transform into the wave vector of a free electron, which is referred to its momentum $\vec{p}$ by the relationship $\vec{p} = \hbar\vec{k}$. When $V_{eff}(\vec{r}) \neq 0$, vector $\vec{p}$ is termed the quasi-momentum of an electron in a lattice and is connected with $\vec{k}$ (2.10) by a similar relationship.

Let us consider the properties of the quasi-momentum that distinguish it from the momentum of a free electron. For this, we take a certain direction in the crystal structure which is given by a unit vector $\vec{n}_j$. Accordingly, let $\vec{a}_j$ and L_j be the minimum period of the crystal structure and its dimension in that direction. Because of the cyclic boundary conditions posed, the wave functions in points $\vec{r}$ and $\vec{r} + L_j\vec{n}_j$ are identical:

$$\psi(\vec{r}) \equiv \psi(\vec{r} + L_j\vec{n}_j)$$

or

$$u(\vec{r})\, e^{i\vec{k}\vec{r}} \equiv u(\vec{r} + L_j\vec{n}_j)\, e^{i\vec{k}(\vec{r}+L_j\vec{n}_j)}$$

[1]) The condition at which the perturbation introduced by an arbitrarily weak pseudo-potential ceases to be low coincides with the Wulff-Bragg condition for diffraction of electron waves on a crystal lattice. It determines the boundaries of energy bands and is considered in detail later in this book.

Hence, the relationship for vector $\vec{k}$ is:

$$e^{i(\vec{k}\vec{n}_j)L_j} = 1 \quad (2.12)$$

which implies that the projection of $\vec{k}$ onto the direction $\vec{n}_j$ must be multiple of $\frac{2\pi}{L_j}$:

$$(\vec{k}\vec{n}_j) = \frac{2\pi}{L_j} n \quad n = 1, 2, 3, \ldots \quad (2.13)$$

Therefore, the projection of the quasi-momentum $\vec{p}$ of an electron onto a certain direction $\vec{n}_j$ in the crystal can take a discrete set of values, of which the minimum one is equal to $\frac{2\pi\hbar}{L_j}$. That the quasi-momentum of an electron in a crystal is a discrete quantity is connected only with that the electron is confined within the potential box of finite dimensions, i.e. with the presence of boundaries in the crystal [1]).

From another property of a crystal, i.e. the translational symmetry of its lattice, there follows the multivaluedness of the quasi-momentum of an electron. The quasi-momentum in a definite direction $\vec{n}_j$ is a vector which can be determined with an accuracy to a whole number of vectors $\frac{2\pi}{a_j}\vec{n}_j$. This already follows from expression (2.12) which determines the projection $\vec{k}$ onto the direction $\vec{n}_j$.

Expression (2.12) will not change if a vector $\pm\frac{2\pi}{a_j}\vec{n}_j m$ with any integer m is added to vector $\vec{k}$. In other words, vectors $\vec{k}$ and $\vec{k}' = \vec{k} \pm \frac{2\pi}{a_j}\vec{n}_j m$ are equivalent to one another in their definition. With $\vec{k}$ being replaced by $\vec{k}'$, the wave function ψ of an electron retains the form of Bloch's function (2.10).

Indeed, let $\psi_{\vec{k}}(\vec{r}) = u(r)\, e^{i\vec{k}\vec{r}}$. We are to find the function $\psi_{\vec{k}'}(\vec{r})$ corresponding to vector $\vec{k}'$:

$$\psi_{\vec{k}'}(\vec{r}) = u(\vec{r})\, e^{i\vec{k}'\vec{r}} = u(\vec{r})\, e^{\pm i\frac{2\pi}{a_j}(\vec{n}_j\vec{r})m}\, e^{i\vec{k}\vec{r}}$$

---

[1]) This is an extra proof that the cyclic boundary conditions only remove the non-equivalence of the positions of various ions in a crystal, but cannot exclude the principal presence of boundaries. In particular, they do not imply the transition from a finite-size crystal to an infinite space filled with a crystal lattice.

The multiplier $e^{\pm i\frac{2\pi}{a_j}(\vec{n}_j\vec{r})m}$ is periodical with the period a_j, and therefore, can be included into the modulating function $u(\vec{r})$. Substituting the functions $\psi_{\vec{k}'}$ and $\psi_{\vec{k}}$ into Schrödinger's equation we can show that they are the solutions corresponding to the same values of the energy of an electron.

Thus, the wave vectors $\vec{k}$ and $k \pm \frac{2\pi}{a_j}\vec{n}_j m$, as also the quasi-momenta $\vec{p}$ and $\vec{p} \pm \frac{2\pi\hbar}{a_j}\vec{n}_j m$ are physically equivalent, i.e. correspond to the same physical state of the electron in the lattice. Hence it follows that the physically different values of the wave vector in the direction $\vec{n}_j$ are confined within an interval of the width $\frac{2\pi}{a_j}$.

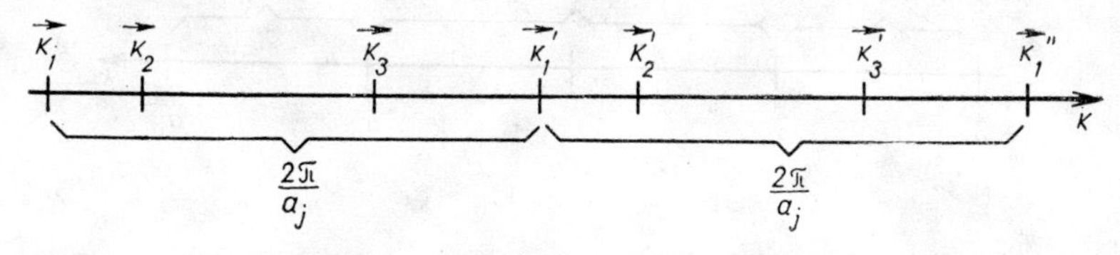

Fig. 41

For instance, vectors $\vec{k}_2$ and $\vec{k}_3$ in Fig. 41 are not equivalent to one another, whereas $\vec{k}_2$ is equivalent to $\vec{k}'_2$ and $\vec{k}_3$ is equivalent to $\vec{k}'_3$.

Since there is no current in the crystal in the absence of electric field, the values of wave vectors $\vec{k}$ and $-\vec{k}$ are equivalent. In this connection the interval including the physically different values of $\vec{k}$ must be symmetrical relative to the origin of coordinates in $\vec{k}$-space, for instance, it may extend from $-\frac{\pi}{a_j}$ to $+\frac{\pi}{a_j}$. This interval (in the case of a one-dimensional lattice) is termed the first Brillouin zone. The second Brillouin zone is located symmetrically with respect to the first zone and to the origin of coordinates and includes two intervals: from $-\frac{2\pi}{a_j}$ to $-\frac{\pi}{a_j}$ and from $+\frac{\pi}{a_j}$ to $+\frac{2\pi}{a_j}$; the third Brillouin zone includes the intervals from $-\frac{3\pi}{a_j}$ to $-\frac{2\pi}{a_j}$ and from $+\frac{2\pi}{a_j}$ to $+\frac{3\pi}{a_j}$, etc.

The total interval corresponding to a Brillouin zone of any number is always equal to $\frac{2\pi}{a_j}$ (Fig. 42).

In the case of a two-dimensional plane lattice with two main translational vectors a_x and a_y and the dimensions L_x and L_y of the crystal, the components of the wave vector along the axes k_x and k_y can acquire only discrete values from the sets:

$$k_x = \frac{2\pi}{L_x} n_x, \quad k_y = \frac{2\pi}{L_y} n_y \text{ (where } n_x \text{ and } n_y \text{ are integers)}$$

The physically different values of the components k_x are located within the interval from $-\frac{\pi}{a_x}$ to $+\frac{\pi}{a_x}$, and those of the components k_y, within the interval from $-\frac{\pi}{a_y}$ to $+\frac{\pi}{a_y}$. These conditions determine the set of vectors plotted from the origin of coordinates

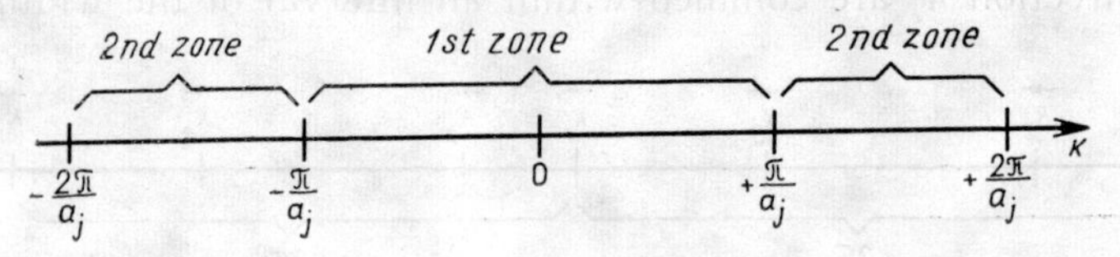

Fig. 42

to the vertices of the cells constructed within a rectangle of the dimensions $\frac{2\pi}{a_x}$ and $\frac{2\pi}{a_y}$ along the axes k_x and k_y respectively and located symmetrically relative to the origin of coordinates (Fig. 43).

Each cell corresponding to one allowed wave vector $\vec{k}$ has the dimensions $\frac{2\pi}{L_x}$ and $\frac{2\pi}{L_y}$ along the axes k_x and k_y.

All physically different values of a wave vector are located within a rectangle, such as shown in Fig. 43. This rectangle is called the first Brillouin zone for a plane two-dimensional lattice.

For a simple rectangular three-dimensional lattice, the first Brillouin zone can be constructed in the form of a parallelepiped bounded by plane surfaces (Fig. 44):

$$k_x = \pm \frac{\pi}{a_x}, \quad k_y = \pm \frac{\pi}{a_y}, \quad k_z = \pm \frac{\pi}{a_z}$$

The volume $v_{\vec{k}}$ of an elementary cell corresponding to a single value of the wave vector in $\vec{k}$-space is equal to

$$v_{\vec{k}} = \frac{2\pi}{L_x} \cdot \frac{2\pi}{L_y} \cdot \frac{2\pi}{L_z} = \frac{(2\pi)^3}{V}$$

where V is the volume of the crystal.

The corresponding volume $v_{\vec{p}}$ of an elementary cell in the space of quasi-momenta ($\vec{p}$-space) is

$$v_{\vec{p}} = \frac{(2\pi\hbar)^3}{V} \tag{2.14}$$

It is the volume of phase space relating to a single electron state in a crystal of volume V (the state is twice degenerated, since two electrons with oppositely directed spins can be present in it according to the Pauli principle).

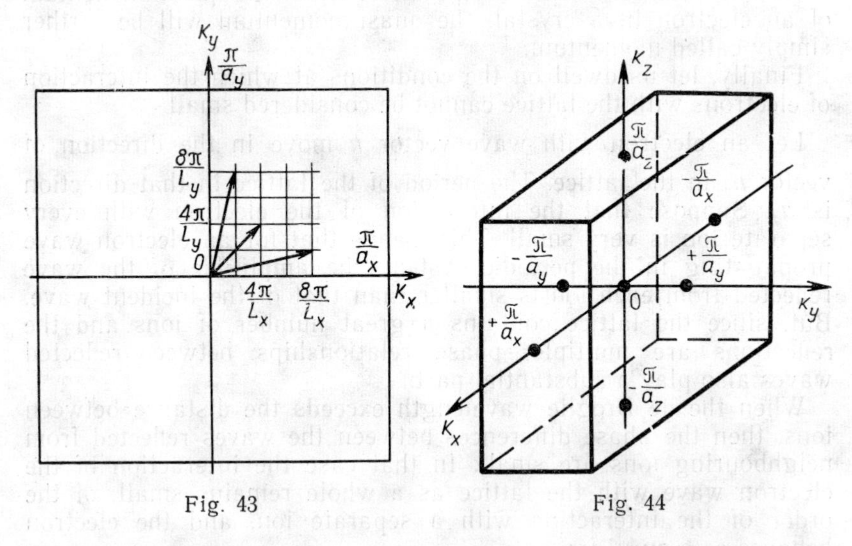

Fig. 43 Fig. 44

It follows from expression (2.14) that electrons are distributed with a constant density in the space of quasi-momenta. The magnitude of volume $v_{\vec{p}}$ is independent of the shape of crystal. Expression (2.14) obtained for a crystal in the form of a parallelepiped is the general one. This statement requires no strict proof and actually follows from that the conditions at the boundaries in a crystal of macroscopic dimensions (i.e. having a large number of elementary cells) must have no effect on the motion of electrons in its volume. The larger the crystal, the greater its volume-to-surface ratio and the greater part of the electrons do not "sense" the size and shape of the crystal. Within a sufficiently large crystal of any shape, we can always mentally separate a parallelepiped and remove the remaining parts. The states of electrons within this parallelepiped will not change noticeably (to an accuracy of the fraction of surface electrons). In order not to adhere to a particular crystal of a definite volume, all relationships are usually written for a crystal of a unit volume $V = 1$. In that case a volume equal to $(2\pi\hbar)^3$ corresponds to one state in $\vec{p}$-space.

Thus, if the electrons occupy a volume Δ in the space of quasi-momenta, their concentration n in the real space is

$$n = 2\frac{\Delta}{(2\pi\hbar)^3} \qquad (2.15)$$

Formula (2.15) takes into account that the spin of each electron state undergoes double degeneration.

Having thus elucidated the peculiarities of the quasi-momentum of an electron in a crystal, the quasi-momentum will be further simply called momentum.

Finally, let us dwell on the conditions at which the interaction of electrons with the lattice cannot be considered small.

Let an electron with wave vector $\vec{k}$ move in the direction of vector $\vec{n}_j$ in the lattice. The period of the lattice in that direction is a_j. Suppose that the interaction of the electron with every separate ion is very small. This means that for an electron wave propagating in the periodic system, the amplitude of the wave reflected from each ion is smaller than that of the incident wave. But, since the lattice contains a great number of ions and the reflections are multiple, phase relationships between reflected waves also play a substantial part.

When the de Broglie wavelength exceeds the distance between ions, then the phase differences between the waves reflected from neighbouring ions are small. In that case the interaction of the electron wave with the lattice as a whole remains small, of the order of the interaction with a separate ion, and the electron behaves as a quasi-free one.

With $m\lambda_B = 2a_j$ (where m is a whole number), the phase difference of reflected waves will be a number multiple of 2π. The waves reflected from various ions are then in phase with each other and their amplitudes add together. As a result, total reflection of the electron from the lattice will occur even with a small amplitude of the wave reflected from one ion. In other words, with $m\lambda_B = 2a_j$, an electron wave cannot propagate in the lattice.

The condition of total reflection coincides with the well-known Wulff-Bragg relationship for diffraction of electron waves in crystals. If the latter is written for the wave vector of an electron $|\vec{k}| = 2\pi/\lambda_B$, we obtain an equation defining the boundaries of Brillouin zones in $\vec{k}$-space

$$(\vec{k}\vec{n}_j) = \pm\frac{\pi}{a_j}m, \quad m = 1, 2, \ldots$$

Thus, the boundaries of Brillouin zones correspond to such values of the wave vectors (or momenta) of an electron at which the electron wave cannot propagate in the lattice. At approaching the boundary of a Brillouin zone (i.e. at an increase of $|k|$) a running electron wave is more and more decelerated by the lattice and becomes a standing wave at the boundary values of $\vec{k}$. This peculiarity of the interaction of an electron and lattice constitutes the principal distinction between the dynamics of an electron in a metal and that of a free electron.

2-4. THE DYNAMICS OF AN ELECTRON IN THE CRYSTAL LATTICE

The quasi-classical approximation for the description of the motion of an electron uses the common Hamilton function $\mathcal{H}(\vec{p}, \vec{r})$ with the classical momentum being replaced by quasi-momentum. Hamilton's equations are then written in the common manner[1]):

$$\vec{v} = \frac{d\vec{r}}{dt} = \frac{\partial \mathcal{H}}{\partial \vec{p}}, \quad \frac{dp}{dt} = -\frac{\partial \mathcal{H}}{\partial \vec{r}} \tag{2.16}$$

The first of them determines the velocity. The Hamilton function $\mathcal{H}(\vec{p}, \vec{r})$ is the sum of kinetic and potential energies:

$$\mathcal{H}(\vec{p}, \vec{r}) = \varepsilon(\vec{p}) + V(\vec{r})$$

This makes it possible to express the velocity of an electron through the dispersion law:

$$\vec{v} = \frac{\partial \varepsilon}{\partial \vec{p}} \quad \text{or} \quad \vec{v} = \text{grad}_{\vec{p}}\, \varepsilon \tag{2.17}$$

Expression (2.17) coincides with the relationship for the group velocity $\vec{v}_g$ of a wave packet in quantum mechanics. Indeed, for a quantum particle, $\varepsilon = \hbar\omega$ and $\vec{h} = \hbar\vec{k}$. Therefore, the group velocity $\vec{v}_g = \frac{\partial \omega}{\partial \vec{k}} = \frac{\partial \varepsilon}{\partial \vec{p}} = v$.

---

[1]) Here and further, differentiation of a scalar quantity over a vector implies the following operation: $\frac{\partial \mathcal{H}}{\partial \vec{p}}$ is a vector with the components $\left\{ \frac{\partial \mathcal{H}}{\partial p_x}, \frac{\partial \mathcal{H}}{\partial p_y}, \frac{\partial \mathcal{H}}{\partial p_z} \right\}$.

Hamilton's second equation (2.16) is an equation of motion and describes the variation of the momentum of an electron $\frac{d\vec{p}}{dt}$ under the action of a force $\vec{F} = -\frac{\partial \mathcal{H}}{\partial \vec{r}} = -\frac{\partial V}{\partial \vec{r}}$:

$$\frac{d\vec{p}}{dt} = \vec{F} \tag{2.18}$$

Dynamics of an electron in the lattice has a specific feature consisting in that the variation of the momentum $\frac{d\vec{p}}{dt}$ of the electron is only determined by the action of the force $\vec{F}_{ext}$ which is external relative to the lattice.

Let us show this for a particular example when an external electric field $\vec{E}$ acts on the electron. We shall calculate the work $\delta\varepsilon$ done by the external force $\vec{F}_{ext} = -|e|\vec{E}$ during time δt to an electron located on the Fermi surface. Since the velocity $v = \frac{\partial \varepsilon}{\partial \vec{p}}$ of the electron on the Fermi surface is very great, then the path travelled by the electron during the time δt can be calculated by neglecting the variation of this velocity.

The work $\delta\varepsilon$ is equal to $|e|\ Ev\delta t$. On the other hand, since ε is a function of the momentum p, then $\delta\varepsilon = \frac{\partial \varepsilon}{\partial p}\, \delta p$ (for simplicity, we consider a case when all the vectors, i.e. $\vec{v}$, $\vec{p}$, $\vec{E}$, and $\delta\vec{p}$, are collinear).

Equating the expressions for $\delta\varepsilon$, we get

$$\frac{\partial \varepsilon}{\partial p}\, \delta p = v\, \delta p = |\, e\, | Ev\, \delta t$$

whence

$$\frac{dp}{dt} = |\, e\, | E$$

Thus, the variation of the momentum is related to the action of only an external force $-|e|E$ and is independent of the internal forces acting on the electron from the lattice. But the variation of the group velocity $\vec{v}$ $\left(\text{kinematic acceleration of the electron } \frac{d\vec{v}}{dt}\right)$ is naturally determined by the sum of all external $\vec{F}_{ext}$ and internal $\vec{F}_{int}$ forces acting on the electron.

The equation of the Newton second law for the kinematic acceleration of the electron in the lattice can be written as

$$m_0 \frac{d\vec{v}}{dt} = \sum \vec{F}_{ext} + \sum \vec{F}_{int} \tag{2.19}$$

The equation for the variation of the momentum, determining the velocity of the electron $\frac{d\vec{p}}{dt}$ in the momentum space, has then the form

$$\frac{d\vec{p}}{dt} = \sum \vec{F}_{ext} \tag{2.20}$$

By comparing equations (2.19) and (2.20), we can see that the momentum $\vec{p}$ of an electron in the lattice and its velocity $\vec{v}$ are not related together by the common equation $\vec{p} = m_0\vec{v}$ which holds for a free electron. The velocity vector $\vec{v}$ is only determined by the dispersion law $\varepsilon(\vec{p})$ through equation (2.17) and in the general case is even not collinear with vector $\vec{p}$.

The kinematic acceleration of the electron $\frac{d\vec{v}}{dt}$ can be formally expressed solely through the external forces. Indeed, for the j-th component of $\frac{d\vec{v}}{dt}$ we have:

$$\left(\frac{d\vec{p}}{dt}\right)_j = \frac{d}{dt}\left(\frac{\partial \varepsilon}{\partial p_j}\right) = \sum_{i=1}^{3} \frac{\partial^2 \varepsilon}{\partial p_j \, \partial p_i} \cdot \frac{\partial p_i}{\partial t}$$

But according to (2.20) $\frac{dp_i}{dt} = \left(\sum F_{ext}\right)_i$. The derivatives $\frac{\partial^2 \varepsilon}{\partial p_j \, \partial p_i}$ form a symmetrical second-rank tensor. Denoting the inverse tensor by m^*_{ji} we can write [1])

$$\sum_{i=1}^{3} m^*_{ji} \frac{dv_i}{dt} = \sum (\vec{F}_{ext})_j \tag{2.21}$$

The tensor $\frac{\partial^2 \varepsilon}{\partial p_j \, \partial p_i}$ is called the tensor of inverse effective masses of electrons in a crystal. The effective masses m^*_{ji} determined by

---

[1]) Note that if the determinant of the matrix $\left\{ \frac{\partial^2 \varepsilon}{\partial p_j \, \partial p_i} \right\}$ in a certain point is equal to zero, then the inverse tensor m^*_{ji} in that point is non-existent.

it, as distinct from the mass of a free electron, depend in general on the momentum and are not constant quantities. Therefore, this tensor of effective masses describes the motion of an electron in the given point of $\vec{p}$-space under the action of the external force applied.

The transition from equation (2.19) to equation (2.21) consists in that the unknown term $\sum \vec{F}_{int}$ in (2.19) is transferred to the left-hand part and the difference vector $m_0 \frac{d\vec{v}}{dt} - \sum \vec{F}_{int}$ is formally expressed through the vector $\frac{d\vec{v}}{dt}$ by means of the second-rank tensor m^*_{ji}.

The unknown internal forces acting on the electron from the lattice are included by this technique into the definition of the tensor of effective mass (which then also remains unknown). But introduction of the effective mass tensor has not only the formal meaning which shows the difference in the laws of motion of a free electron and an electron in the lattice.

The inclusion of the internal forces into the definition of the effective mass tensor has also the physical meaning, since under certain conditions the components of tensor m^*_{ji} become constant quantities having like signs (in the system of the main axes of tensor m^*_{ji} in which its matrix is diagonal). The three main components of tensor m^*_{ji} in such cases become the important dynamic characteristics of the electron in the crystal and determine the nature of motion of the electron in external electric and magnetic fields.

Measurements of the components of m^*_{ji} by different independent experimental methods have shown their good coincidence. This proves that the components of m^*_{ji} are physical parameters of an electron in the crystal, similarly as m_0 is a physical parameter of a free electron.

Knowledge of the components of the effective mass tensor makes it possible to describe the motion of an electron under an assumption that it is acted upon only by external forces. Many formulae derived for an electron in free space are used in the lattice, with replacing the mass of the free electron by the respective component (or combination of components) of the effective mass tensor. When using the effective mass, one has, however, to take into account that the analogy between the behaviour of an electron in the lattice and that of an electron in free space, which is useful owing to its clear evidence, has only a restricted field of applicability. The use of the effective mass tensor in various formulae every time requires proper preliminary substantiation.

By the method of its introduction, the effective mass tensor in the given point of the momentum space is determined by the law of dispersion of electrons in metal $\varepsilon = \varepsilon(\vec{p})$. The question of the magnitude of the effective mass in some or other metal, as also the question of calculation of the effective mass under account of the characteristics of the metal, will be considered after description of the Harrison method of constructing the Fermi surfaces. We shall consider in the same section the influence of electron-phonon interaction on the magnitude of the effective mass of the electrons located at the Fermi level.

The use of the effective masses as physical parameters characterizing the electron in the metal is evidently expedient for those regions of $\vec{p}$-space in which the main components of tensor m^*_{ji} are practically independent of $\vec{p}$ and are constant (or nearly constant) quantities. Such conditions are observed, for instance, in the vicinity of points $\vec{p}_0$ in which the energy ε of an electron attains a relative maximum or minimum. These points of extrema of the energy play a substantial part in the description of the energy spectrum of electrons, since they correspond to the boundaries of the energy regions.

Let ε_0 be the energy of the electron in an extremum point. In the vicinity of this point, $\varepsilon(\vec{p})$ may be expanded into a series of powers of $(\vec{p}-\vec{p}_0)_j$. Since $\frac{\partial \varepsilon}{\partial p_j} = 0$ in extremum points, the terms of the first power of $(\vec{p}-\vec{p}_0)_j$ will be absent in the expansion. To an accuracy of quadratic terms, the expansion will then have the form:

$$\varepsilon(\vec{p}) = \varepsilon(\vec{p}_0) + \frac{1}{2}\sum_{ij} \alpha_{ij}(p_i - p_{0i})(p_j - p_{0j})$$

where $\alpha_{ij} = \frac{\partial^2 \varepsilon}{\partial p_i\, \partial p_j}$ at $\vec{p} = \vec{p}_0$.

Passing over to a system of coordinates related to the main axes of the tensor α_{ij}, we have

$$\varepsilon(\vec{p}) = \varepsilon_0 + \frac{1}{2}\sum_{i=1}^{3} \alpha_{ii}(p_i - p_{0i})^2 \tag{2.22}$$

In the vicinity of point $\vec{p}_0$, where we can limit ourselves to the quadratic terms in $(\vec{p}-\vec{p}_0)_i$, the components of the effective mass tensor are constant and equal to

$$m_{ii}^* = \frac{1}{\alpha_{ii}} = \left(\frac{\partial^2 \varepsilon}{\partial p_i^2} \bigg|_{\vec{p}=\vec{p}_0} \right)^{-1} \tag{2.23}$$

The values of m_{ii}^* are positive near a point of minimum energy and negative near a point of maximum. In the first case an electron in the lattice behaves qualitatively like a free electron: its kinematic velocity in an electric field increases in the direction of the acting force $-|e|\vec{E}$. In the second case the electron is accelerated in the direction opposite to that of the external force, i.e. behaves as a negative mass particle.

The behaviour of a charged particle in electric and magnetic fields is determined by the sign of its charge-to-mass ratio, rather than by that of its mass. For that reason, an electron of a negative mass behaves dynamically as a particle of a positive charge and positive mass in the vicinity of points of maximum energy.

If the origin of coordinates of $\vec{p}$-space is placed into a point of an extremum of energy $\vec{p}_0$ and the energy ε is calculated from ε_0 then equation (2.22) can be written as:

$$\varepsilon' = \sum_{i=1}^{3} \frac{p_i'^2}{2m_{ii}^*} \tag{2.24}$$

where

$$\varepsilon' = \varepsilon - \varepsilon_0 \text{ and } p_i' = p_i - p_{0i}$$

Expression (2.24) is an equation of a three-axial ellipsoid

$$\frac{p_1'^2}{2m_{11}^*\varepsilon'} + \frac{p_2'^2}{2m_{22}^*\varepsilon'} + \frac{p_3'^2}{2m_{33}^*\varepsilon'} = 1 \tag{2.25}$$

whose semi-axes are respectively equal to $\sqrt{2m_{11}^*\varepsilon'}$, $\sqrt{2m_{22}^*\varepsilon'}$, and $\sqrt{2m_{33}^*\varepsilon'}$ (Fig. 45).

Thus, the surfaces of constant energy $\varepsilon =$ const near the extremum points for electrons in a crystal are ellipsoidal surfaces. In a particular case, when $m_{11}^* = m_{22}^* = m_{33}^*$, the ellipsoid degenerates into a sphere. The components m_{11}^*, m_{22}^*, and m_{33}^* determine the effective masses of the electron at its motion along the axes $\vec{p}_1'$, $\vec{p}_2'$, and $\vec{p}_3'$ respectively.

If the momentum $\vec{p}$ of the electron is directed along the unit vector $\vec{n}_j$, then its effective mass $m_{\vec{n}_j}^*$ is proportional to the square of the segment of the straight line OA from the origin of coordinates O to the surface of the ellipsoid in the direction $\vec{n}_j$ (Fig. 45). The magnitude of $m^*_{\vec{n}_j}$ is equal to $\frac{(OA)^2}{2\varepsilon'}$.

Let us see what is the physical meaning of the effective mass. The difference between the effective mass m^*_{ji} and the mass of a free electron is evidently caused by the interaction of the electron with the crystal lattice (note that for a free electron $\varepsilon = \frac{p^2}{2m_0}$ and the main values of the tensor $m^*_{ii} = 1 \Big/ \frac{\partial^2 \varepsilon}{\partial p_i^2}$ simply coincide with m_0).

As has been shown in the previous section, the strongest perturbation of the motion of an electron is observed when its wave vector passes to the boundary of the Brillouin zone. In that case the electron wave in the direction of the normal to the zone boundary is completely braked by the lattice and becomes a standing wave, while the normal component of the wave group velocity $\vec{v}_g = \frac{\partial \varepsilon}{\partial \vec{p}}$ becomes zero.

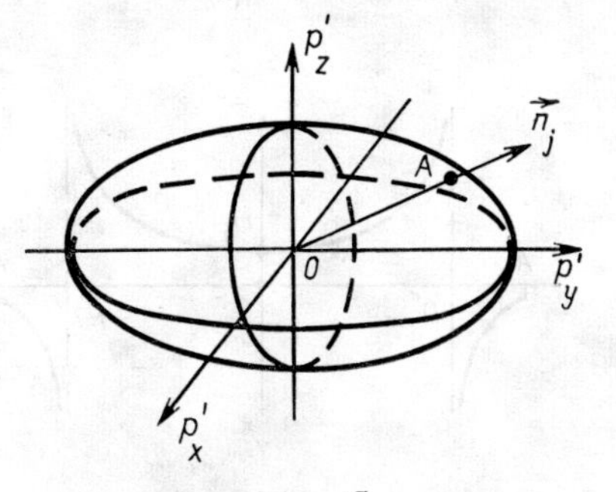

Fig. 45

The physical meaning of the effective mass can be understood by considering the process of acceleration of an electron in an external electric field at various values of its momentum.

If the momentum of the electron is initially small (the electron is near the centre of the first Brillouin zone), it behaves practically as a free electron: in an electric field the electron undergoes a common acceleration at which its velocity v increases in the direction of the external force. But at the same time, its momentum p increases continuously [see (2.20)] and de Broglie wavelength λ_B decreases. The electron is not only accelerated kinematically, but also approaches the boundary of the Brillouin zone in the momentum (phase) space.

This approach to the boundary results in that the reaction of the lattice to the electron wave becomes stronger, or in other words, the amplitude of the wave reflected from the lattice increases. The velocity of the electron placed into an external field $\vec{E}$ then increases not so rapidly as would the velocity of a free electron, since part of the accelerating external force is spent to withstand the reaction of the lattice. This means that as the momentum is being increased, the effective mass of the electron becomes greater than that of a free electron.

If the action of the electric field E lasts longer and no scattering of the electron still occurs, the momentum of the electron increases continuously in time [see (2.20)]. Since the reaction of the lattice also increases, the growth of the group velocity of the electron is decelerated more and more and is stopped at a certain value p' (Fig. 46a).

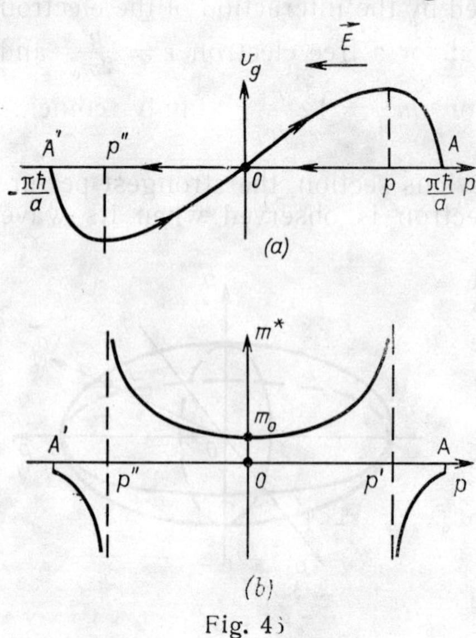

Fig. 4ͅ

The effective mass of the electron in this point inside the Brillouin zone turns to infinity (Fig. 46b) and its velocity attains a maximum. The sum of external and internal forces in equation (2.19) actually turns to zero at that moment and

$$\left.\frac{dv}{dt}\right|_{p=p'}=0.$$

The infinite magnitude of m^* shows that in equation (2.21) a finite external force $+|e|E$ causes no further growth of the velocity v in the positive direction of the axis p (m^* tends to infinity as $1\Big/\frac{dv}{dt}$, so that the product $m^*\frac{dv}{dt}$ remains equal to a finite value $+|e|E$). Note that the external field $\vec{E}$ in Fig. 46a and b has the direction $-\vec{p}$, so that the projection of the external force onto the $+\vec{p}$ axis is equal to $+|e|E$, where E is the magnitude of $\vec{E}$.

A further increase of the momentum at $p>p'$ results in the reaction of the lattice becoming greater than the external force. At this stage of acceleration of the electron in the external field the component of the electron wave reflected from the lattice increases most effectively. Under the action of the reaction of the lattice, the kinematic velocity of the electron begins to decrease and the sum of forces in equation (2.19) changes sign. In this region, the electron continues to move by inertia in the former direction and is braked in the external field (actually, it is braked by the reaction of the lattice). With respect to the external electric field, the electron behaves like a negative-mass particle (Fig. 46*b*).

At the boundary of the Brillouin zone (point A in Fig. 46*a*) the momentum acquires the magnitude $p=\frac{\pi\hbar}{a}$, and the kinematic velocity turns to zero: $v=\frac{\partial\varepsilon}{\partial p}=0$. If the action of the electric field continues more, the electron begins to be accelerated in the opposite direction. The point A in Fig. 46*a* is equivalent to point A'. The motion of the imaging point in phase space is now continued from point A' in the former direction. It is common to say that the electron is reflected from the boundary of the Brillouin zone, since its momentum at point A changes stepwise to a reverse one. The growth of velocity v in the inverse direction occurs up to the point $p = p''$ in which $\frac{dv}{dt}$ again becomes zero and the effective mass turns to minus infinity. Further, the mass of the electron changes sign and the electron begins again to be accelerated "normally" (as a positive-mass particle) up to the point p'.

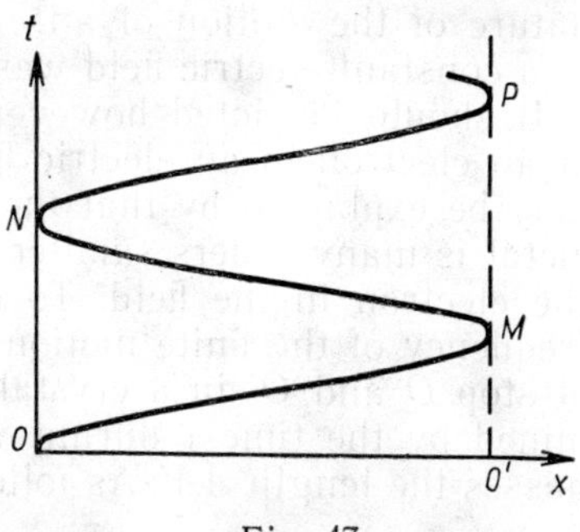

Fig. 47

Thus, under the action of a constant electric field E, the electron in the crystal lattice performs a periodic motion in $\vec{p}$-space, each time passing the Brillouin zone in the same direction which coincides with the direction of the force $-|e|\vec{E}$ acting on it. The electron is also in a periodic motion in the real coordinate space.

Indeed, having begun its motion in a definite point O in the crystal (Fig. 47) which corresponds to the point O in the phase space of Fig. 46*a*, the electron first moves in the positive direction of the x axis (in the forward direction) up to the point of stop O' which corresponds to the point A (or A') in the phase space, in which its velocity becomes zero. Then the electron moves in the reverse direction up to the next stop, when in the point O of the phase space its velocity again becomes zero.

When moving in the reverse direction, the electron is first accelerated against the action of the force $-|e|\vec{E}$ as a particle of a negative charge and negative mass, and is then braked by the same force $-|e|\vec{E}$. As a result of the reverse motion, it stops in the initial point O in the crystal. After that, the whole cycle of motion is repeated again from points O in the coordinate and phase spaces.

Thus, the motion of an electron in the coordinate space is an oscillatory motion in the direction collinear with the electric field vector between the two points O and O' in the crystal. The time pattern of this motion is shown in Fig. 47.

Thus, we have come to a paradoxical conclusion that the motion of an electron in a constant electric field is finite (i.e. cyclic). The electron in the lattice periodically passes through one and the same point of the crystal. Such a motion differs qualitatively from the motion of a free electron in a constant electric field. The motion of the free electron in the field is uniformly accelerated and infinite and the electron never returns to its initial point. The periodic nature of the motion of an electron in a lattice under the action of a constant electric field was first mentioned by I. M. Lifshits.

It should be noted however that the paradoxical cyclic motion of an electron in an electric field is practically not observed. This may be explained by that the free-path length of an electron in a metal is many orders smaller than the amplitude of oscillations of the electron in the field. To demonstrate it, let us determine the frequency of the finite motion and the distance between the points of stop O and O' in a crystal. The period of oscillations is determined by the time τ during which the imaging point in Fig. 46*a* passes the length AA'. As follows from equation (2.20) the time τ is

$$\tau = \int_{-\frac{\pi\hbar}{a}}^{+\frac{\pi\hbar}{a}} \frac{dp}{|e|E} = \frac{2\pi\hbar}{a|e|E}$$

The frequency of oscillations ν_E in an electric field is

$$\nu_E = \frac{1}{\tau} = \frac{a|e|E}{2\pi\hbar}$$

The amplitude of oscillations X is then equal to $\frac{\varepsilon_A - \varepsilon_0}{|e|E}$ where ε_A and ε_0 are the values of the energy of the electron in points A and 0 of phase space. The difference $\varepsilon_A - \varepsilon_0$ is the width of the energy band and is of an order of several electron-volts in metals. At feasible values of the electric field $E \sim 10^{-6}$ V/cm, the amplitude X attains approximately 10^6 cm, that is hundreds of millions times greater than the free-path length of an electron[1]). Oscillations

[1]) An electric field of $E \sim 10^{-6}$ V/cm corresponds to the current density in metal $j \sim 10^2$ A/cm² at unit resistance $\rho \sim 10^{-8}$ ohm·cm. The numerical example has been taken from the book: Lifshits I. M., Azbel M. Ya., Kaganov M. I., *Electron Theory of Metals* (in Russian). Moscow, "Nauka", 1971.

of such a gigantic amplitude would occur with an extremely low frequency

$$\nu_E = \frac{a\,|e|\,E}{2\pi\hbar} \sim 50 \text{ Hz}$$

With each act scattering, the momentum of the electron changes irregularly. For that reason a finite motion along a path exceeding the free-path length is impossible. At small sections of the path between successive acts of scattering, the electron naturally moves progressively. We have come to another paradoxical conclusion that the flow of a direct current in metals is caused by scattering of electrons. With no scattering, an electric current would not grow infinitely, as in free space, but an alternating current would be formed with a frequency depending on the magnitude of the electric field applied (close to the industrial frequency of 50 Hz).

2-5. DEPENDENCE OF THE ENERGY OF AN ELECTRON IN THE LATTICE ON QUASI-MOMENTUM

The dependence of the energy on the momentum of an electron at application of a weak periodic potential $V_{eff}(\vec{r})$ may be considered in the simplest way on a one-dimensional model in which the energy ε depends only on one component of the momentum $\vec{p}$. We shall proceed from the law of dispersion for a free electron: $\varepsilon = \frac{p^2}{2m_0}$.

An application of a weak periodic potential $V_{eff}(x)$ cannot change substantially the law of dispersion at small momenta at large distances from the boundaries of the Brillouin zone. An electron then moves as a free particle and its law of dispersion practically does not differ from the parabolic relationship $\varepsilon = p^2/2m_0$.

As the momentum approaches the boundary values $\pm\frac{\pi\hbar}{a}$, the electron wave is braked by the lattice, its velocity $v_g = \frac{\partial\varepsilon}{\partial p}$ decreases, and the effective mass of the electron increases. This results in that the relationship $\varepsilon(p)$ deviates from the parabola $p^2/2m_0$ towards smaller values of ε (Fig. 48).

At $p = p'$ (see Fig. 46*a* and *b*) where the effective mass turns to infinity, the curve $\varepsilon(p)$ has a bend point $\left.\frac{\partial^2\varepsilon}{\partial p^2}\right|_{p=p'} = 0$.

The relationship $\varepsilon(p)$ undergoes the greatest perturbation in the vicinity of the boundary momentum. Since at $p = \pm\frac{\pi\hbar}{a}$ the velocity of the electron becomes zero, the curve $\varepsilon(p)$ must approach the boundary with the zero slope of the tangent (Fig. 48):

$$\left.\frac{\partial \varepsilon}{\partial p}\right|_{p=\pm\frac{\pi\hbar}{a}}=0$$

Let us consider in more detail the behaviour of the energy of an electron near the boundary of the Brillouin zone.

The energy levels of an electron in a crystal at the momenta sufficiently distant from the boundary of the Brillouin zone can be calculated by the theory of perturbations; with the motion of a free electron taken as the motion not disturbed by a periodic potential $V_{eff}(\vec{r})$. The theory of perturbations makes it possible to determine qualitatively the dependence of energy on momentum also in direct vicinity of the boundary of the Brillouin zone. We shall make a brief calculation by the theory of perturbations for the one-dimensional model selected.

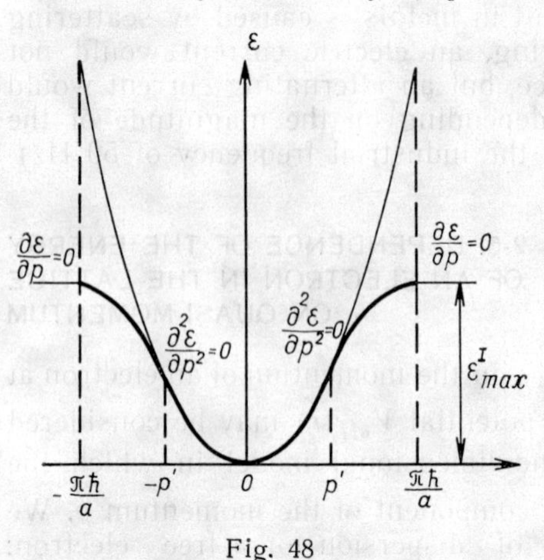

Fig. 48

With a zero approximation, the perturbation $V_{eff}(x)=0$. The wave functions of the zero approximation $\varphi_k^{(0)}(x)$ obey Schrödinger's equation for a free particle:

$$-\frac{\hbar^2}{2m_0}\cdot\frac{d^3}{dx^2}\psi_k^{(0)}(x)=\varepsilon^{(0)}\psi_k^{(0)}(x) \tag{2.26}$$

and are plane waves: $\psi_k^{(0)}(x)=Ae^{ikx}$. The constant A is found from the normalization condition

$$\int_0^L \psi_k^{(0)*}(x)\,\psi_k^{(0)}(x)\,dx=A^2\int_0^L dx=1$$

where L is the length of the "crystal" in the selected one-dimensional model. The wave function $\psi_k^{(0)}(x)$ is of the form:

$$\psi_k^{(0)}(x)=\frac{1}{\sqrt{L}}\,e^{ikx} \tag{2.27}$$

With the zero approximation, the energy spectrum is quasi-continuous:

$$\varepsilon^{(0)}=\frac{\hbar^2k^2}{2m_0}=\frac{p^2}{2m_0} \tag{2.28}$$

Let us recall that k can take a discrete set of values [see (2.7)]:

$$k = \frac{2\pi}{L} n, \text{ where } n = \pm 1, \pm 2, \ldots$$

According to the theory of perturbations, in order to find the energy and eigenfunctions in the first approximation, we have to calculate the matrix elements of the pseudo-potential:

$$(V_{eff})_{k'k} = \int_0^L \psi_{k'}^{(0)*}(x) V_{eff}(x) \psi_k^{(0)}(x)\, dx \tag{2.29}$$

It can be demonstrated that the matrix elements $(V_{eff})_{k'k}$ are other than zero only at definite values of k' and k. Since $V_{eff}(x)$ is a periodic function with the lattice period a, it can be expanded into a Fourier series of the form

$$V_{eff}(x) = \sum_n C_n e^{i \frac{2\pi}{a} nx} \tag{2.30}$$

Substituting (2.30) into (2.29) gives:

$$(V_{eff})_{k'k} = \frac{1}{L} \int_0^L e^{-ik'x} \sum_n C_n e^{i \frac{2\pi}{a} nx} e^{ikx}\, dx =$$

$$= \frac{1}{L} \sum_n C_n \int_0^L e^{i\left(k + \frac{2\pi}{a} n - k'\right) x} dx = \sum_n C_n \delta_{k',\, k + \frac{2\pi}{a} n} =$$

$$= \begin{cases} 0 \text{ at } k' \neq k + \dfrac{2\pi}{a} n \\ C_{n'} \text{ at } k' = k + \dfrac{2\pi}{a} n' \end{cases} \tag{2.31}$$

Thus, the matrix elements $(V_{eff})_{k'k}$ differ from zero only at $k' = k + \frac{2\pi}{a} n'$. The correction $\varepsilon^{(1)}(k)$ to the energy $\varepsilon^{(0)}(k)$ in the first approximation is equal to the diagonal matrix element:

$$\varepsilon^{(1)}(k) = (V_{eff})_{kk} = \frac{1}{L} \int_0^L V_{eff}(x)\, dx = \overline{V}_{eff} \tag{2.32}$$

i.e. to the magnitude of the pseudo-potential averaged over the crystal. This correction is independent of wave vector. The energy spectrum of the electron in the first approximation of the theory of perturbations changes by a constant value $\overline{V}_{eff}$. This variation is not a principal one, since the constant shift of the energy levels $\overline{V}_{eff}$ can be accounted for by varying the origin of calculation of the energy.

According to the theory of perturbations, the value of energy at the second approximation can be written in the form

$$\varepsilon^2(k) = \varepsilon^{(0)}(k) + \varepsilon^{(\mathrm{I})}(k) + \varepsilon^{(\mathrm{II})}(k) =$$

$$= \varepsilon^{(0)}(k) + \bar{V}_{eff} + \sum_{k'} \frac{|(V_{eff})_{k'k}|^2}{\varepsilon^{(0)}(k) - \varepsilon^{(0)}(k')} =$$

$$= \varepsilon^{(0)}(k) + \bar{V}_{eff} + \sum_n \frac{|C_n|^2}{\varepsilon^{(0)}(k) - \varepsilon^{(0)}\left(k + \frac{2\pi}{a} n\right)} \qquad (2.33)$$

The use of the theory of perturbations is only feasible when the disturbing potential $V_{eff}(x)$ is sufficiently low. The values $|C_n|^2$ then are of the second order of smallness, so that the correction to the energy in the second approximation is inessential, provided that the difference $\varepsilon^{(0)}(k) - \varepsilon^{(0)}\left(k + \frac{2\pi}{a} n\right)$ is substantially greater than $|C_n|^2$.

Under the same condition, the wave function in the first approximation $\psi_k^{(\mathrm{I})}(x)$ differs only slightly from $\psi_k^{(0)}(x)$:

$$\psi_k^{(\mathrm{I})}(x) = \psi_k^{(0)}(x) + \sum_n \frac{C_n}{\varepsilon^{(0)}(k) - \varepsilon^{(0)}\left(k + \frac{2\pi}{a} n\right)} \psi^{(0)}_{k + \frac{2\pi}{a} n}(x) \qquad (2.34)$$

The states are of interest for which the difference $\varepsilon^{(0)}(k) - \varepsilon^{(0)}\left(k + \frac{2\pi}{a} n\right)$ is comparable with $|C_n|^2$. Namely these states correspond to the wave vectors which are close to the boundaries of the Brillouin zone.

Indeed, let $\varepsilon^{(0)}(k) = \varepsilon^{(0)}\left(k + \frac{2\pi}{a} n\right)$. This equality implies that $\frac{\hbar^2 k^2}{2m_0} = \frac{\hbar^2}{2m_0}\left(k + \frac{2\pi}{a} n\right)^2$, whence $k^2 = k^2 + \frac{4\pi}{a} nk + \frac{4\pi^2 n^2}{a^2}$ or $k = -\frac{\pi}{a} n$. Thus, since $n = \pm 1, \pm 2, \ldots$, the equality of the energies $\varepsilon^{(0)}(k)$ and $\varepsilon^{(0)}\left(k + \frac{2\pi}{a} n\right)$ is observed for the momenta corresponding to the boundaries of the Brillouin zone. With a small difference between these energies the correction to the wave function $\psi_k^{(0)}(x)$ in equation (2.34) sharply increases. The free motion of the electron then undergoes a strong perturbation.

The maximum perturbation is observed when the denominator of the correction to the wave function $\psi_k^{(0)}(x)$ becomes zero. Expressions (2.33) and (2.34) then become meaningless, since the condition of the applicability of the theory of perturbations is violated.

With $\varepsilon^{(0)}(k) \to \varepsilon^{(0)}\left(k + \frac{2\pi}{a} n\right)$, the coefficient at $\psi^{(0)}_{k+\frac{2\pi}{a}n}(x)$ in (2.34) increases. This implies that the state $\psi^{(1)}_k(x)$ becomes a mixed one, the proportion of the state $\psi^{(0)}_{k+\frac{2\pi}{a}n'}(x)$ in it being not less than the proportion of the state $\psi^{(0)}_k(x)$.

With the exact equality $\varepsilon^{(0)}(k) = \varepsilon^{(0)}\left(k + \frac{2\pi}{a} n'\right)$ which corresponds to $k = -\frac{\pi}{a} n'$, the level $\varepsilon^{(0)}(k)$ is degenerated, since two different functions: $\psi^{(0)}_k(x)$ and $\psi^{(0)}_{k+\frac{2\pi}{a}n'}(x)$ correspond to one and the same energy. Therefore, near the boundaries of the Brillouin zone, degeneration must be already taken into account in the zero approximation, and the wave function $\psi^{(0)}(x)$ must have the form:

$$\psi^{(0)}(x) = \alpha\psi^{(0)}_k(x) + \beta\psi^{(0)}_{k+\frac{2\pi}{a}n'}(x) \qquad (2.35)$$

where α and β are unknown coefficients.

With no degeneration, $\beta \ll \alpha$; for degenerated states, α and β are values of the same order.

Let the wave function (2.35) be substituted into Schrödinger's equation for a disturbed system

$$\left(-\frac{\hbar^2}{2m_0} \cdot \frac{d^2}{dx^2} + V_{eff}(x)\right)\psi^{(0)} = \varepsilon\psi^{(0)} \qquad (2.36)$$

We also take into account that $\psi^{(0)}_k(x)$ and $\psi^{(0)}_{k+\frac{2\pi}{a}n'}(x)$ are the solutions of the equation for a free electron (2.26). This gives the following relationship:

$$\alpha\varepsilon^{(0)}(k)\,\psi^{(0)}_k(x) + \beta\varepsilon^{(0)}\left(k + \frac{2\pi}{a} n'\right)\psi^{(0)}_{k+\frac{2\pi}{a}n'}(x) + V_{eff}(x)\,\psi^{(0)}(x) = \varepsilon\psi^{(0)}(x) \qquad (2.37)$$

Let us denote for convenience:

$$\psi^{(0)}_k(x) = \psi_1, \quad \psi^{(0)}_{k+\frac{2\pi}{a}n'}(x) = \psi_2$$

$$\varepsilon^{(0)}(k) = \varepsilon_1, \quad \varepsilon^{(0)}\left(k + \frac{2\pi}{a} n'\right) = \varepsilon_2$$

Equation (2.37) will then take the form

$$\alpha\varepsilon_1\psi_1 + \beta\varepsilon_2\psi_2 + V_{eff}(\alpha\psi_1 + \beta\psi_2) = \varepsilon(\alpha\psi_1 + \beta\psi_2)$$

Multiplying it from the left-hand side by ψ_1^* and ψ_2^* respectively and integrating by x, we get

$$\alpha\varepsilon_1 + \alpha(V_{eff})_{11} + \beta(V_{eff})_{12} = \varepsilon\alpha$$
$$\beta\varepsilon_2 + \alpha(V_{eff})_{21} + \beta(V_{eff})_{22} = \varepsilon\beta$$

or

$$\begin{aligned} &[\varepsilon_1 + (V_{eff})_{11} - \varepsilon]\,\alpha + (V_{eff})_{12}\,\beta = 0 \\ &(V_{eff})_{21}\,\alpha + [\varepsilon_2 + (V_{eff})_{22} - \varepsilon]\,\beta = 0 \end{aligned} \tag{2.38}$$

The system of linear homogeneous equations (2.38) relative to α and β has a non-trivial solution only when the determinant of the system turns to zero:

$$\begin{vmatrix} \varepsilon_1 + (V_{eff})_{11} - \varepsilon, & (V_{eff})_{12} \\ (V_{eff})_{21}, & \varepsilon_2 + (V_{eff})_{22} - \varepsilon \end{vmatrix} = 0 \tag{2.39}$$

This, in turn, gives an equation for determining the energy. Noting that

$$(V_{eff})_{11} = (V_{eff})_{22} = \overline{V}_{eff}$$

and

$$(V_{eff})_{12}\,(V_{eff})_{21} = |\,C_{n'}\,|^2 = \left|\frac{1}{L}\int_0^L V_{eff}(x)\, e^{-i\frac{2\pi}{a}n'x}\,dx\right|^2 = \left|(V_{eff})_{\frac{2\pi}{a}n'}\right|^2$$

we have

$$\varepsilon = \frac{\varepsilon^{(0)}(k) + \varepsilon^{(0)}\left(k + \frac{2\pi}{a}n'\right)}{2} + \overline{V}_{eff} \pm$$
$$\pm\sqrt{\frac{\left[\varepsilon^{(0)}(k) - \varepsilon^{(0)}\left(k + \frac{2\pi}{a}n'\right)\right]^2}{4} + \left|(V_{eff})_{\frac{2\pi}{a}n'}\right|^2} \tag{2.40}$$

It can be seen from (2.40) that at application of a perturbation $V_{eff}(x)$, the energy of an electron has a discontinuity in a point where $\varepsilon^{(0)}(k) = \varepsilon^{(0)}\left(k + \frac{2\pi}{a}n'\right)$, i.e. at the boundaries of Brillouin zones. With approaching the boundary from the left- or right-hand side, the energy tends respectively to

$$\varepsilon^{(0)}\left(\frac{2\pi}{a}n'\right) - \left|(V_{eff})_{\frac{2\pi}{a}n'}\right| \quad \text{and} \quad \varepsilon^{(0)}\left(\frac{2\pi}{a}n'\right) + \left|(V_{eff})_{\frac{2\pi}{a}n'}\right|$$

The magnitude of discontinuity of the energy at the zone boundary is $2\left|(V_{eff})_{\frac{2\pi}{a}n'}\right|$ and becomes zero only at $V_{eff}(x) \equiv 0$. Energy discontinuities at boundaries of Brillouin zones are equivalent to that the energy spectrum of an electron in a metal has forbidden states (and the corresponding forbidden values of the energy). The interval of forbidden energies coincides with the magnitude of the discontinuity $2\left|(V_{eff})_{\frac{2\pi}{a}n'}\right|$.

The dependence of the energy ε of an electron on momentum under account of energy discontinuities at boundaries of Brillouin zones is shown in Fig. 49. The dotted lines in the figure show the relationship $\varepsilon^{(0)}(p)$ for a free electron.

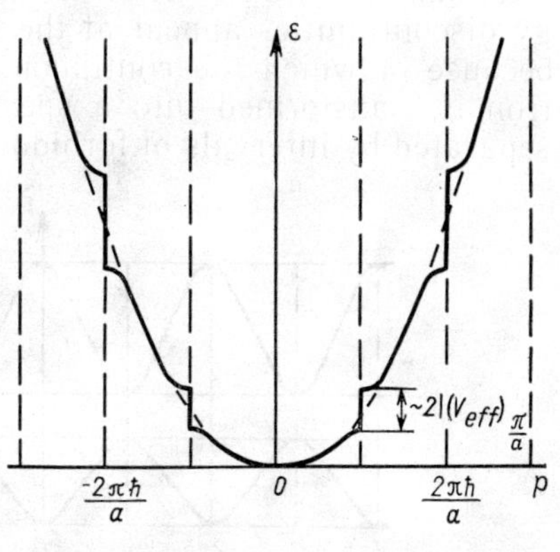

Fig. 49

As has been noted in Sec. **2-3**, Chapter Two, the values of momentum in the first, second, etc. Brillouin zones are physically equivalent. But physically equivalent values of momentum must have in correspondence the same values of the energy. In order to satisfy this requirement, we translate (shift) the equivalent halves of the zones $\left(\text{the intervals } 0 \div \frac{\pi\hbar}{a};\ \frac{\pi\hbar}{a} \div \frac{2\pi\hbar}{a}, \text{ etc.}\right)$ and the respective sections of the curve $\varepsilon(p)$ with a period $\frac{2\pi\hbar}{a}$.

As a result of this translation, the picture shown in Fig. 49 will be transformed into the one illustrated in Fig. 50. The energy then becomes a periodic $\left(\text{with a period of } \frac{2\pi\hbar}{a}\right)$ and multiple-valued function of momentum. Bands of allowed states of energy are formed in the energy spectrum, which are termed energy bands (the first, the second, etc.), separated by intervals of forbidden energy values. The dependence of energy on momentum then remains single-valued only within the limits of each energy band.

What has been done above with the momentum-energy relationship should be regarded as the method of transition from the continuous energy spectrum of a free electron to a banded energy spectrum of an electron in a weak periodic field of an effective potential. The following aspect then requires special attention. If the Brillouin zones are considered from the standpoint of possible values of the momentum, then one can limit oneself to the first zone, since in that sense all the zones are equivalent and the first zone includes all the physically different values of momenta.

But in order to construct energy bands (Fig. 50) we have to consider a number of Brillouin zones. The relationship $\varepsilon(p)$ in the first Brillouin zone has resulted, after translation, in the construction of the first energy band. The same relationship in the second Brillouin zone has made it possible to construct the second energy band, etc.

The one-dimensional model considered makes it possible to elucidate the main qualitative peculiarities of the momentum-energy relationship for an electron in a crystal. They consist in that energy discontinuities appear at the boundaries of the Brillouin zones, because of which the continuous energy spectrum of a free electron is transformed into a spectrum consisting of energy bands separated by intervals of forbidden values of energy.

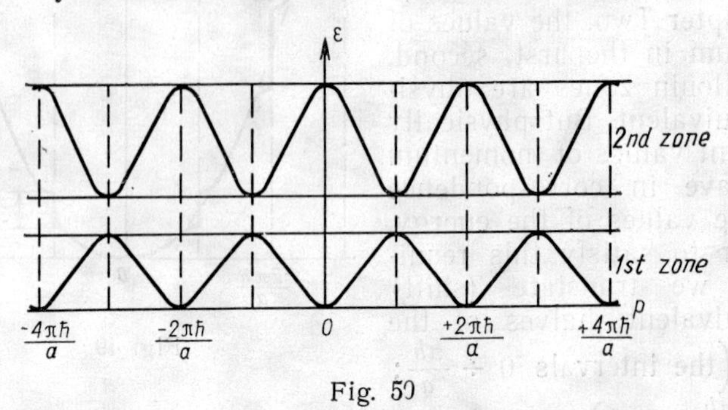

Fig. 50

This picture of band spectrum becomes substantially more complicated in cases of two or three dimensions. This is connected with the distances to the boundaries of Brillouin zones becoming dependent on the direction of momentum, and therefore, the relation between energy and momentum becomes different for different directions in the lattice.

The boundaries of the Brillouin zones for a two- or three-dimensional crystal are respectively the straight lines or planes in $\vec{p}$-space on which the energy of an electron undergoes a discontinuity. The interference between the incident electron wave and that reflected from atomic planes results in that standing waves are formed in directions perpendicular to the atomic planes; running waves can only propagate along atomic planes.

For that reason, the normal component of the velocity of the electron $(\vec{v}_g)_n = \left(\frac{\partial \varepsilon}{\partial \vec{p}}\right)_n$ turns to zero at the zone boundaries. This results in that the constant-energy surfaces $\varepsilon(\vec{p}) = \text{const}$ cannot touch the boundaries of the Brillouin zones and must intersect the latter in the direction of a normal to the boundary. In other words, a normal to a constant-energy surface $\varepsilon(\vec{p})$ = const coinciding in direction with the velocity of electron $\vec{v}_g = \text{grad}_{\vec{p}}\,\varepsilon$ at the boundary of a Brillouin zone must be located in the plane of that boundary.

In a two-dimensional case, the boundaries of a Brillouin zone intersect each other in certain points. The components of the vector $\text{grad}_{\vec{p}}\,\varepsilon$ must evidently turn to zero in those points, and the energy of the electron attains a relative maximum (or minimum).

In a three-dimensional case, two boundaries of a Brillouin zone intersect along a certain straight line. The vector $\vec{v}_g = \text{grad}_{\vec{p}}\,\varepsilon$ can have a component other than zero only along this line. In the common point belonging to the three boundaries (at the vertex of the zone), $\text{grad}_{\vec{p}}\,\varepsilon = 0$ and the energy also attains a relative maximum (or minimum).

When calculating the energy of an electron in the vicinity of the vertices of the zones by means of the theory of perturbations, we have to take into account that the levels at the vertices undegro quadruple degeneration. The wave function of the electron at the zero approximation near a vertex must be the sum of four items with the wave vectors entering the equation of the three boundaries intersecting at the vertex. This gives a fourth-order equation [similar to equation (2.39)] to determine the energy of the electron.

The energy of an electron is a periodic multiple-valued function of momentum, irrespective of the number of dimensions of the lattice for any direction in the crystal. The general pattern of the momentum-energy relationship for a selected direction is shown in Figs. 49 and 50.

2-6. CONSTRUCTION OF BRILLOUIN ZONES

In Sec. **2-3** of this chapter we have defined the Brillouin zones for the simplest one-dimensional model of crystal (see Fig. 42) and also the first Brillouin zone for a plane square lattice in a two-dimensional case and for a cubic lattice in a three-dimensional case. Let us pass over to the construction of Brillouin zones of any number for cases of two and three dimensions.

In order to demonstrate the method of geometrical construction of Brillouin zones, we will consider a two-dimensional crystal with a simple square lattice having the period a (Fig. 51). The boundaries of the Brillouin zones are the straight lines in $\vec{k}$-space on which the energy undergoes discontinuities. Therefore, to construct the general pattern of the zones, we have first to find all these straight lines in $\vec{k}$-space.

Let us turn to the two-dimensional lattice chosen. For a two-dimensional periodic structure we can plot various families of equidistant parallel straight lines, what are called "network" lines, on which all atoms of the structure are located (Fig. 52). Generally speaking, the number of such families may be infinite. The construction is convenient to be started from the families of "network" lines which pass at the maximum distances from each other and

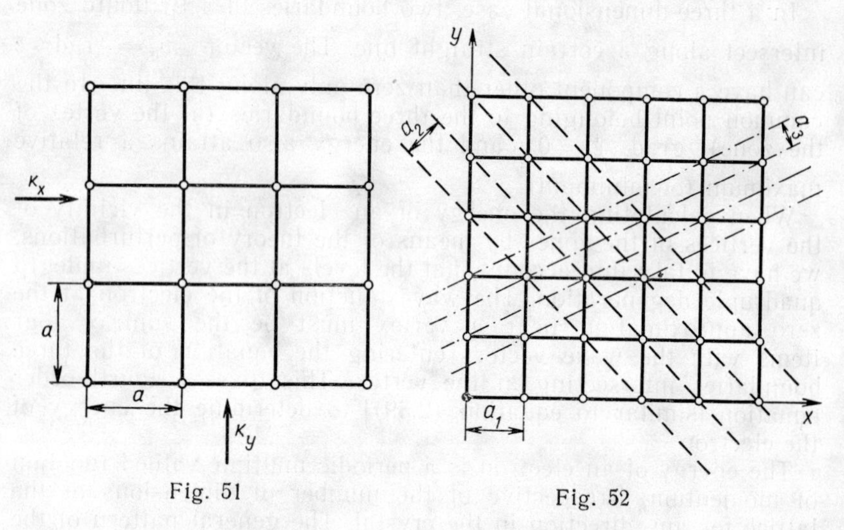

Fig. 51 Fig. 52

then go over to families of lines passing closer and closer to each other. In the case considered, two families of the first kind are possible: the one consists of lines perpendicular to the x axis (the distance between the lines $a_1 = a$), and the other is formed by lines perpendicular to the y axis (solid lines in Fig. 52). It can be easily seen that no other families with the period a can exist.

Then follow two families of lines located at distances $a_2 = \frac{a}{\sqrt{2}}$ from each other (diagonal dotted lines in Fig. 52).

Next follow four equivalent families of straight lines. One of these families is shown in Fig. 52 by dot-and-dash lines. The straight lines constituting this family pass at distances of $a_3 = \frac{a}{\sqrt{5}}$ from each other.

Let us consider one of the families plotted in which the network lines pass at a distance a_i from each other (Fig. 53). Let an electron with the wave vector $\vec{k}_i$ move perpendicular to the lines of this family. At the values of the vector $k_i = \pm \frac{\pi}{a_i} n$, the energy undergoes discontinuity (the electron is reflected by the lattice). These values of the wave vector determine the position of the boundaries of the Brillouin zones in the given direction.

We choose the origin of coordinates in $\vec{k}$-space and draw a line through it which is parallel to vector $\vec{k}_i$. We mark the points $\pm \frac{\pi}{a_i} n$ on that line and draw perpendiculars through them (Fig. 54). These lines are the lines on which the energy undergoes discontinuity. (For all the vectors passed from the origin of coordinates in $\vec{k}$-space to any point on these lines, their projection into the chosen direction of

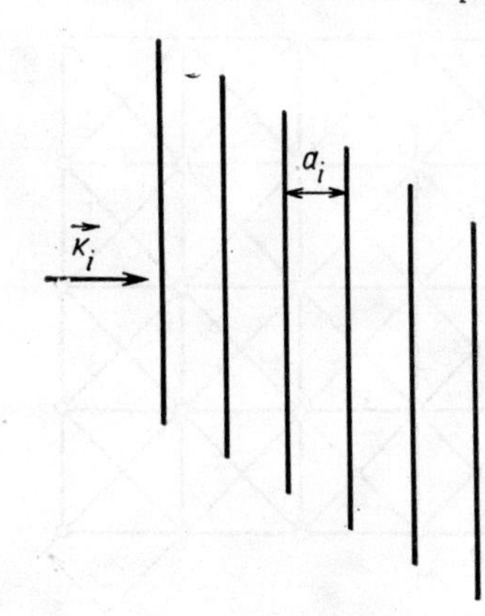

Fig. 53

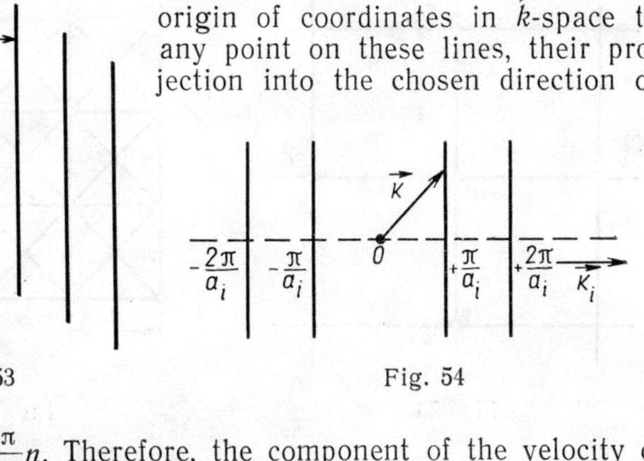

Fig. 54

$\vec{k}_i$ is equal to $\pm \frac{\pi}{a_i} n$. Therefore, the component of the velocity of motion of an electron along the direction $\vec{k}_i$ turns to zero and the electron wave is reflected by the lattice.)

The greater the distance a_i between the equi-distant "network" lines, the closer the lines of discontinuity of energy are located to the origin of coordinates in $\vec{k}$-space. Since the first Brillouin zone is the area (or the volume in a three-dimensional case) in $\vec{k}$-space bounded by a combination of straight lines (or planes) of energy discontinuities located most closely to the origin of coordinates, then it is clear that in order to construct this zone, we have first to consider the families of equi-distant straight lines in the initial lattice which are at the greatest distance from each other.

Thus, the construction of the first Brillouin zone for the lattice shown in Fig. 51 must be started by using two mutually perpendicular families of lines with the maximum distances between them equal to the lattice period a. Let us consider the wave vectors of the electron which are orthogonal to each of these families (vectors k_x and k_y in Fig. 51).

We draw straight lines parallel to vectors k_x and k_y through the origin of coordinates in $\vec{k}$-space. We draw lines through the points corresponding to the values of wave vector $k_i = \pm \frac{\pi}{a} n$ (where $n = 1, 2, 3, \ldots$), perpendicular to these lines. Note that these latter lines are parallel to the chosen families of "network" lines in the crystal lattice, but the distance between them is inversely proportional to a (Fig. 55).

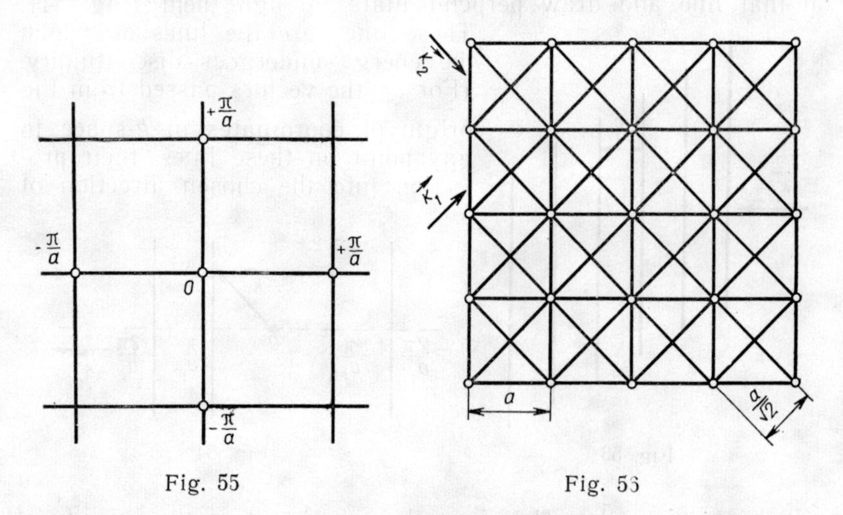

Fig. 55

Fig. 56

The next two families of "network" lines with the period equal to $a/\sqrt{2}$ (diagonal families in Figs. 52 and 56) will give the lines of energy discontinuity in $\vec{k}$-space located at an angle of 45 degrees to those constructed earlier and passing at distances $\pm \frac{\sqrt{2}\,\pi}{a} n$ from the origin of coordinates (Fig. 57).

Similarly, we can construct a set of lines of energy discontinuity corresponding to four families of "network" lines having the period $a/\sqrt{5}$, etc.

Continuing the construction further, we can plot in $\vec{k}$-space all the possible lines of energy discontinuity that are more and more distant from the origin of coordinates as the period of engendering "network" lines becomes smaller. But, as will be clear later, plotting of a large number of various families of lines of energy discontinuity has no practical meaning.

We can now construct the Brillouin zones since the boundaries separating these zones have been defined. As has been indicated, the first Brillouin zone is an area bounded by a set of energy discontinuity lines that are located most closely to the origin of coordinates. The second Brillouin zone is a set of areas adjoining the boundaries of the first zone and equal in area to the first zone. Similarly, the third zone consists of areas adjoining the boundaries of the second zone and is equal in area to the second zone.

It can be taken conditionally that the first Brillouin zone adjoins a zero zone which has degenerated into a point at the origin of

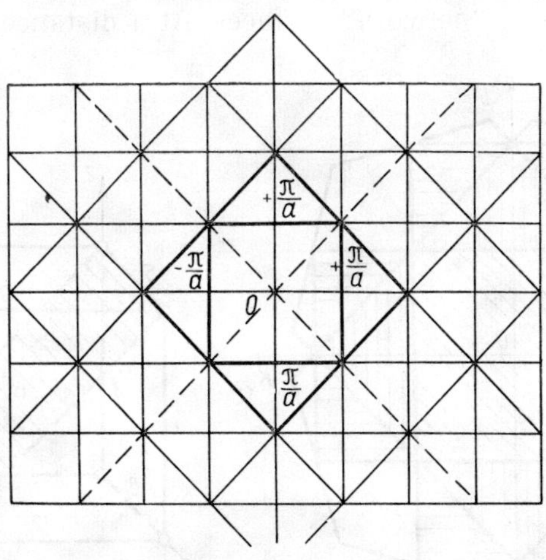

Fig. 57

coordinates. Under such an assumption, the methods of constructing each zone become similar. The first four Brillouin zones for a two-dimensional square lattice are shown in Fig. 58. The first zone is shown shaded at 45 degrees to the axes, the second, by vertical shading, the third, by horizontal, and the fourth is left unshaded. As can be seen from the figure, all the zones beginning from the second become multi-bonded.

The construction of Brillouin zones for a three-dimensional case is similar to that just described and reduces to finding the set of planes of energy discontinuity in $\vec{k}$-space and to classifying (by zones) the regions bounded by these planes.

Let us construct, as an example, the first Brillouin zone for a face-centered cubic lattice [such a lattice have alkali metals at room temperature (Fig. 3)]. For this, we first find the families of "network" surfaces located at the greatest distance from each other. Such families are formed by diagonal planes, one of which is shown in Fig. 59. The distances between these planes are equal to $a/\sqrt{2}$. In the case considered, there may be six equivalent families of this type (half the number of edges of the cube). The surfaces of one of these families are parallel to the plane passing through edges *5* and *7* of the cube (Fig. 59). The planes of the second family are parallel to the plane passing through edges *6* and *8*, those of the third, to the plane passing through edges *10* and *12*, etc.

We draw straight lines through the origin of coordinates in $\vec{k}$-space in the direction of wave vectors perpendicular to each of the six families of "network" surfaces. At a distance $\sqrt{2}\ \pi/a$ from

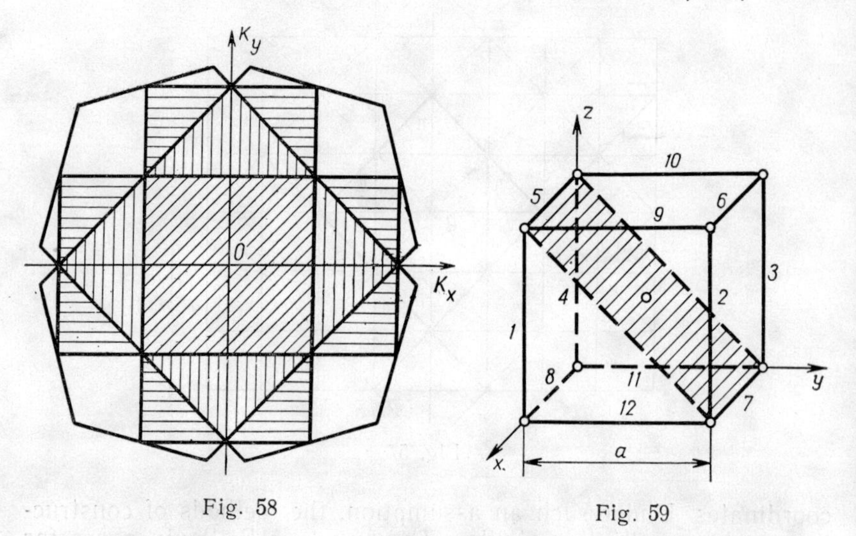

Fig. 58

Fig. 59

the origin of coordinates, we plot the planes perpendicular to these straight lines (each pair of these planes is parallel to one of the families of "network" surfaces, or in other words, each family of "network" surfaces with the period $a/\sqrt{2}$ engenders in $\vec{k}$-space parallel planes of energy discontinuity passing at distances $\frac{2\sqrt{2}}{a}\pi$ from each other).

The intersections of twelve such planes confine the volume of the first Brillouin zone which has the form of a rhombic dodecahedron (Fig. 60). Two families of diagonal planes perpendicular to the *xy* plane (in Fig. 59, these are the families passing through edges *1-3* and *2-4*) engender four faces which are shown shaded in Fig. 60. The remaining four families of planes located at an angle of 45 degrees to the *xy* plane engender eight non-shaded faces. Two planes of energy discontinuity located symmetrically relative to the origin of coordinates correspond to each family. The centres of all faces of the rhombic dodecahedron of the first zone are located at an equal distance $\frac{\sqrt{2}\,\pi}{a}$ from the origin of coordinates.

Let us now construct the first Brillouin zone for a face-centered cubic lattice (Fig. 4). Such a lattice is characteristic, for instance, of gold, silver, and copper. In this case the families of "network" surfaces passing at the greatest distances from each other are the planes located at distances $a/\sqrt{3}$ apart (two such planes are shown shaded in Fig. 61). Each of such planes includes the diagonals of three faces of the cube with the common vertex. There are

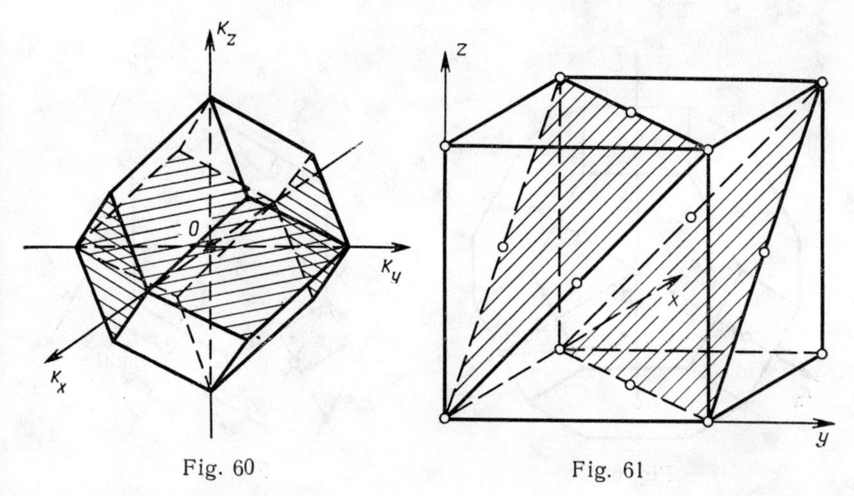

Fig. 60 Fig. 61

four families with the period $a/\sqrt{3}$. In the wave vector space, they will give eight planes of energy discontinuity spaced at $\frac{\sqrt{3}\,\pi}{a}$ from the origin of coordinates.

The second group of families of "network" surfaces is characterized by the period $\frac{a}{2}\left(\frac{a}{2} < \frac{a}{\sqrt{3}}\right)$ and consists of the faces of cubes and parallel planes passing through the central atom of a face. In $\vec{k}$-space, these families will give six planes perpendicular to the axes k_x, k_y, k_z and spaced at $\frac{2\pi}{a}$ from the origin of coordinates.

Thus, eight planes of the first type $\left(\text{with the period } \frac{2\sqrt{3}\,\pi}{a}\right)$ and six planes of the second kind $\left(\text{with the period } \frac{4\pi}{a}\right)$, by intersecting with each other, confine the volume of the first Brillouin zone of the face-centered cubic lattice in the form of a cubo-octahedron

(Fig. 62). By selecting the families of "network" surfaces with a period smaller than $a/2$ and constructing the corresponding surfaces of energy discontinuity in $\vec{k}$-space, we can plot the second zone adjoining the first one, etc.

Here, as in the two-dimensional case, all the zones beginning from the second become multi-bonded. But the sum of all the volumes relating to each zone is constant and equal to the volume of the first zone.

To conclude this section, we shall construct the first Brillouin zone of a close-packed hexagonal lattice (Fig. 5). Let us recall

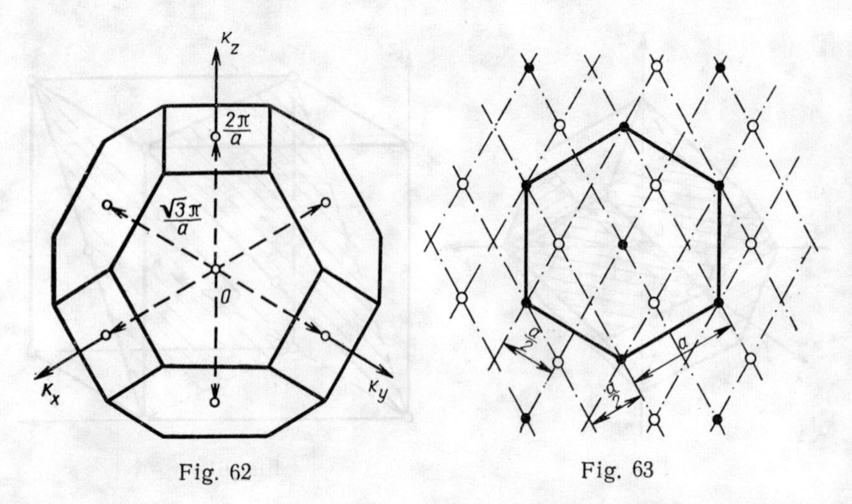

Fig. 62 Fig. 63

that this lattice is characterized by the ratio c/a equal to $\sqrt{\frac{8}{3}} \approx$ ≈ 1.633 (see Sec. **1-2,** Chapter One).

Hexagonal crystalline structure is found in many bivalent metals (beryllium, magnesium, zinc, cadmium) and some trivalent metals (yttrium and thallium). The lattices of these metals have a c/a ratio close to that indicated for the close-packed hexagonal lattice. For instance, according to Barrett [37], for thallium at the temperature of liquid helium, $c/a = 1.593$. Thus, the ideal close-packed hexagonal lattice is a sufficiently good model of the crystalline structure of the metals indicated.

Let us consider the geometry of the close-packed hexagonal lattice. The axis of symmetry coinciding with the axis of any of the hexagonal prisms in the lattice (Fig. 5) is a spiral axis 6_3, which is usually denoted as the c axis. The planes perpendicular to this axis are termed the basal ones. The "network" surfaces parallel to one of the basal planes are located at distances $c/2$ from each other. In $\vec{k}$-space, these surfaces engender the planet of energy discontinuities which are orthogonal to the wave vector $\vec{k}_c$ directed along the c axis. The family of such planes of energy discontinuities has the period $\frac{4\pi}{c}$. The two planes of the family, which are the closest to the origin of coordinates of $\vec{k}$-space, bound the first Brillouin zone in the $\vec{k}_c$-direction.

In order to construct the boundaries of the first Brillouin zone in directions perpendicular to $\vec{k}_c$, let us consider one of the "network" surfaces which is parallel to the basal plane (Fig. 63). The atoms located directly on the given "network" surface are shown in dark circles here, while the bright circles show the projections onto this surface of the atoms located on the neighbouring "network" surfaces shifted a distance $\pm c/2$ along the c axis. Shown separated in the figure is one of the hexagons in the basal plane in whose vertices and centre are atoms of the lattice (shown as dark circles).

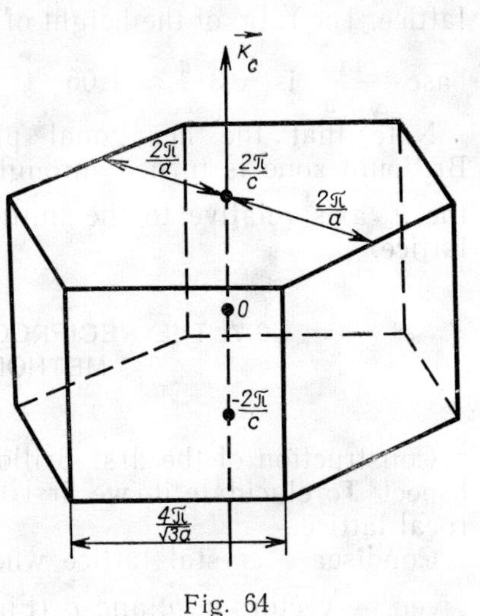

Fig. 64

Let us consider various families of "network" surfaces parallel to the c axis. Among them there are six families whose surfaces are located at the greatest distance from each other. One of these families passes over into any of the remaining five families at rotation through an angle $\pm\frac{2\pi}{6}n$, where $n = 1, 2, 3, 4, 5$, about the c axis. The period of a family is equal to $a/2$. The dot-and-dash line in the figure shows the lines of intersection of the basal plane with the "network" surfaces of two of the indicated families, which are directed at an angle of 60 degrees to one another.

In the wave vector space, the "network" surfaces of each of the families described with the period $a/2$ engender planes of energy discontinuity located at distances of $\frac{4\pi}{a}$. Six such planes, which are symmetrical relative to the $\vec{k}_c$ axis and the closest to the origin of coordinates of $\vec{k}$-space, together with the two closest planes perpendicular to $\vec{k}_c$, form a hexagonal prism (Fig. 64).

The sides of the right hexagon in the base of the prism are equal to $\frac{4\pi}{\sqrt{3}\,a}$ and the height of the prism is $\frac{4\pi}{c}$.

When considering various families of "network" surfaces directed at different angles to the c axis (other than zero or 90 degrees), we can easily see that there are no planes of energy discontinuity in $\vec{k}$-space that intersect the hexagonal prism constructed. Thus, the surface of the prism consists of the planes of energy discontinuity that are closest to the origin of coordinates, and the prism itself is the first Brillouin zone of a close-packed hexagonal lattice. The ratio of the height of the prism $\frac{4\pi}{c}$ to the side of the base $\frac{4\pi}{\sqrt{3}\,a}$ is $\sqrt{3}\,\frac{a}{c} \approx 1.06$.

Note that the hexagonal prism corresponding to the first Brillouin zone is turned through an angle of ± 30 degrees about the $\vec{k}_c$ axis relative to the initial hexagonal prism in the crystal lattice.

2-7. THE RECIPROCAL LATTICE. THE WIGNER-SEITZ METHOD FOR CONSTRUCTING THE FIRST BRILLOUIN ZONE

Construction of the first Brillouin zone has another geometrical aspect. To elucidate it, we first introduce the concept of the reciprocal lattice.

Condiser a crystal lattice where the principal translations are given by vectors $\vec{a}$, $\vec{b}$, and $\vec{c}$ (Fig. 65). A reciprocal lattice will be the one with the vectors of the principal translations $\vec{a}^*$, $\vec{b}^*$, and $\vec{c}^*$ which are determined as follows:

$$\vec{a}^* = \frac{[\vec{b} \cdot \vec{c}]}{(\vec{a}\,[\vec{b} \cdot \vec{c}])}, \quad \vec{b}^* = \frac{[\vec{a} \cdot \vec{c}]}{(\vec{b}\,[\vec{a} \cdot \vec{c}])}, \quad \vec{c}^* = \frac{[\vec{a} \cdot \vec{b}]}{(\vec{c}\,[\vec{a} \cdot \vec{b}])} \tag{2.41}$$

As is known from analytical geometry, the magnitude of a scalar triple product $(\vec{a}\,[\vec{b} \cdot \vec{c}])$ is equal to the volume of a parallelepiped constructed on the vectors $\vec{a}$, $\vec{b}$, and $\vec{c}$. The magnitude of vector product $|[\vec{b} \cdot \vec{c}]|$ gives the area of the base of the parallelepiped, i.e. the area of the face which contains the vectors $\vec{b}$ and $\vec{c}$ (see Fig. 65). Therefore, the vector $\vec{a}^* = \frac{[\vec{b} \cdot \vec{c}]}{(\vec{a}\,[\vec{b} \cdot \vec{c}])}$, which is perpendicular to the base, is equal in magnitude to $1/h$, where h is the height of the parallelepiped.

The faces of the parallelepiped containing the vector pairs $\vec{a}$ and $\vec{b}$, $\vec{b}$ and $\vec{c}$, or $\vec{a}$ and $\vec{c}$, as also their parallel faces, coincide with the planes of the three families of "network" surfaces of the crystal. Each of the vectors $\vec{a}^*$, $\vec{b}^*$, and $\vec{c}^*$ determining the reciprocal lattice, is orthogonal to one of those families and is equal in magnitude to the inverse distance between the "network" planes of the family. This property of the vector of a reciprocal lattice makes it possible to use the following technique for constructing this lattice. Let it be explained by an example of constructing the reciprocal lattice for a face-centered cubic lattice.

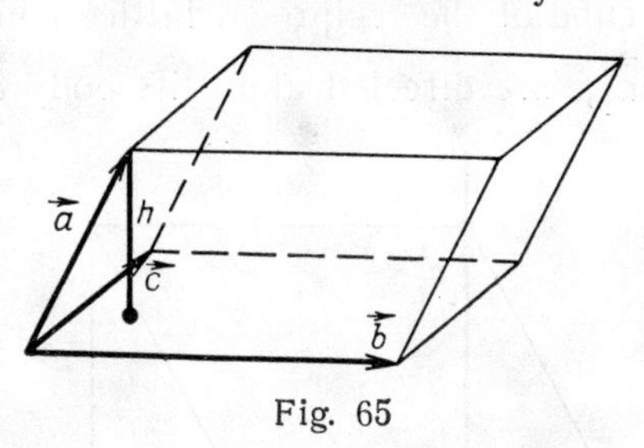

Fig. 65

We consider the atoms located on the faces of a single elementary cell (Fig. 66). Two types of families of "network" surfaces

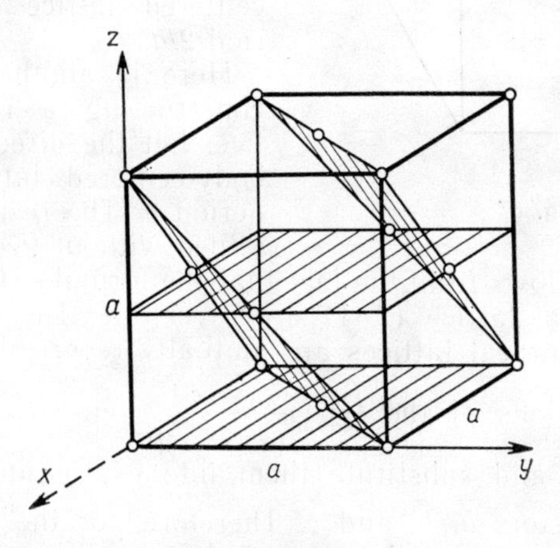

Fig. 66

with the periods $a/\sqrt{3}$ and $a/2$ pass through these atoms. Three families of planes with a period $a/2$, which are perpendicular to the Cartesian axes, define the three vectors of the reciprocal lattice coinciding in direction with the Cartesian axes. Each of these vectors, $\vec{a}^*_x$, $\vec{a}^*_y$, or $\vec{a}^*_z$, is equal in magnitude to the inverse distance $2/a$ between the planes (Fig. 67).

The four families of "network" surfaces with the period $a/\sqrt{3}$, which are perpendicular to the four body diagonals of the cube, define the four orthogonal vectors of the reciprocal lattice $\vec{b}_1^*$, $\vec{b}_2^*$, $\vec{b}_3^*$, and $\vec{b}_4^*$. These vectors start at the vertices of the elementary cube of the reciprocal lattice constructed on vectors $\vec{a}_x^*$, $\vec{a}_y^*$, and $\vec{a}_z^*$, are directed along its body diagonals, and end in the centre of this cube. Since they thus define only one node of the reciprocal lattice, i.e. the centre of the elementary cube, it is sufficient to use only one of the four vectors $\vec{b}_1^*$, $\vec{b}_2^*$, $\vec{b}_3^*$, and $\vec{b}_4^*$, for instance, $\vec{b}_1^* = \vec{b}^*$ (Fig. 67).

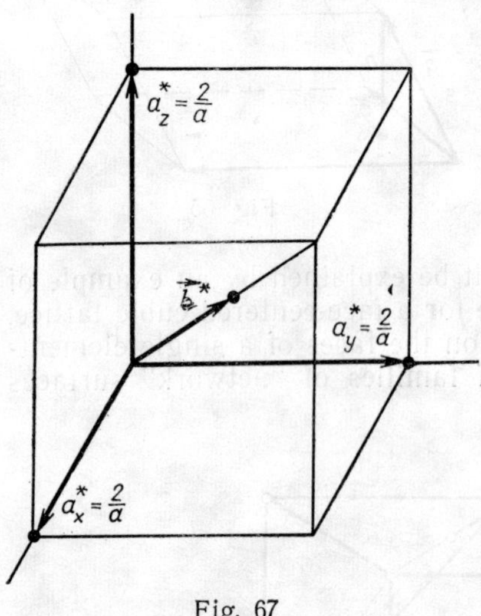

Fig. 67

The construction we have made shows that for a face-centered lattice with the period a the reciprocal lattice is a body-centered lattice with the period $2/a$.

Here is another example of constructing a reciprocal lattice. Let the direct lattice be a body-centered lattice with the period a. The result can be obtained without geometrical construction. It follows from the fact that the formulae for the vectors of the reciprocal lattice (2.41) are reversible. The vectors of the direct and reciprocal lattices are mutually reversible: if we form three combinations of the type $\frac{[\vec{b}^* \cdot c^*]}{(\vec{a}^* [\vec{b}^* \cdot \vec{c}^*])}$ etc. from the vectors $\vec{a}^*$, $\vec{b}^*$, and c^* and substitute them into expressions (2.41), we then obtain vectors $\vec{a}$, $\vec{b}$, and $\vec{c}$. Therefore, for the body-centered lattice with the period a, the reciprocal one will be a face-centered lattice with the period $2/a$.

The same result will now be obtained by geometrical construction. For this, we consider, as before, the atoms located in the vertices of an elementary cube and in its centre. One cell defines only two types of families of "network" surfaces with the periods $a/\sqrt{2}$ and $a/2$. The planes of the families with the period $a/2$ are perpendicular to the Cartesian coordinates. One of these planes is shown in Fig. 68 as the shaded horizontal plane. The diagonal shaded plane in the figure belongs to one of the families with the period $a/\sqrt{2}$.

The vectors $\vec{a}_1^*$, $\vec{a}_2^*$, and $\vec{a}_3^*$ corresponding to the three first families with the period $a/2$ are directed along the Cartesian axes of the space of the reciprocal lattice. The magnitude of these vectors is $2/a$; these vectors define an elementary cube of the reciprocal lattice (Fig. 69). The six diagonal families of "network" surfaces with the period $a/\sqrt{2}$ have respectively three different

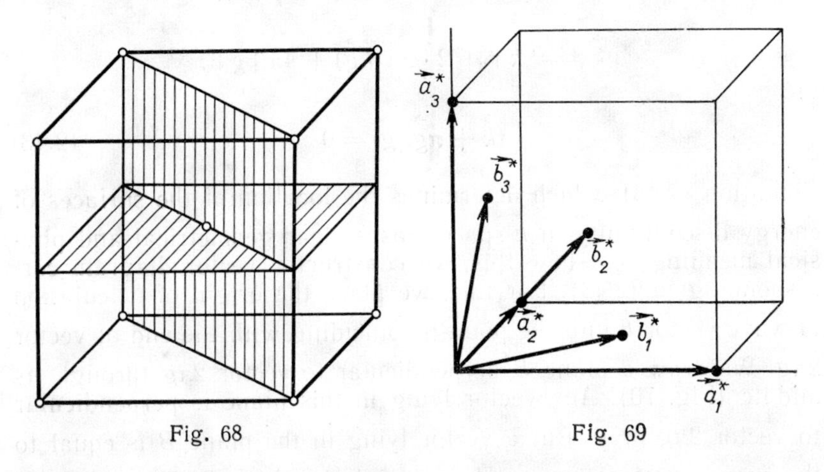

Fig. 68 Fig. 69

vectors of the reciprocal lattice $\vec{b}_1^*$, $\vec{b}_2^*$, and $\vec{b}_3^*$. These vectors are directed from the vertex of the elementary cube to the centres of the three faces for which the given vertex is the common one. The six vectors $\vec{a}_1^*$, $\vec{a}_2^*$, $\vec{a}_3^*$ and $\vec{b}_1^*$, $\vec{b}_2^*$, $\vec{b}_3^*$ define the nodes of a face-centered lattice with the period $2/a$.

These geometrical constructions of reciprocal lattices show that the space of a reciprocal lattice is closely linked with the phase space or wave vector space. (The space of a reciprocal lattice in its dimension is identical to $\vec{k}$-space).

By comparing the construction of reciprocal lattice with that of Brillouin zones, which has been described in Sec. **2-6**, Chapter Two, we can conclude that the surfaces of energy discontinuities in the wave vector space are orthogonal to the vectors of the reciprocal lattice. It has been found for the one-dimensional model of a crystal (see Sec. **2-5**, Chapter Two) that the location of energy discontinuities in $\vec{k}$-space is determined by the equality $\varepsilon^{(0)}(k) = \varepsilon^{(0)}\left(k + \frac{2\pi}{a} n\right)$. By a similar calculation of the wave function and energy of an electron for a three-dimensional case, it may be shown that the surfaces in $\vec{k}$-space on which the energy undergoes a discontinuity are also determined by the equalities

$$\varepsilon^{(0)}(\vec{k}) = \varepsilon^{(0)}(\vec{k} + 2\pi\vec{g}) \tag{2.42}$$

where $\vec{g}$ is one of the vectors of the reciprocal lattice. Since $\varepsilon^{(0)}(\vec{k}) = \frac{\hbar^2 |\vec{k}|^2}{2m_0}$, it then follows from (2.42) that

$$|\vec{k}|^2 = |\vec{k}|^2 + 2 \cdot 2\pi(\vec{k}\vec{g}) + 4\pi^2 |\vec{g}|^2$$

or

$$(\vec{k} + \pi\vec{g}, \vec{g}) = 0 \tag{2.43}$$

Condition (2.43), which determines the location of the surfaces of energy discontinuity in $\vec{k}$-space, has a clear geometrical and physical meaning. To show this, we construct a vector diagram corresponding to (2.43). For this, we place the origin of calculation of wave vector $\vec{k}$ into the point 0 coinciding with the end of vector $2\pi\vec{g}$. We draw a plane B perpendicular to vector $2\pi\vec{g}$ through its middle (Fig. 70). Any vector lying in this plane is perpendicular to vector $2\pi\vec{g}$ or $\vec{g}$. But a vector lying in the plane B is equal to the sum of vectors $\pi\vec{g}$ and $\vec{k}$, if $\vec{k}$ ends in plane B. Thus, the end of vector $\vec{k}$ satisfying equation (2.43) gives all the points of plane B. In order to find all the possible values of $\vec{k}$ satisfying (2.43), it is necessary to give all the vectors $\vec{g}$ of the reciprocal lattice. Vectors $\vec{g}$ are expressed through the basis of the reciprocal lattice $\vec{a}^*$, $\vec{b}^*$, and $\vec{c}^*$ as follows:

$$\vec{g} = n_1\vec{a}^* + n_2\vec{b}^* + n_3\vec{c}^* \tag{2.44}$$

where n_1, n_2, n_3 are whole numbers. The combination of the planes passing through the middles of vectors $2\pi\vec{g}$ and perpendicular to them gives, therefore, all the planes of energy discontinuities in $\vec{k}$-space. Note that the space of the reciprocal lattice differs from k-space only in that the scale of axes is changed by a factor of 2π.

These considerations make it possible to suggest the following simple method for constructing the first Brillouin zone. For the given lattice, we construct the reciprocal one, after which the scale of space of the reciprocal lattice is changed by a factor of 2π, which allows us to pass over into $\vec{k}$-space. Any selected node of the reciprocal lattice is then connected by straight lines with the nearest neighbouring nodes, and perpendicular planes are drawn through the middles of the sections obtained. The volume confined within these planes will be the first Brillouin zone.

This construction of the first Brillouin zone, which is essentially a unit cell in $\vec{k}$-space bounded by the planes of energy discontinuity that are the nearest to the given node of the reciprocal lattice, was first proposed by Wigner and Seitz. Because of this the first Brillouin zone is sometimes called the Wigner-Seitz cell in reciprocal space. Wigner-Seitz cells constructed near each node of a reciprocal lattice tightly fill the whole reciprocal space.

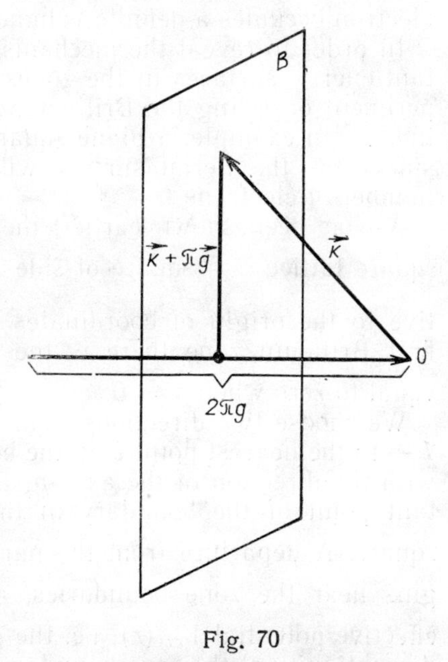

Fig. 70

As will be shown later, the geometric frame obtained in such manner plays the most important part in the construction of Fermi surfaces by Harrison's method in the scheme of repeating zones.

2-8. FILLING THE BRILLOUIN ZONE WITH ELECTRONS

Having determined the position of surfaces of energy discontinuity in $\vec{k}$- or $\vec{p}$-space and the Brillouin zones for lattices of various kinds, we can begin to construct the energy spectrum of electrons [1]. Our aim is to find the boundaries of Fermi distribution of electrons, i.e. of the Fermi surface, since only the electrons located sufficiently close to this surface can interact with external fields.

---

[1]) $\vec{k}$-and $\vec{p}$-spaces differ from one another only in the scale factor $\hbar$, since $\vec{p} = \hbar\vec{k}$.

For construction of the Fermi surface in $\vec{p}$-space, it is immaterial whether electrons in the metal are a Fermi gas or a Fermi liquid, since in both cases they obey the Fermi-Dirac statistics and each electron occupies a definite volume in $\vec{p}$-space.

In order to reveal the mechanism of formation of complex constant-energy surfaces in the zones, we will carry out a mental experiment on filling the Brillouin zones with electrons by considering, as an example, a plane square lattice with the period a. Consider how the Fermi surface will change at an increase of the number of electrons.

As has been shown earlier, the first Brillouin zone for a plane square lattice is a square of side $\frac{2\pi\hbar}{a}$ located symmetrically relative to the origin of coordinates in $\vec{p}$-space. In the centre of the first Brillouin zone there is the minimum of energy, since ε is equal to zero when $\vec{p} = 0$.

We choose two directions from the centre of the zone (Fig. 71): *1* — to the nearest point c on the boundary (this direction coincides with the direction of the axes p_x and p_y) and *2* — to the most distant point of the boundary in the direction of diagonals of the square. A departure from the parabolic relationship $\varepsilon = \frac{p^2}{2m_0}$ begins near the zone boundaries, and the earlier, the greater the effective potential $V_{eff}(\vec{r})$, i.e. the greater the magnitude of energy discontinuity at the zone boundary (see Fig. 49).

On the other hand, at small momenta, near the centre of the first zone, the law of dispersion is sufficiently close to the quadratic law $\varepsilon = \frac{p^2}{2m_0}$, so that the constant energy surfaces (constant-energy curves in the two-dimensional case considered) $\varepsilon(\vec{p}) =$ $=$ const can be approximated with circles of a radius

$$p = \sqrt{2m_0\varepsilon}$$

The distances from the origin of coordinates to the boundaries of the first Brillouin zone along the two directions selected, *1* and *2* (Fig. 71), are respectively equal to $\frac{\pi\hbar}{a}$ and $\frac{\sqrt{2}\,\pi\hbar}{a}$. The relationships for $\varepsilon(p)$ along each of these directions are shown in Fig. 72 (curves *1* and *2*).

In the direction *1*, the boundary of the zone is located closer by a factor $\sqrt{2} \approx 1.41$ than that in the direction *2*, owing to which the departure of the function $\varepsilon(p)$ from parabola occurs at smaller values of momentum in the first case. In the region of small momenta (when p is smaller than a certain value p' corresponding to the energy ε' of an electron), curves *1* and *2* in Fig. 72 practically coincide with one another and with the parabola $\varepsilon = \frac{p^2}{2m_0}$. With $p > p'$, curve *1* passes below curve *2*, so that for the same energy (for instance $\varepsilon = \varepsilon''$) the momenta along direction *1* (p_2'') are greater than those along direction *2* (p_1'').

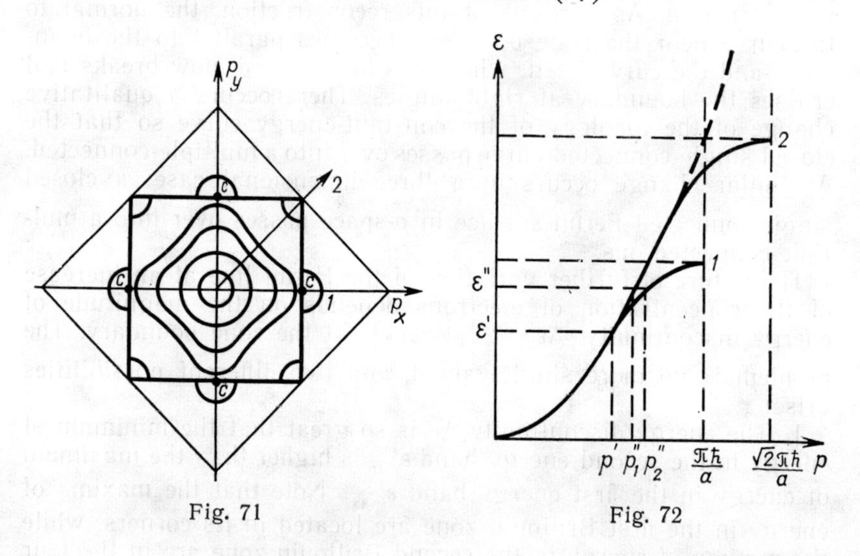

Fig. 71

Fig. 72

This means that the constant-energy surfaces $\varepsilon(\vec{p}) = \text{const}$, which were concentric circles of radius $p = \sqrt{2m_0\varepsilon}$ at $\varepsilon < \varepsilon'$, begin to be attracted, as it were, to the middles of the boundaries of the Brillouin zone that are located most closely to the centre of the zone. The degree of distortion of constant-energy curves is determined by how near they are to the boundary of the zone, and also by the magnitude of the pseudo-potential $V_{eff}(\vec{r})$. It is evident that with $V_{eff}(\vec{r}) = 0$ there will be no distortion, and the boundaries of the Brillouin zones will vanish.

The size of a constant-energy surface depends on the number of electrons. With an increase of this number, the boundary energy of Fermi-distribution increases. At the same time, the volume of $\vec{p}$-space bounded by the Fermi surface $\varepsilon(\vec{p}) = \varepsilon_F$ also increases, since the volume of $\vec{p}$-space per each two electron states is a constant equal to $\frac{(2\pi\hbar)^3}{V}$. Therefore, when mentally filling the

Brillouin zone with electrons, we bring the constant energy surface nearer to its boundaries [1]).

At an increase of the concentration of electrons, the "projections" of the distorted Fermi circle come nearer to the boundaries along the four equivalent directions *1*. Since a Fermi surface cannot touch the boundary of the Brillouin zone, then, with a definite small distance between the "projections" and the boundary in points *c*, the constant-energy curve should undergo qualitative reconstruction. As a result of this reconstruction, the normal to the curve near the zone boundary becomes parallel to the boundary, and the curve itself, which has been closed, now breaks and crosses the boundary at right angles. There occurs a qualitative change of the topology of the constant-energy curve so that the closed single-connected curve passes over into a multiple-connected. A similar change occurs in a three-dimensional case: a closed single-connected Fermi surface in $\vec{p}$-space passes over into a multiple-connected one.

The nature of further variation of the Fermi line at an increase of the concentration of electrons depends on the magnitude of energy discontinuity $\Delta\varepsilon = 2\left|(V_{eff})_{\vec{g}}\right|$ at the zone boundary. The problem is no more single-valued, and two different possibilities arise:

1. The energy discontinuity $\Delta\varepsilon$ is so great that the minimum of energy in the second energy band ε^{II}_{min} is higher than the maximum of energy in the first energy band ε^{I}_{max}. Note that the maxima of energy in the first Brillouin zone are located in its corners, while the minima of energy in the second Brillouin zone are in the four equivalent points *c* corresponding to the following values of the momentum: $p_x = \pm\frac{\pi\hbar}{a}$ and $p_y = 0$; $p_x = 0$ and $p_y = \pm\frac{\pi\hbar}{a}$. In that case, with an increase of the concentration *n* of electrons, it is profitable from the standpoint of energy that all the permitted states in the first zone were filled.

The total number of permitted states in the zone is twice the number of atoms $2N$. This can be easily shown, for instance, for a simple cubic lattice with the period *a*. Indeed, the volume of each Brillouin zone $V_{\vec{p}}$ in $\vec{p}$-space is equal to the volume of a cube with the side $\frac{2\pi\hbar}{a}$, i.e. $V_{\vec{p}} = \left(\frac{2\pi\hbar}{a}\right)^3$. The volume of one electron state twice degenerated by the spin is $\frac{(2\pi\hbar)^3}{V}$, where V is

[1]) An experiment on filling the zone with electrons is principally feasible. The number of valence electrons may be varied by introducing some or other amount of an impurity of a greater valence (donor impurities) into the metal. For semi-metals Bi, As and Sb, such an impurity may be, Te or Se.

the volume of crystal. Thus, the number of electrons filling the zone is $2\frac{V_{\vec{p}}}{(2\pi\hbar)^3/V}$, or $2\frac{V}{a^3}=2N$.

Passing to a unit-volume crystal, it may be concluded that the first Brillouin zone must be completely filled with the concentration of electrons twice the concentration of atoms in the lattice. This condition is satisfied for bivalent substances. In the case considered, bivalent substances are dielectrics or semiconductors, depending on the magnitude of the interval of forbidden energies between the bands, which is equal to the difference $\varepsilon^{II}_{min}-\varepsilon^{I}_{max}$. Location of the bands at $\varepsilon^{II}_{min}>\varepsilon^{I}_{max}$ is shown schematically in Fig. 73.

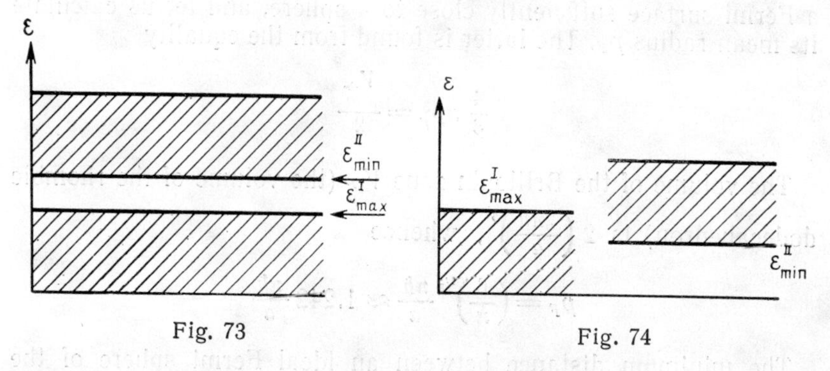

Fig. 73

Fig. 74

2. The energy discontinuity $\Delta\varepsilon$ is not large and the minimum of energy in the second band lies below the maximum of energy in the first zone ($\varepsilon^{II}_{min}<\varepsilon^{I}_{max}$, Fig. 74).

In that case filling of the first band proceeds only to the energy level ε equal to ε^{II}_{min}. At greater values of ε, it becomes energetically profitable to fill the free states in the second band near the points c corresponding to the minimum values of energy (Fig. 71). The corners of the first zone then remain unfilled. With such a situation, bivalent substances are metals. These conditions are realized in all the elements of the group II of the Periodic Table. Elementary semiconductors or dielectrics with a valency of 2 are non-existent in nature.

2-9. FERMI SURFACES OF THE ELEMENTS OF GROUP I OF THE PERIODIC TABLE

At filling the first Brillouin zone with electrons, the analysis of variation of constant-energy surfaces which has been made in the previous section, is sufficient to predict the most probable shape of Fermi surfaces for the elements of group I of the Periodic Table. These include alkali metals (Li, Na, K, Rb) having a body-centered lattice at room temperature, and also the metals Cu, Ag, and Au having a face-centered lattice.

Let us start from alkali metals. The first Brillouin zone for a body-centered lattice was constructed in Sec. 2-6 of this chapter and has the form of a rhombic dodecahedron (Fig. 60) in which all the faces are equal and located at the same distance $\frac{\sqrt{2}\,\pi\hbar}{a}$ from the centre of the zone, where a is the period of the lattice.

In alkali metals, N valence electrons fill one half of the volume of the first zone. Let us assume that the part of the space of the first Brillouin zone which is filled with electrons is bounded by a Fermi surface sufficiently close to a sphere, and let us calculate its mean radius p_F. The latter is found from the equality

$$\frac{4}{3}\pi p_F^3 = \frac{V_{\vec{p}}}{2}$$

The volume of the Brillouin zone $V_{\vec{p}}$ (the volume of the rhombic dodecahedron) is $2\left(\frac{2\pi\hbar}{a}\right)^3$, whence

$$p_F = \left(\frac{6}{\pi}\right)^{1/3} \frac{\pi\hbar}{a} \approx 1.242\,\frac{\pi\hbar}{a}$$

The minimum distance between an ideal Fermi sphere of the same radius and the boundary of the first zone is

$$\sqrt{2}\,\frac{\pi\hbar}{a} - p_F \approx 0.172\,\frac{\pi\hbar}{a}$$

In other words, the radius of the Fermi sphere p_F constitutes approximately 88 per cent of the minimum distance from the origin of coordinates to the boundary of the first Brillouin zone. Since the magnitude of effective potential, generally speaking, is not great, we can expect that with such a relatively large distance from the Fermi sphere to the faces of the Brillouin zone their distorting effect on the shape of the Fermi surface will also be not great.

But the radius of the Fermi sphere is not so small as to make completely unnoticeable the distortion of the real constant-energy surface in alkali metals at close distances to the zone boundary. The expected small deviations of the Fermi surface from the sphere should evidently have the form of twelve projections located in the directions of the shortest distances from the centre of the zone to its faces (Fig. 75).

A small deviation of the Fermi surface from the sphere can be characterized qualitatively by the anisotropy of this surface

$$\frac{\Delta S}{S} = 2\,\frac{(S_{max} - S_{min})}{(S_{max} + S_{min})}$$

where S_{max} and S_{min} are the maximum and the minimum areas of the section of the Fermi surface with the planes passing through the centre of the Brillouin zone. For a sphere, evidently, $\frac{\Delta S}{S} = 0$.

The values of anisotropy of Fermi surfaces for alkali metals, found experimentally by Shoenberg [55], are given in the table below

Metal	Na	K	Rb	Cs
$\frac{\Delta S}{S}$, %	0.2	0.6	0.7	1.4

As will be seen from the table, anisotropy of Fermi surfaces does not exceed 1.5 per cent, so that replacement of a real Fermi surfacc with a sphere gives only an insignificant error. An increase of the anisotropy $\Delta S/S$ together with the atomic number in the alkali metal row indicates to an increase of the pseudo-potential $V_{eff}(\vec{r})$ as a result of more complicated shape of atomic structures. The effective mass m^* of electrons in alkali metals, as should be expected, must be close to the mass of a free electron. Measurements have shown that for all alkali metals [1]) $m^* \approx (1.0\text{-}1.2)\, m_0$.

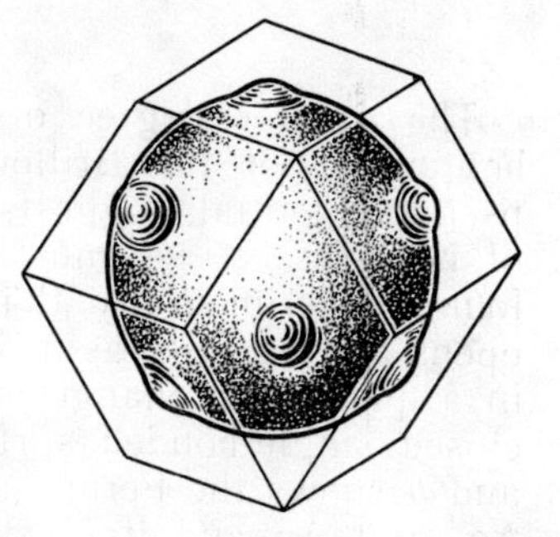

Fig. 75

We now consider the metals of group I having a face-centered lattice (Cu, Ag, Au). In this case the first Brillouin zone is a cubo-octahedron (Fig. 62) which is less symmetrical than the rhombic dodecahedron. The eight hexagonal faces of the zones which are the nearest to the centre extend from the latter at a distance of $\frac{\sqrt{3}\,\pi\hbar}{a}$. Six square faces are at a distance $\frac{2\pi\hbar}{a}$

[1]) The cause of the slight increase of the effective mass compared with the mass of a free electron will be discussed later.

from the centre. The volume of the cubo-octahedron $V_{\vec{p}}$ is equal to $4\left(\frac{2\pi\hbar}{a}\right)^3$. It also is half filled with electrons. The radius p_F of an ideal Fermi sphere in this case is $\left(\frac{3}{2\pi}\right)^{1/3}\frac{2\pi\hbar}{a} \approx 1.56\frac{\pi\hbar}{a}$ and differs by only 9.8 per cent from the minimum distance from the centre to the zone boundary, which is approximately equal to $1.73\frac{\pi\hbar}{a}$. In addition, because of the more complex shape of ionic structures, the effective potential $V_{eff}(\vec{r})$ of copper, silver and gold exceeds substantially $V_{eff}(\vec{r})$ of alkali metals.

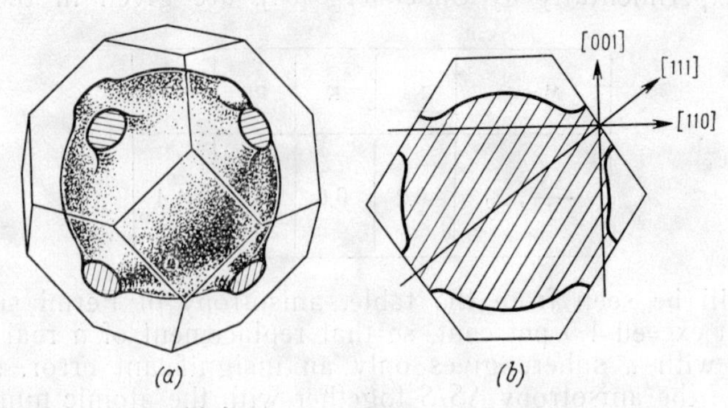

Fig. 76

The greater degree of closeness of the Fermi sphere to the boundaries of the Brillouin zone, in combination with greater pseudo-potential, suggests that the Fermi surface near the centres of hexagonal faces must be substantially distorted. It has been found experimentally that the Fermi surface near these points opens and adheres, as it were, to the zone boundaries. This results in a qualitative change of the topology of Fermi surface, i.e. the closed single-bonded surface becomes an open one. Figure 76 *a* and *b* shows the Fermi surface in the first Brillouin zone for Cu, Ag, and Au and its section with a vertical plane of symmetry passing through the axis p_z and the bisector of the angle between the axes p_x and p_y.

Since energy is a periodic function of momentum, the Fermi surface in first Brillouin zone which has been obtained must be translated for the periods of the principal translations in $\vec{p}$-space. This operation gives a surface which is open in four directions and passes through the whole $\vec{p}$-space. The shape of this surface and of its section with a vertical plane of symmetry passing through the bisector of the coordinate angle are shown in Figs. 77 and 78 respectively.

Thus, the Fermi surfaces of copper, silver and gold form a system of spheres connected by thin "necks" in the direction of

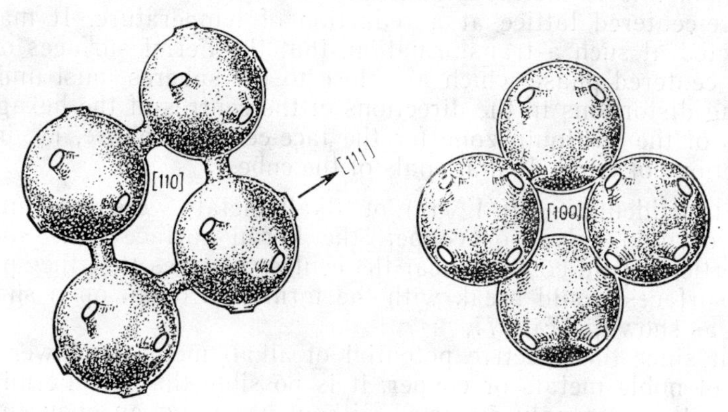

Fig. 77

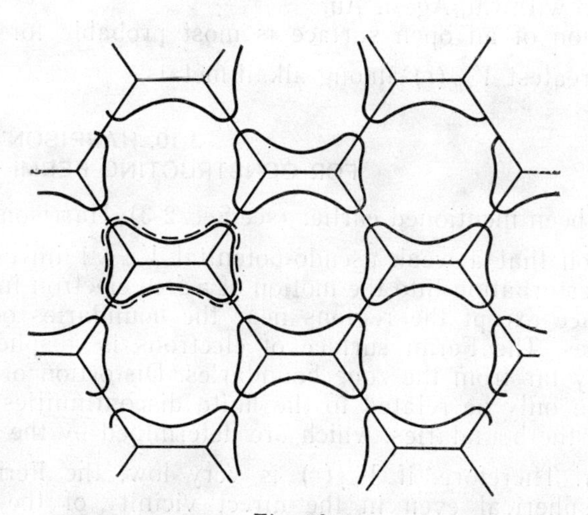

Fig. 78

body diagonals. This structure of the Fermi surface was first found experimentally by Pippard [28]. The spherical portion of the surface is called the "belly" in the specialist literature. The diameter of the necks between spheres strongly depends on pseudo-potential. With the growth of the atomic number in the Cu, Ag, Au series, the ionic structure of the metal becomes more complicated, which results in an increase of $V_{eff}(\vec{r})$ and diameter of the "neck" of the Fermi surface.

It has been indicated earlier that the body-centered lattice of alkali metals is thermodynamically unstable and transforms into a face-centered lattice at a reduction of temperature. It may be expected at such a transformation, that the Fermi surfaces of the body-centered phase which are close to the spheres must undergo strong distortions in the directions of the centres of the hexagonal faces of the Brillouin zone for the face-centered phase, i.e. in the directions of the body diagonals of the cube.

It is evident that if $V_{eff}(\vec{r})$ of alkali metals were the same as that of gold, silver or copper, the Fermi surface after such a distortion would come so near the centres of faces that the sphere-like surfaces would break with the formation of an open surface, such as shown in Fig. 77.

But since the effective potential of alkali metals is lower than that of noble metals or copper, it is possible that the Fermi surface will be strongly distorted without becoming an open surface or with the formation of open surfaces with substantially thinner necks than with Cu, Ag, or Au.

Formation of an open surface is most probable for Cs, which has the greatest $V_{eff}(\vec{r})$ among alkali metals.

2-10. HARRISON'S METHOD FOR CONSTRUCTING FERMI SURFACES

As has been mentioned earlier (see Sec. **2-3**), Harrisons's method is based on that a weak pseudo-potential $V_{eff}(\vec{r})$ introduces only a small perturbation into the motion of a free electron in the whole phase space except the regions near the boundaries of the Brillouin zones. The Fermi surface of electrons is a sphere if it is sufficiently far from the zone boundaries. Distortion of the Fermi sphere can only be related to the finite discontinuities $\Delta\varepsilon$ of the energy at the boundaries, which are determined by the magnitude of $V_{eff}(\vec{r})$. Therefore, if $V_{eff}(\vec{r})$ is very low, the Fermi surface remains spherical even in the direct vicinity of the planes of energy discontinuity.

In the final result, if $V_{eff}(\vec{r})$ is made arbitrarily low, a sphere filled with electrons will not practically be distorted at the zone boundaries and at an increase of the concentration of electrons will grow in its volume without changing the shape. It means physically that if an energy discontinuity $\Delta\varepsilon$ at the boundary of the zones becomes substantially smaller than, for example, the energy of thermal oscillations $\sim kT$, then the band nature of the energy spectrum will have no effect on the motion of electrons. Because of thermal oscillations, electrons in that case can freely overcome the interval of forbidden energies, which actually is equivalent to the absence of the forbidden interval. Consequently, with an arbitrarily low $V_{eff}(\vec{r})$ the Fermi surface can cross the boundaries of a number of Brillouin zones without "noticing" the structure formed by the planes of energy discontinuities.

We shall proceed as follows. First we assume that $V_{eff}(\vec{r})$ is negligibly low. In that case we can describe a Fermi sphere of any radius p_F from the centre of the first Brillouin zone (i.e. a Fermi sphere comprising any number of electrons) which will cross without distortion a whole number of zone boundaries.

Now let $V_{eff}(\vec{r})$ be other than zero. This will result in energy discontinuities appearing at the zone boundaries and the surface of the sphere will be distorted in such a manner that will cross the boundaries at right angles. The sphere will then break into a number of pieces. Noting the periodicity of energy and translating these pieces of the surface, we can construct constant-energy surfaces in various zones. This exactly is the principal idea behind Harrisons's method for constructing constant-energy surfaces.

The main assumption in Harrison's method is that a Fermi sphere constructed at $V_{eff}(\vec{r}) \equiv 0$ undergoes distortion at finite values of the pseudo-potential only near the boundaries of the zones proper, these distortions being the smaller, the lower $V_{eff}(\vec{r})$ is.

In other words, it is assumed that we can continuously pass over from the state of free electrons to a new qualitative state, i.e. the interaction of electrons with the periodic lattice which manifests itself in the appearance of Bragg reflection of electrons from the lattice with the momenta corresponding to the boundaries of the zones.

Notwithstanding this coarse assumption, Harrisons's method is practically reasonable for all monovalent and polyvalent simple metals, i.e. for metals in which inner electron shells of ionic structures are filled. For these metals, a simple approximation of the model of free electrons can be used, which makes it possible to determine qualitatively the Fermi surface. This approximation is essentially based on the possibility of separation of electrons into atomic (i.e. electrons of inner shells which do not contribute to metal conductivity) and collective electrons (i.e. those which fill the outer shell in a separate atom of metal and are current carriers in metal crystals).

The metals for which this approximation is valid (simple metals) relate to *s*- and *p*-elements, since their main properties are determined by *s*- and *p*-electrons filling the outer shells. In addition, the Periodic Table contains three distinct periods of transition metals beginning with the elements Sc, Y, and La and ending with the elements Ni, Pd, and Pt. These periods correspond to the filling of the 3*d*-, 4*d*-, and 5*d*-shells respectively. The free-electron approximation is inapplicable for the metals of these periods. Some special properties of the transition metals indicated are to a large extent determined by *d*-electrons.

Note, however, that the group of metals with unfilled *d*-shells also includes such monovalent metals as Cu, Ag, Au, for which the Fermi surfaces can be qualitatively constructed within the frames of the almost free electron approximation, and bivalent metals Zn, Cd, and Hg for which this approximation is also fairly applicable. The point is that inapplicability of the model of almost free electrons is determined not only by the existence of an unfilled *d*-shell (or of a deeper *f*-shell, as is the case with the series of the rare-earth metals of the lanthanum group), but is also related to the metal conductivity of *d*- (*f*-) electrons [or else of the conductivity over the *d*-(*f*-) bands].

In mono- and bivalent metals with an unfilled *d*-shell, the Fermi level passes in the bands corresponding to *s*- and *p*-electrons; *d*-electrons in such metals are fairly well localized, so that the wave functions of these electrons belonging to neighbouring atoms practically do not overlap and there is no conduction in the *d*-zone. Because of this, there are no specific features which are characteristic of polyvalent transition metals and related to the peculiarities of conduction in the *d*-band.

The properties of mono- and bivalent metals with an unfilled *d*-shell are mainly determined by *s*- and *p*-electrons, as also the properties of simple metals. This explains why the model of almost free electrons is applicable to them.

In the case of polyvalent transition metals, the Fermi level is almost in the middle of the energy band corresponding to *d*-shell electrons. Their electronic properties (conductivity, electronic fraction of heat capacity, magnetic susceptibility, etc.) are practically defined completely by *d*-electrons near the Fermi level. The data on complex Fermi surfaces of transition metals cannot be explained by means of the almost free energy zone corresponding to *s*- and *p*-electrons. The latter indicates to that *s*- and *p*-electrons play no substantial part in the electronic properties of these metals.

The electronic *d*-states of transition metals, though ensuring metallic conductivity over the *d*-band, cannot be related to almost free states. The orbits of *d*-electrons are rather small compared with the orbits of other (*s*- and *p*-) valence electrons. As a result, the *d*-states behave to a large extent as localized ones, being not greatly perturbed by the lattice potential. The conductivity over a d-band corresponds more closely to quantum tunnelling junctions of d-electrons between neighbouring atoms owing to a small overlap of the d-wave functions of these atoms, rather than to the quasi-free motion of s- and p-electrons which is characteristic of simple metals.

Having made preliminary remarks on the idea behind Harrison's method and on the causes by which the method is inapplicable for polyvalent transition metals, we shall now discuss the particular techniques for constructing Fermi surfaces of transition metals.

First of all, we determine the radius p_F of Harrison's sphere which surrounds the volume occupied by electrons in the zero approximation: $V_{eff}(\vec{r}) \equiv 0$. For this, we express p_F through the volume $V_{\vec{p}}$ of the Brillouin zone. The volume in the zone per one electron is equal to $\frac{V_{\vec{p}}}{2N}$.

Hence the volume of $\Delta\vec{p}$-space occupied by electrons in a metal of valence $\varkappa$ is

$$\Delta = \frac{V_{\vec{p}}}{2N}\varkappa N = \frac{\varkappa}{2} V_{\vec{p}} \tag{2.45}$$

The volume of the Harrison sphere $\left(\frac{3}{4}\pi p_F^3\right)$ coincides with Δ, and therefore, p_F is only determined by the type and parameters of the lattice (through $V_{\vec{p}}$) and the valence $\varkappa$ of the metal:

$$p_F = \left(\frac{3\varkappa V_{\vec{p}}}{8\pi}\right)^{1/3} \tag{2.46}$$

In a two-dimensional case, Harrison's sphere degenerates into Harrison's circle whose radius is

$$p_F = \sqrt{\frac{\varkappa S_{\vec{p}}}{2\pi}} \tag{2.47}$$

where $S_{\vec{p}}$ is the area of the two-dimensional Brillouin zone in $\vec{p}$-space.

For more simplicity, let us first construct the Fermi "surface" for a hypothetical trivalent metal ($\varkappa = 3$) having a plane square lattice of period a. The area $S_{\vec{p}}$ of the Brillouin zone for such a lattice is $\left(\frac{2\pi\hbar}{a}\right)^2$. According to (2.47):

$$p_F = \sqrt{\frac{3}{2\pi}}\left(\frac{2\pi\hbar}{a}\right) \approx 1.386\,\frac{\pi\hbar}{a} \tag{2.48}$$

From the centre of the first Brillouin zone (Fig. 55), we draw Harrison's circle. Since the radius $p_F \approx 1.386 \frac{\pi\hbar}{a}$ is smaller than the distances to the corners of the first zone, which are equal to $\sqrt{2}\frac{\pi\hbar}{a}$, the part of the surface of the first zone near its corners will be outside the Harrison circle (Fig. 79). On the other hand, a certain portion of the surface in the second Brillouin zone will be inside that circle.

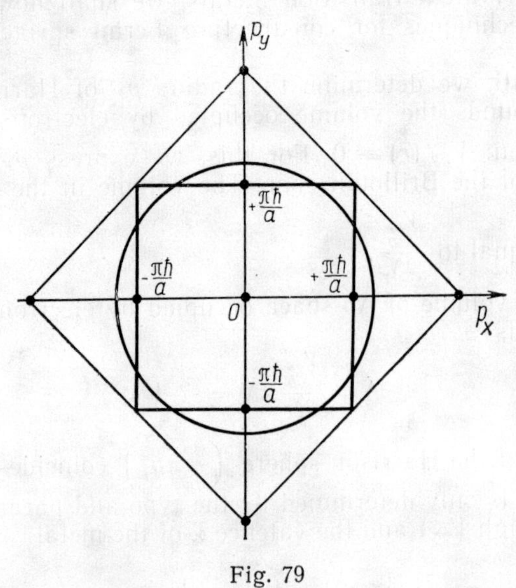

Fig. 79

It then follows that the first zone of the "metal" considered is not completely filled with electrons, and in addition, that the second zone is filled to a low extent. Let us recall that drawing Harrison's circle in such a manner we assume that $V_{eff}(\vec{r}) \equiv 0$. Now, we shall assume $V_{eff}(\vec{r})$ to be low but not zero. Then, as in the one-dimensional case (see Fig. 49), the following will occur:

(1) energy discontinuities appear at the boundaries of the zones [strictly speaking, at $V_{eff}(\vec{r}) \equiv 0$ there were no zones, i.e. surfaces of Bragg reflection, and they only appear at $V_{eff}(\vec{r}) \neq 0$];

(2) energy becomes a periodic function in $\vec{p}$-space with the period $\frac{2\pi\hbar}{a}$ of main translations along the axes p_x and p_y.

As a result of appearance of energy discontinuities at the boundaries of the zones, Harrison's circle breaks at the boundaries and forms separate "pieces" of constant-energy curves in the zones, as shown in Fig. 80.

We translate these "pieces" into periods $2\pi\hbar/a$ along the axes p_x and p_y. From the "pieces" of constant-energy curves in the first zone, after translation, we may construct closed constant-energy curves around the corners of the zone (Fig. 81). Similarly, from the "pieces" of curves in the second zone there are formed closed constant-energy curves on the sides of the square (Fig. 82).

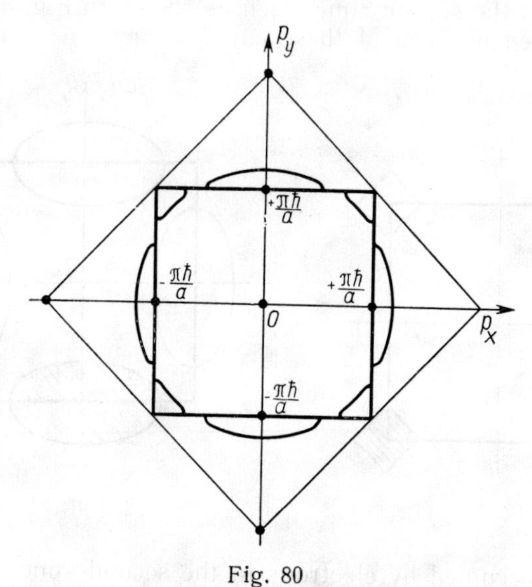

Fig. 80

The set of all constant-energy curves in the first and second zones forms a two-dimensional Fermi surface.

Note that from the pieces of Harrison's circle in the first zone we can construct only a single closed curve, such as shown in Fig. 81 (because of the translational symmetry, this curve is naturally repeated in $\vec{p}$-space with the period $2\pi\hbar/a$ in the direction of the Cartesian coordinates). If the area $S'(\varepsilon)$ bounded by this curve is divided by the elementary area $\left(\frac{2\pi\hbar}{a}\right)^2/2N$ occupied by one electron state, then the total number of vacant places n_1' (unfilled states) in the first energy band will be:

$$n_1' = 2\,\frac{S'(\varepsilon)}{(2\pi\hbar)^2}\,a^2N \tag{2.49}$$

where N is the number of atoms in the lattice. Since a^2N gives the total area of a two-dimensional lattice, the number of vacant places per unit area of the lattice is:

$$n_1 = \frac{n_1'}{a^2 N}$$

or

$$n_1 = \frac{2S'(\varepsilon)}{(2\pi\hbar)^2} \tag{2.50}$$

We can construct two closed curves from the pieces of Harrison's circle in the second zone, such as shown in Fig. 82. The area $S''(\varepsilon)$ bounded by each of these curves constitutes only one half

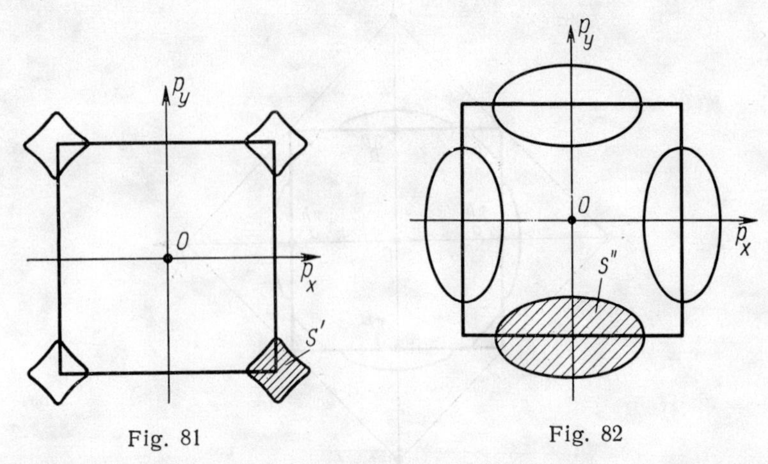

Fig. 81 Fig. 82

of the area occupied by electrons in the second zone. The number of electron states n'' in the second zone will be:

$$n'' = \frac{2S''(\varepsilon)}{\left(\frac{2\pi\hbar}{a^2}\right)^2 / 2N} = 2\,\frac{2S''(\varepsilon)}{(2\pi\hbar)^2}\,a^2 N \tag{2.51}$$

whence the concentration n_2 of electrons in the second energy band is:

$$n_2 = 2\,\frac{2S''(\varepsilon)}{(2\pi\hbar)^2} \tag{2.52}$$

Let us see how the shape of constant-energy curves will vary in the zones if the valence $\varkappa$ of the metal considered is changed from three to four. In that case the radius of Harrison's circle will be:

$$p_F = \sqrt{\frac{2}{\pi}}\left(\frac{2\pi\hbar}{a}\right) \approx 1.6\,\frac{\pi\hbar}{a} \tag{2.53}$$

The Harrison circle of such a radius completely confines the first Brillouin zone and passes through the second, third, and fourth zones (Fig. 83). Transition from $V_{eff}(\vec{r}) \equiv 0$ to a finite pseudo-potential causes breaks in the Harrison circle at the boun-

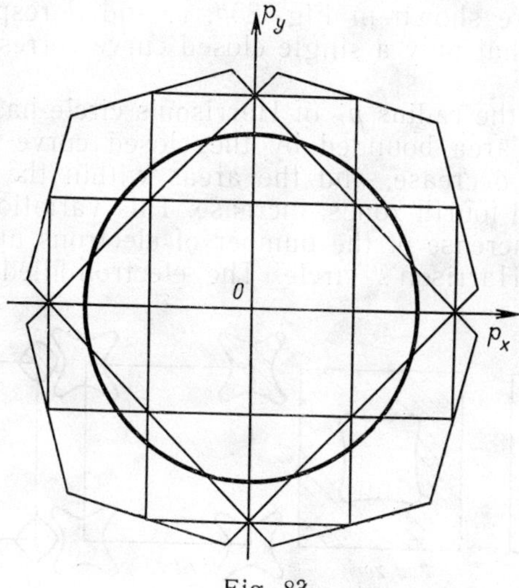

Fig. 83

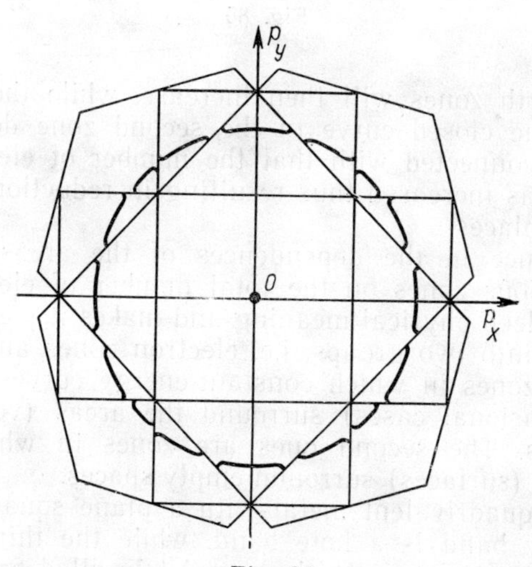

Fig. 84

daries of the zones and a variation of the direction of the normals to constant-energy curves in boundary regions (Fig. 84).

For constructing closed constant-energy curves in various zones, we translate the pieces of Harrison's circle, as before, over the periods $2\pi\hbar/a$ along the axes p_x and p_y. Since the first zone has been completely filled, there are no constant-energy curves in it (Fig. 85*a*). The constant-energy curves in the second, third and fourth zones are shown in Fig. 85*b*, *c*, and *d* respectively. It is easy to verify that only a single closed curve corresponds to each zone.

Assume that the radius p_F of Harrison's circle has been slightly increased. The area bounded by the closed curve in the second zone will then decrease, and the areas within the closed curves of the third and fourth zones, increase. This variation of the areas results in an increase of the number of electrons upon increasing the radius of Harrison's circle. The electron-filled areas in the

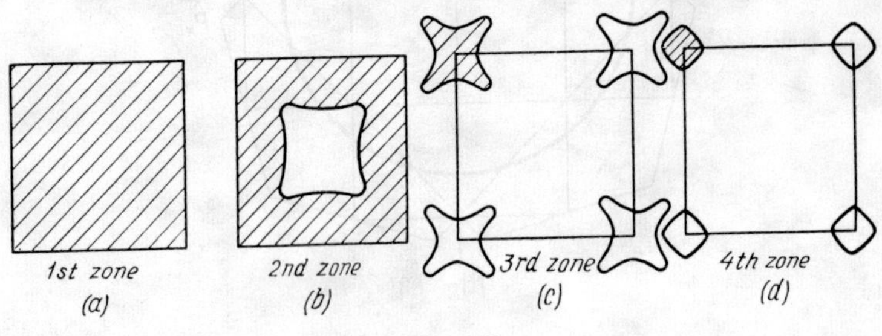

Fig. 85

third and fourth zones will then increase, while the empty area bounded by the closed curve in the second zone decreases. The latter fact is connected with that the number of electrons in the second zone has increased thus resulting in reduction of the number of vacant places.

This difference in the dependences of the areas bounded by curves in various zones on the total number of electrons in the metal is of a deep physical meaning and makes it possible to class energy zones into two groups, i.e. electron zones and hole zones. The first are zones in which constant-energy curves (or surfaces in three-dimensional cases) surround the areas (volumes) filled with electrons. The second ones are zones in which constant-energy curves (surfaces) surround empty spaces.

Thus, in a quadrivalent metal with a plane square lattice, the second energy band is a hole band, while the third and fourth bands are electron zones. In the completely filled first band, electrons do not participate in the physical processes and this band may be excluded from discussion.

Using this classification of zones, electrons will be defined as quasi-particles located on constant-energy surfaces of the electron type. Quasi-particles located on constant-energy surfaces of the hole type will be called holes. The physical differences in the behaviour of quasi-particles in external electric and magnetic fields is connected with the difference in the constant-energy surfaces on which they are located. Electron and hole surfaces are located in different energy bands, with an energy barrier being between them. As a result of this, the quasi-particles located on a surface of one type cannot pass into surfaces of another type under common conditions.

The construction of constant-energy curves by Harrison's method described above on the example of a two-dimensional metal can be generalized for a three-dimensional case. But determination of constant-energy surfaces in real three-dimensional metals encounters a large difficulty, since it is required to construct preliminarily a number of Brillouin zones and find pieces of Harrison's sphere in these zones. A method exists, however, which makes it possible to simplify the construction of a Fermi surface in various zones. To explain this method, let us dwell once more on the sequence of the main stages of the construction discussed above.

First of all, the construction was based on an aperiodic picture of the so-called extended Brillouin zones in $\vec{p}$-space, i.e. the zones built up to each other around the first Brillouin zone. Proceeding from such construction of the zones, there were found pieces of Harrison's circle (sphere) in various zones. Transition to a periodic picture in p-space was made only at the final stage, when closed curves (surfaces) were formed in the zones by means of translations of separate pieces. This final stage is of principal importance, since the energy becomes a multiple-valued and periodic function of momentum only owing to this stage. But the transition to a periodic picture can be made preliminarily. Such an operation, as will be shown later, makes it possible to substantially simplify the construction of the Fermi surface.

Instead of considering the initial aperiodic picture of extended zones, we can preliminarily translate the boundaries of energy discontinuities in the second, third, etc. zones into the first Brillouin zone. The result of such a procedure for the first three Brillouin zones for a case of a plane square lattice with the period a, is shown in Fig. 86.

After this procedure, the first Brillouin zone contains not only a set of physically different values of quasi-momentum, but also all the possible straight lines (planes) on which the energy undergoes discontinuity. Thus we pass over in advance to a periodic picture in $\vec{p}$-space: a periodic repetition of the first Brillouin zone with the surfaces of energy discontinuities contained inside it fills the whole $\vec{p}$-space.

Let us imagine a mosaic composed of the first Brillouin zones placed side by side in $\vec{p}$-space. This structure has been termed the pattern of repeating zones. All neighbouring zones are equivalent between themselves. This means that Harrison's circles (spheres) can be described from the centre of each zone.

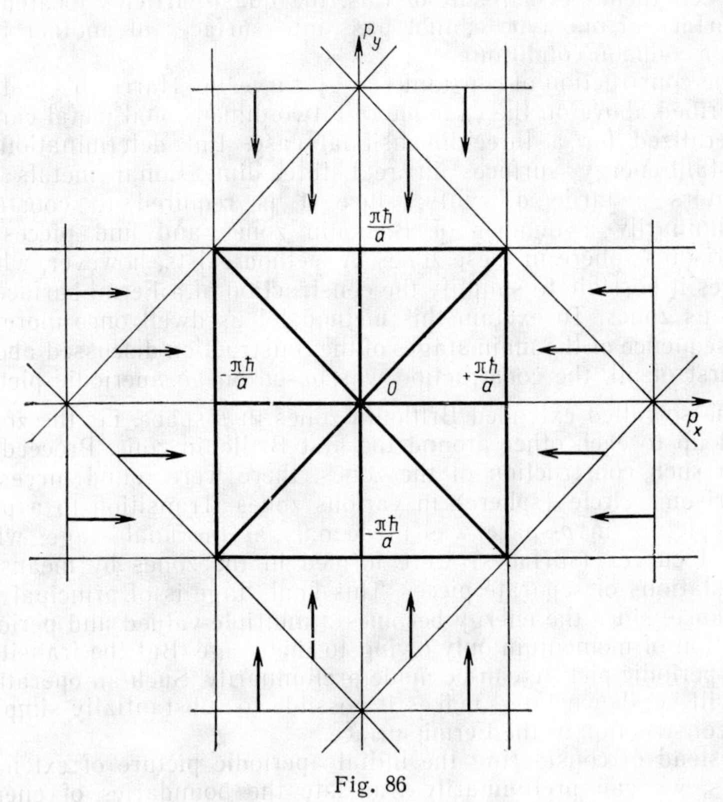

Fig. 86

The result of such an operation for a case of a quadrivalent metal with a plane square lattice is shown in Fig. 87. Harrison's circles described from the centres of neighbouring zones intersect each other and form closed constant-energy curves. This automatically gives the same picture as that constructed earlier for the same case.

An analysis of the picture obtained will reveal the following regularities:

(1) the regions bounded by pieces of lines of negative (relative to the inner areas) curvature (concave lines) and belonging simultaneously to k or more circles represent the regions bounded

by hole-type constant-energy curves in the $(k+1)$-th energy band;

(2) the regions bounded by pieces of lines of positive curvature (convex curves) and belonging simultaneously to k or more circles represent the regions bounded by electron-type constant-energy curves in the k-th energy band.

Let us illustrate the use of these rules for various curves in Fig. 87. Indeed, the region bounded by the closed constant-energy curve in the second zone belongs only to one circle described from the centre of the first zone. It is bounded by the constant-energy

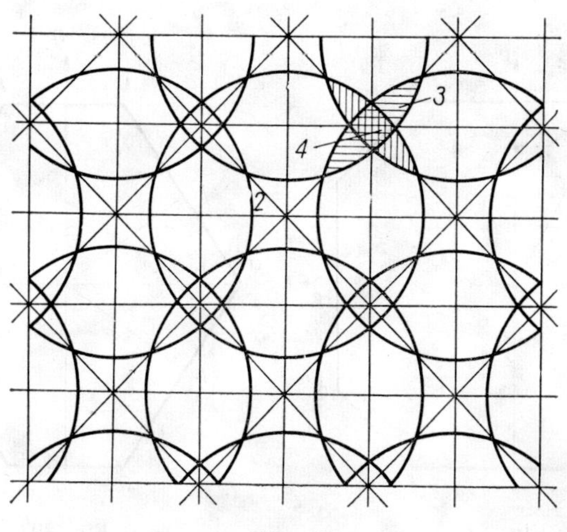

Fig. 87

curve of the hole type (of a negative curvature relative to the internal region). The regions shaded in oblique and cross-hatched simultaneously belong to three circles and are called "rosettes". They are bounded by lines of positive curvature and include the areas belonging to four circles. Their boundary is therefore an electron-type constant-energy curve in the third zone. The central portion of the "rosette", which is shown cross-hatched, belongs simultaneously to four circles. Its boundary consists of "pieces" of curves of positive curvature, and therefore, is an electron-type constant-energy curve in the fourth zone.

It may be clearly seen from Fig. 87 that energy is a multiple-valued function of momentum. Indeed, the central portion of the "rosette" belongs to the fourth energy band. At the same time, it is included into the whole "rosette" and relates to the third energy zone. It also belongs to the second and first bands. Therefore, to each quasi-momentum from the central portion of the "rosette" there correspond the values of energy of electrons from the first, second, third, and fourth bands. Similarly, to each quasi-momentum from the petals of the "rosette" there correspond the values of energy of electrons from the first, second, and third bands, etc.

In order to generalize these rules of classification of constant-energy lines for three-dimensional cases, it is sufficient to change the words "lines" and "areas" to "surfaces" and "volumes" respectively. But in three-dimensional cases the classification becomes more complicated because the open surfaces formed can have regions of both positive and negative curvature relative to the volume confined.

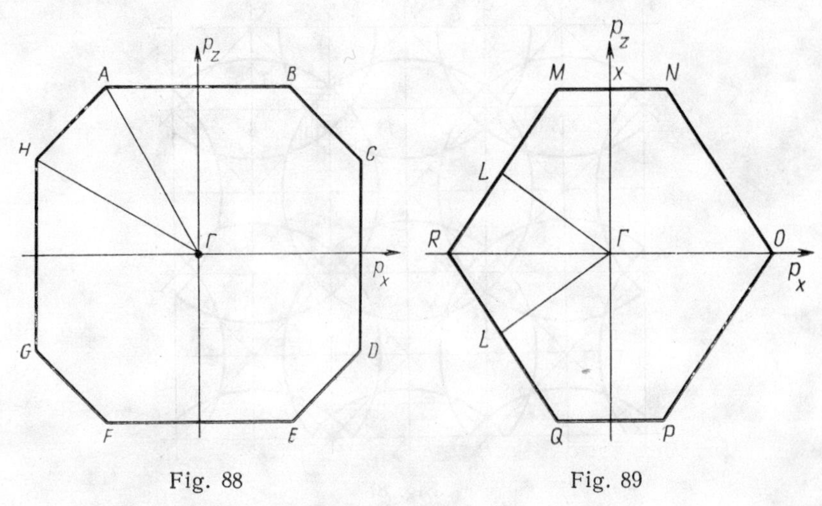

Fig. 88 Fig. 89

Thus, an open surface cannot be fully referred either to the electron or hole type and the above classification is to a large extent conditional for it. In that connection it is only adopted conditionally that if an open surface surrounds the volume belonging to k Harrison's spheres, it is related to the k-th band. At this stage, classification of open surfaces is finished.

Let us now discuss the construction of Fermi surfaces by Harrison's method for real metals. Consider a particular case of a trivalent metal having a face-centered lattice with the period a. An example of such a metal may be aluminium.

The first Brillouin zone for a face-centered lattice is a cubo-octahedron (see Sec. **2-6**, Chapter Two, Fig. 62), which is a body bounded by eight hexagonal and six square faces.

The families of "network" surfaces that have engendered the hexagonal faces of the Brillouin zone are the families of planes perpendicular to the body diagonals of an elementary cube of the crystal lattice. They are at distances $a/\sqrt{3}$ from each other and make an angle arc cos $(1/\sqrt{3})$ with the plane xy. The hexagonal faces of the Brillouin zone in $\vec{p}$-space are, therefore, at a distance $\sqrt{3}\pi\hbar/a$ from the origin of coordinates and make the same angle with the plane $p_x p_y$.

The families of "network" surfaces forming the square faces of the Brillouin zone are families of planes perpendicular to the axes x, y, and z and removed a distance $a/2$ from each other. Thus, the square faces of the Brillouin zone in $\vec{p}$-space are at a distance $2\pi\hbar/a$ from the origin of coordinates and perpendicular to the axes p_x, p_y and p_z.

For clarity, let us plot the sections of the cubo-octahedron by two vertical planes: a coordinate plane, for instance, $p_x p_y$ (Fig. 88), and a plane passing through the axis p_z and the bisector of the coordinate angle in the plane $p_x p_y$ (diagonal plane) (Fig. 89).

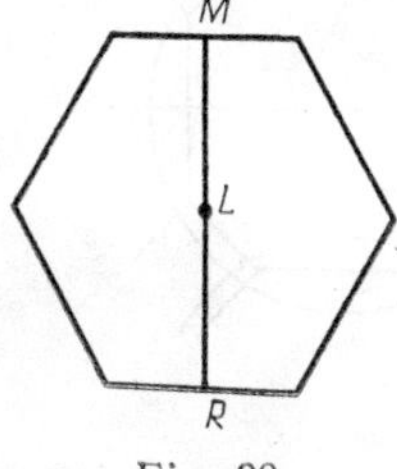

Fig. 90

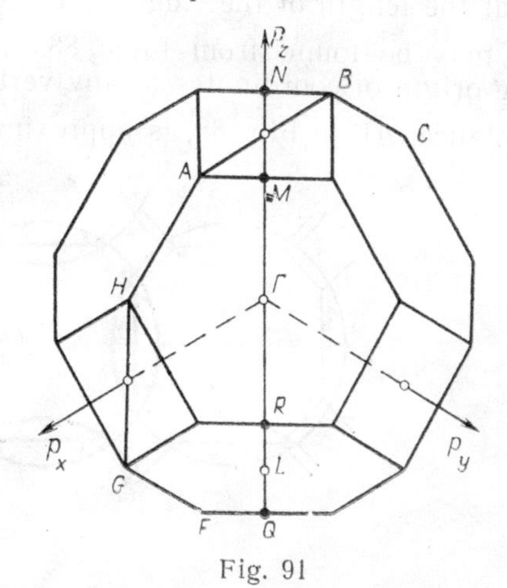

Fig. 91

Because of the symmetry of the cubo-octahedron, the sections of any of the coordinate planes coincide, as well as all the sections by any diagonal plane passing through the axes p_x, p_y, or p_z. The straight lines AB, CD, EF, and GH in Fig. 88 are diagonal of the four square faces: two horizontal and two vertical, while the lines BC, DE, FG, and HA are the sides of the four hexagons. Since the side of the square is equal in length to the side of the hexagon, then, denoting it by d, we can write

$$AB = CD = EF = GH = \sqrt{2}d$$

$$BC = DE = FG = HA = d$$

The straight lines MN and QP in Fig. 89 are the lines of intersection of the square faces with the plane passing through their centre and parallel to the sides of the square: $MN = QP = d$. The lines RM, NO, OP, and QR are lines of intersection of the hexagons with the plane passing through their centres and perpendicular to their horizontal sides (Fig. 90). It is evident that $RM = NO = OP = QP = \sqrt{3}\,d$. The points R and O are the traces of the side which is common for two hexagonal faces. Location of the characteristic points of sections on the cubo-octahedron is shown in Fig. 91.

Knowing the distance $\Gamma L = \frac{\sqrt{3}\,\pi\hbar}{a}$ in $\vec{p}$-space, it is easy to show that the length of the side d is equal to $\frac{\sqrt{2}\,\pi\hbar}{a}$ and $RL = \frac{\sqrt{3}}{2}\,d$. It may be found from Figs. 88 and 89 that the distance from the origin of coordinates to any vertex of the cubo-octahedron, for instance, $A\Gamma$ in Fig. 88, is approximately equal to $2.24\frac{\pi\hbar}{a}$, while

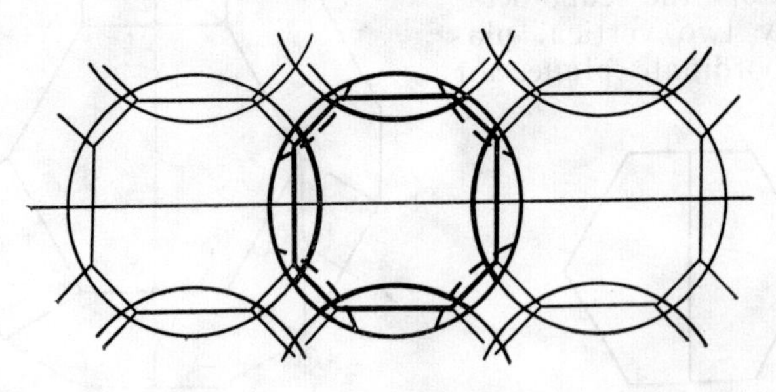

Fig. 92

the distance from the origin of coordinates to the side which is common for two hexagonal faces in Fig. 89 is approximately equal to $2.12\frac{\pi\hbar}{a}$.

The volume of the cubo-octahedron $V_{\vec{p}} = 8\sqrt{2}\,d^3$ or

$$V_{\vec{p}} = 4\left(\frac{2\pi\hbar}{a}\right)^3$$

To calculate the radius p_F of Harrison's sphere, we use formula (2.46) with the valence $\varkappa = 3$:

$$p_F = \left(\frac{9}{2\pi}\right)^{1/3} \frac{2\pi\hbar}{a} \approx 2.26\frac{\pi\hbar}{a} \qquad (2.54)$$

This value of the radius p_F exceeds the distances from the vertices of the first Brillouin zone, which means that the first energy zone of a trivalent metal with a face-centered lattice is completely filled with electrons.

To construct constant-energy surfaces in the second and third zones, we use the sections through the system of repeating zones by a coordinate and a diagonal plane (Figs. 92 and 93). Since the radius p_F of Harrison's sphere only slightly exceeds the distances from the origin of coordinates to the boundary points of the zone, the spheres described from the centres of neighbouring zones pass inside the central zone and cut lens-like volumes (in the form of plane-convex lenses) adjoining its faces from it.

The empty space left inside the central band resembles a cubo-octahedron but with concave faces and edges. This space belongs to the single central Harrison's sphere. The closed single-con-

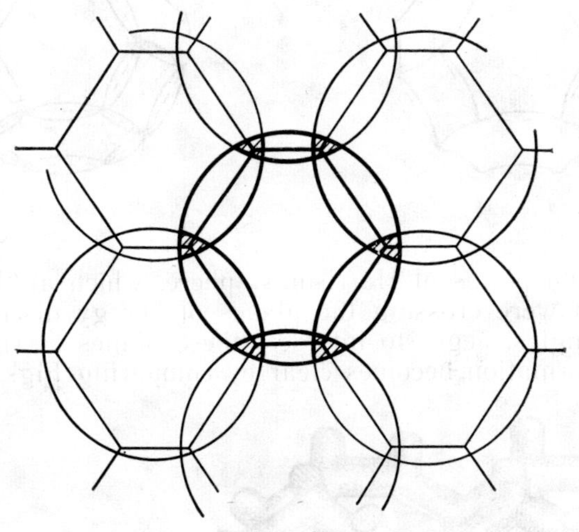

Fig. 93

nected constant-energy surface surrounding it is of the hole type and belongs to the second energy band. The general shape of this surface is shown in Fig. 94.

In Figure 93, the sections of those volumes which belong simultaneously to three spheres are shown hatched. They have the shape of triangular "tubes" of a variable cross-section located along all edges of the cubo-octahedron. To the ends of the edges, the tubes become narrower. The surface of these "tubes" is an electron surface in the third energy band (Fig. 95).

In the zero approximation, the "tubes" on the edges passing from each vertex of the cubo-octahedron are connected with each other. Thus the whole network of "tubes" is a single closed system in $\vec{p}$-space. Since the volumes of each "tube" are comparatively small, we have to discuss the problem of what changes will occur in separate portions of the Fermi surface constructed by Harrison's method in the zero approximation at passing to a finite value of the effective potential and also at subsequent increase of this potential. It is evident that such changes are most essential for those portions of the Fermi surface which confine small volumes.

The nature of variations of the Fermi surface is determined qualitatively by such a transformation of its portions at which

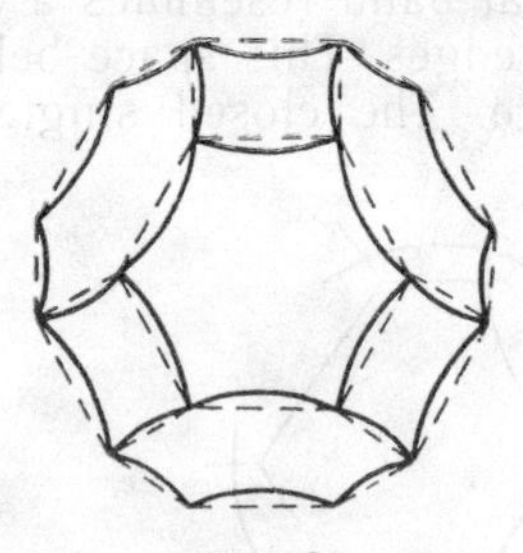

Fig. 94

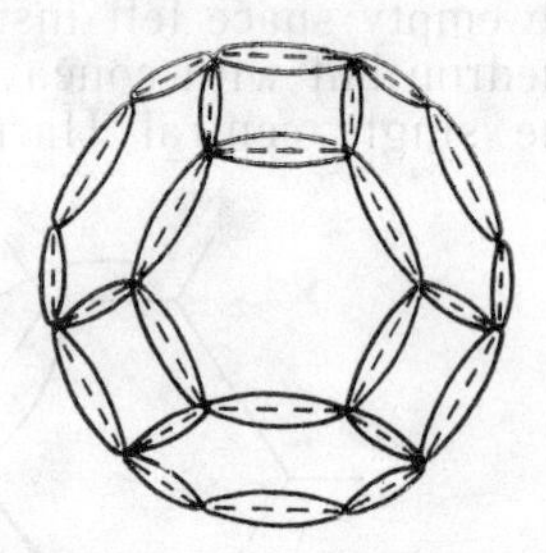

Fig. 95

the individual pieces of Harrison's sphere, which at the zero approximation were crossing the planes of energy discontinuity at arbitrary angles, begin to intersect these planes at right angles. This transformation becomes clear by comparing Figs. 79 and 80

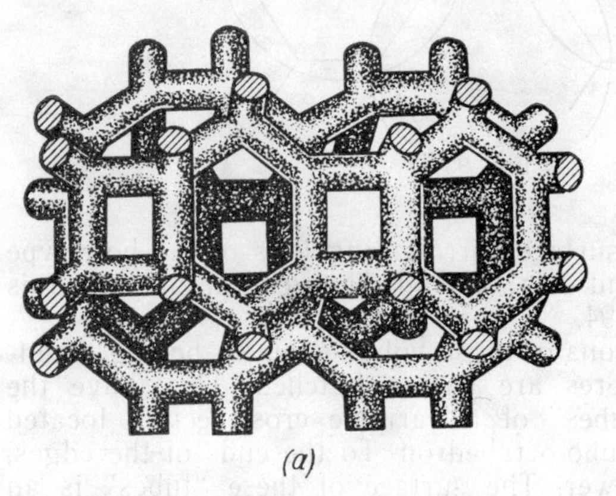

(a)

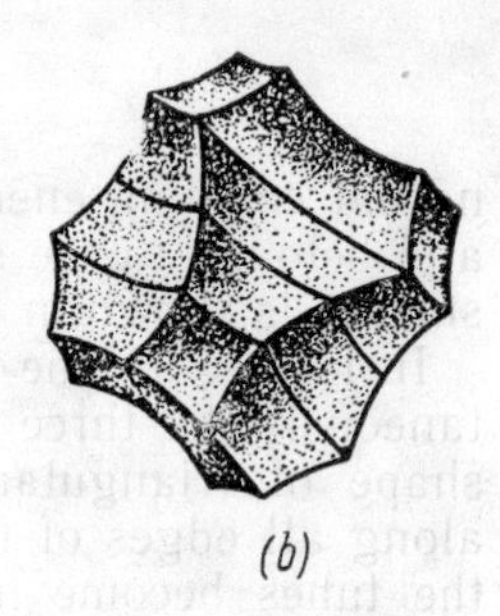

(b)

Fig. 96

and also Figs. 83 and 84. As a result of this process, electron-type surfaces in the zones diminish, and the hole-type ones, increase. The quantitative changes of the surfaces are the greater, the larger $V_{eff}(\vec{r})$ is, since the pieces of Harrison's sphere begin to be distorted at greater distances from the planes of energy discontinuity in that case.

With a sufficiently large $V_{eff}(\vec{r})$, small electron-type surfaces can break into separate smaller pieces or entirely vanish in the zones. Thus, in the case of the electron-type surface of aluminium, it can be expected for the third zone that at a finite value of effective potential the closed network of "tubes" will be broken near the vertices of the cubo-octahedron and will be transformed into elongated "loafs" isolated from each other at each edge.

With an increase of the number of electrons, the cross-sectional area of electron "tubes" in the third zone increases, while the size of the hole-type surface in the second zone decreases. This may be shown by passing over from a trivalent metal with a face-centered lattice to a quadrivalent one, for instance, lead. The electron surface of lead in the third zone is open and consists of thick tubes located around the edges of all the cubo-octahedrons filling the whole $\vec{p}$-space. A part of such a surface is shown in Fig. 96*a*. The hole-type surface of lead in the second zone, which is noticeably smaller compared with the corresponding surface of aluminium, is shown in Fig. 96*b*.

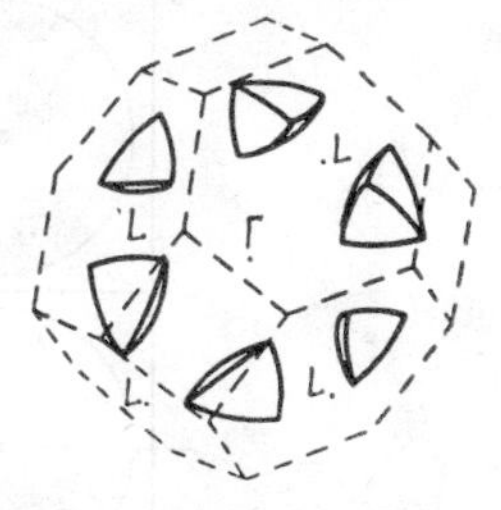

Fig. 97

The electron surface of a trivalent metal with a face-centered lattice is not confined solely to the surface in the third zone. An accurate construction shows that around the vertices of the square faces of the cubo-octahedron (around the points *A*, *B*, *N*, *C*, etc. in Fig. 91) there are located small regions belonging simultaneously to four Harrison's spheres. These regions, resembling tetrahedrons in shape, are bounded by the electron-type surface in the fourth energy band. Their volume is extremely small compared with that of the first Brillouin zone (of the order of 0.001 $V_{\vec{p}}$).

When passing to a finite value of effective potential, the electron-type surfaces in the fourth zone reduce still more and can even vanish completely. In the case of a quadrivalent metal, these surfaces are substantially greater. They remain at the value of $V_{eff}(\vec{r})$ corresponding to a real metal. For instance, the existence of electron-type surfaces in the fourth zone in lead has been proved experimentally by Khaikin and Mina [56].

Figure 97 shows electron-type surfaces in the fourth zone for a quadrivalent metal with a face-centered lattice at $V_{eff}(\vec{r}) \equiv 0$. For clarity, these surfaces are shown displaced half the distance between the centres of the opposite hexagonal faces of the first Brillouin zone along one of the directions of the type ΓL (see Fig. 91).

2-11. THE MODERN SCHEME OF CONSTRUCTING FERMI SURFACES. TOPOLOGICAL CLASSIFICATION OF FERMI SURFACES

In the earlier sections we have discussed various methods for constructing the surfaces of energy discontinuity and constant-energy surfaces in phase space. Using them as the basis, it is possible to formulate a convenient scheme of constructing Fermi surfaces for a given metal. This scheme includes four successive operations:

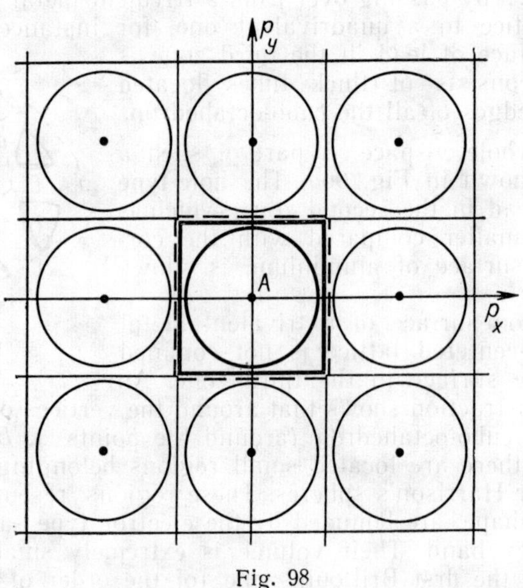

Fig. 98

1. For the given metal lattice, we construct the reciprocal lattice.
2. Near each node of the reciprocal lattice, we construct a unit cell by the Wigner-Seitz method. By varying the scale of the axes $2\pi\hbar$ times, we pass to a phase space filled with periodically repeating first Brillouin zones.
3. For the given parameters of the Brillouin zone and the valence of the metal, we determine the radius p_F of the Harrison sphere and describe spheres of this radius from the centres of the periodically repeating zones.
4. The constant-energy surfaces, formed by intersections of the Harrison spheres, are classified by the rules given in Sec. **2-10** of this chapter, with determination of the type of surface (either

electron or hole), its shape and dimensions, and also of the number of the energy band to which it belongs.

The scheme of constructing Fermi surfaces may be illustrated on an example of a monovalent metal with a simple cubic lattice.

The vectors of main translations of the reciprocal lattice are perpendicular to three families of equivalent "network" planes with the period a. An elementary cell of the reciprocal lattice is a cube of side $1/a$. Let the origin of coordinates be placed in one of the nodes of the reciprocal lattice and be connected by straight lines with the nearest nodes. There may evidently be six such lines of the length $1/a$.

The space scale of the reciprocal lattice is now changed by a factor $2\pi\hbar$. The first Brillouin zone will be obtained in the form of a cube with the side $2\pi\hbar/a$ bounded by six planes orthogonal to these six straight lines and passing through their middles. The zone is symmetrical about the origin of coordinates. The volume of the zone (in $\vec{p}$-space) is $\left(\frac{2\pi\hbar}{a}\right)^3$. The radius of the Harrison sphere (2.46) is:

$$p_F = \left(\frac{3}{8\pi}\right)^{1/3}\left(\frac{2\pi\hbar}{a}\right) \approx 0.986\,\frac{\pi\hbar}{a} \tag{2.55}$$

Since the shortest distance from the origin of coordinates to the boundaries of the Brillouin zone is $\pi\hbar/a$, Harrison's sphere almost touches the boundaries of the zone. We describe spheres of radius p_F about the centres of the periodically repeating first Brillouin zones and examine the picture in the plane $p_x p_y$ (Fig. 98). The points in the figure denote the nodes of the reciprocal lattice. One of the cells of the reciprocal lattice with the centre at point A is marked separately in the figure (A is the origin of coordinates of reciprocal space).

Taking into account that the effective potential $V_{eff}(\vec{r})$ is other than zero, the Harrison spheres should "stick" to the centres of faces of the Brillouin zone. The pictures in the planes $p_x p_y$, $p_x p_z$, and $p_y p_z$ are evidently similar to each other. The constant-energy surface in the first zone passes through the whole $\vec{p}$-space and consists of spheres joined in the directions of the axes p_x, p_y, and p_z. A surface of this type has been called "monster" in the specialist literature (Fig. 99).

In our example, the second and subsequent energy bands contain no electrons.

For a bivalent metal with the same type of crystalline lattice, the radius p_F of Harrison's sphere is $2^{1/3} \approx 1.26$ times greater. It is approximately equal to $1.24\frac{\pi\hbar}{a}$ and exceeds the shortest dis-

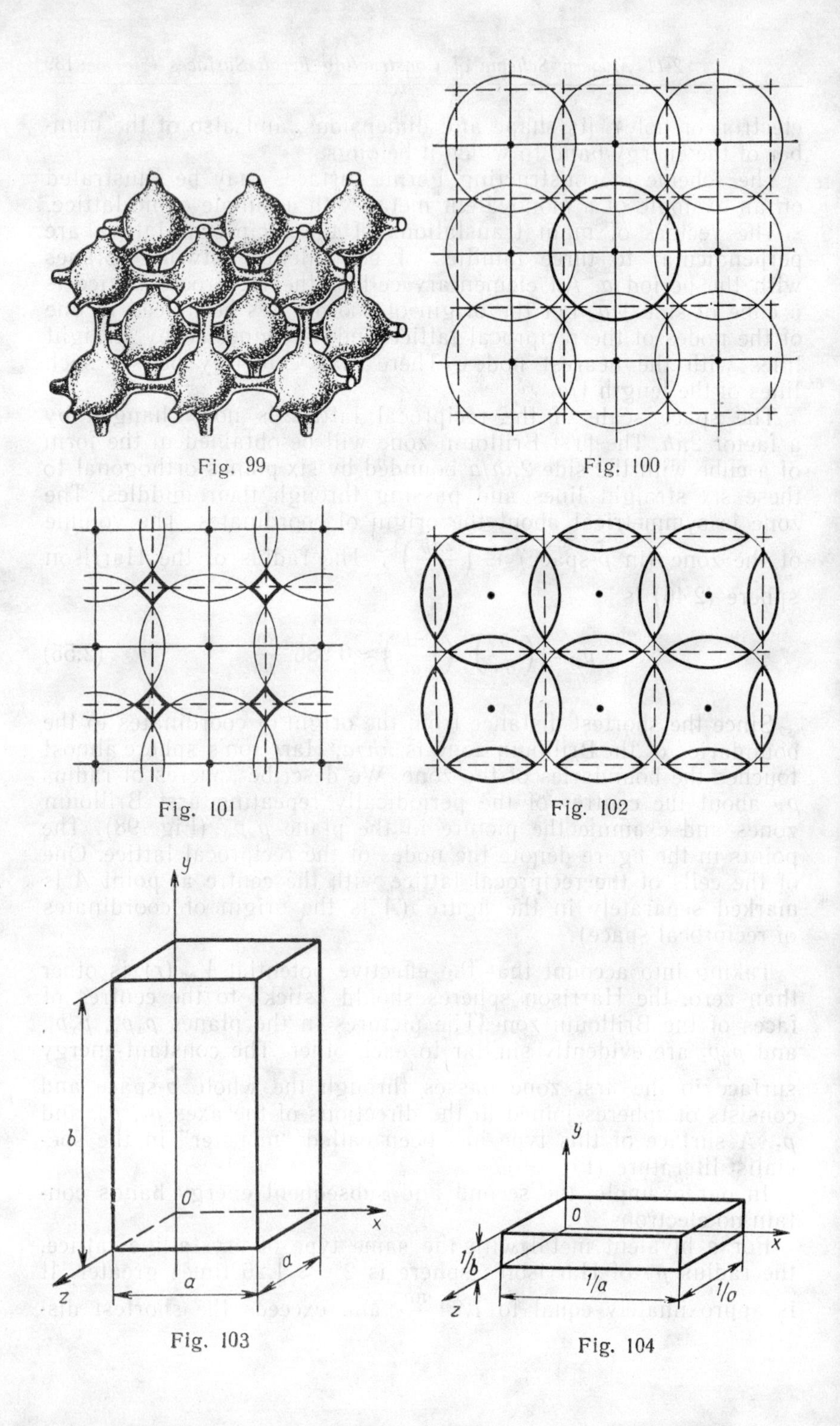

Fig. 99

Fig. 100

Fig. 101

Fig. 102

Fig. 103

Fig. 104

tance to the boundary of the Brillouin zone (Fig. 100). The constant-energy surface in the first zone is the same "monster", but with considerably thicker "necks" between individual spheres. The section of this surface by the plane p_xp_y is shown in Fig. 101.

The constant-energy surface in the second zone is of the electron type. It consists of three isolated closed "pockets" resembling lenses in shape. Each "lens" is formed of two halves on the opposite faces of the zone. The section of the surface in the second zone by the plane p_xp_y is shown in Fig. 102.

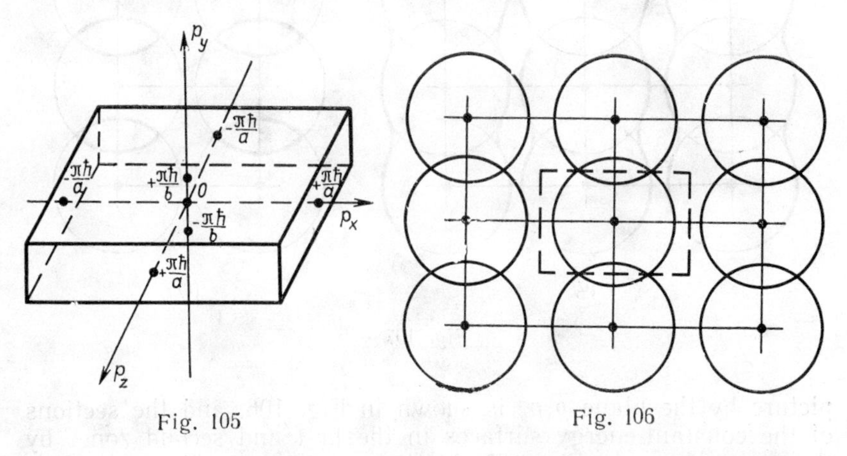

Fig. 105

Fig. 106

To conclude with, let us consider a monovalent metal with the lattice in which an elementary cell has the form of a right-angled parallelepiped (Fig. 103). The vectors of the reciprocal lattice are

$$|\vec{a}_x^*| = \frac{1}{a}, \quad |\vec{a}_y^*| = \frac{1}{b}, \quad |\vec{a}_z^*| = \frac{1}{a}$$

A cell of the reciprocal lattice is shown in Fig. 104.

The first Brillouin zone constructed by the Wigner-Seitz method is a parallelepiped with the sides $2\pi\hbar/a$, $2\pi\hbar/b$, and $2\pi\hbar/a$ having the volume $V_{\vec{p}} = \frac{8(\pi\hbar)^3}{a^2b}$ (Fig. 105).

The radius of Harrison's sphere (at $\varkappa = 1$) is

$$p_F = \left(\frac{3}{8\pi}\right)^{1/3} \frac{2\pi\hbar}{(a^2b)^{1/3}} \simeq 0.984 \frac{\pi\hbar}{(a^2b)^{1/3}} \tag{2.56}$$

With $\frac{a}{b} < \sqrt{\frac{3}{\pi}} \approx 0.98$, the following inequality holds true:

$$\frac{\pi\hbar}{b} < p_F < \frac{\pi\hbar}{a}$$

In this case, Harrison's sphere crosses the upper and lower boundaries of the Brillouin zone, but does not reach the side boundaries.

We now describe Harrison's spheres from the centres of all periodically extended first Brillouin zones. The section of this

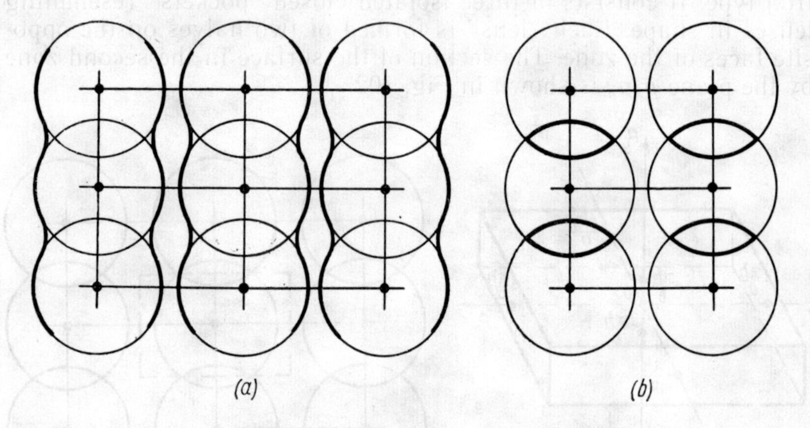

Fig. 107

picture by the plane $p_x p_y$ is shown in Fig. 106, and the sections of the constant-energy surfaces in the first and second zones by the same plane, in Fig. 107. In the case considered, the surface of the "monster" type in the first zone has transformed into a surface of the type of "warped" cylinder (Fig. 108).

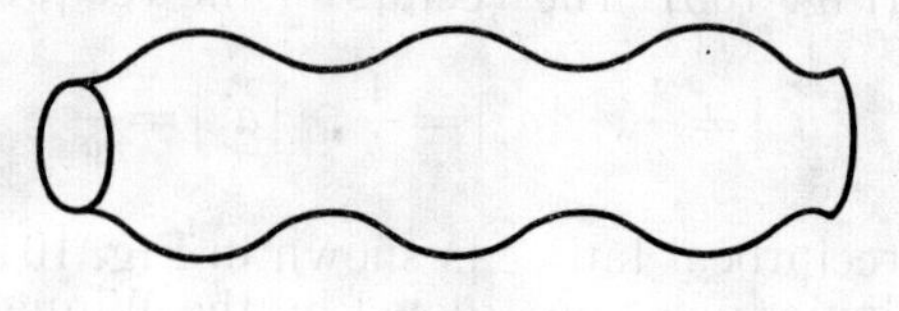

Fig. 108

The examples considered can show that Fermi surfaces can be divided by their topology into two main classes: closed and open surfaces.

A surface is called closed if the lines of its intersection with any planes are closed. Examples are the constant-energy surfaces in the second zone in Figs. 102 and 107.

A surface is called open if there exists at least one plane section giving an unclosed curve. Examples are the "monster" (Figs. 99 and 101) and "warped" cylinder (Figs. 107 and 108).

Conclusions.

1. The energy spectrum of electrons in a metal consists of a number of energy zones (bands). It means that different energy states may correspond to one and the same value of the quasi-momentum of an electron.

2. The properties of electrons in each band are determined by the shape of the constant-energy surface in that zone. The energy in the bands is usually counted from the points in which it is a minimum (for electron-type surfaces) or a maximum (for hole-type surfaces).

3. A metal may have both open and closed constant-energy surfaces.

4. The concentrations of current carriers in the zones can sharply differ from each other.

The concentration n_i in the i-th zone is determined by the formula

$$n_i = \frac{2\Delta_i}{(2\pi\hbar)^3} \tag{2.57}$$

where Δ_i is the total volume of $\vec{p}$-space bounded by all constant-energy surfaces in the i-th zone.

If the Fermi surface is an open one, then Δ_i implies the volume bounded by this surface within one Brillouin zone (the boundary planes of the Brillouin zone are part of the boundaries of this volume).

5. The volumes bounded by Fermi surfaces in various zones can overlap each other, but the individuality of each zone is then completely retained.

6. On the basis of the theorem stating that isothermal lines cannot intersect, constant-energy surfaces cannot intersect but may have common points.

7. All physical processes involve electrons located on the whole set of constant-energy surfaces.

2-12. MOTION OF QUASI-PARTICLES IN A CONSTANT MAGNETIC FIELD

The motion of an electron in a constant magnetic field $\vec{H}$ in the quasi-classical approximation can be described by the equation

$$\frac{d\vec{p}}{dt} = \frac{e}{c}[\vec{v}_g \vec{H}] \tag{2.58}$$

where $\vec{p}$ is the classical momentum of the electron. The Lorentz force $F_H = \frac{e}{c}[\vec{v}_g \cdot \vec{H}]$ exerted by the magnetic field $\vec{H}$ on the electron is directed perpendicular to its velocity $\vec{v}_g$. Because of this, the electron moves in free space along a spiral around a magnetic line of force. The radius of its orbit r_H projected onto the plane perpendicular to the magnetic field $\vec{H}$ is equal to $\frac{cp_\perp}{|e|H}$, where $p_\perp$ is the component of the momentum which is perpendicular to the magnetic field. Applicability of the quasi-classical approximation (see Sec. **2-1** of this chapter) is determined by the inequality

$$\lambda_B \ll r_H \tag{2.59}$$

Since $\lambda_B = \frac{2\pi}{k} = \frac{2\pi\hbar}{p}$, inequality (2.59) for an electron in free space is equivalent to the inequality

$$\hbar\omega \ll \varepsilon_F \tag{2.60}$$

where ω is, as before, the cyclotron frequency of precession of the electron.

Variations of the quasi-momentum of an electron in a crystal (see Sec. **2-4** of this chapter) are only determined by the action of external forces. The equation of motion of an electron in a metal will then have the form coinciding with equation (2.58), but $\vec{p}$ now denoting the quasi-momentum of the electron.

By scalar multiplication of both parts of equation (2.58) by $\vec{v}_g$ and $\vec{H}$, we obtain two equalities as follows:

$$\left(\vec{v}_g \frac{d\vec{p}}{dt}\right) = 0 \text{ and } \left(\vec{H} \frac{d\vec{p}}{dt}\right) = 0 \tag{2.61}$$

The first equality is exactly the law of conservation of energy. Indeed:

$$\vec{v}_g = \frac{\partial \varepsilon}{\partial \vec{p}} \text{ and } \left(\vec{v}_g \frac{d\vec{p}}{dt}\right) = \left(\frac{\partial \varepsilon}{\partial \vec{p}} \frac{d\vec{p}}{dt}\right) = \frac{d\varepsilon}{dt} = 0$$

Hence it follows that the electron moves in the magnetic field $\vec{H}$ along a constant-energy surface:

$$\varepsilon(\vec{p}) = \text{const} \tag{2.62}$$

Let vector $\frac{d\vec{p}}{dt}$ be resolved into two components: $\left(\frac{d\vec{p}}{dt}\right)_\parallel$, which is parallel to $\vec{H}$, and $\left(\frac{d\vec{p}}{dt}\right)_\perp$, perpendicular to $\vec{H}$. The second equality in (2.61) implies that $\left(\left(\frac{d\vec{p}}{dt}\right)_\parallel \vec{H}\right) = 0$, or, in other words,

$$\left|\left(\frac{d\vec{p}}{dt}\right)_\parallel\right| \cdot |\vec{H}| = 0, \text{ i.e. } \left|\left(\frac{d\vec{p}}{dt}\right)_\parallel\right| = 0$$

It then follows that the projection of the momentum of electron onto the magnetic field is retained in motion, i.e.

$$p_{\parallel} = \text{const} \tag{2.63}$$

Relationships (2.62) and (2.63) describe the path of an electron in momentum space. This path is a curve along which a constant-energy surface (2.62) is intersected by a plane perpendicular to the magnetic field (Fig. 109). Equality (2.63) is the equation of secant plane.

Depending on the topology of constant-energy surface and the direction of magnetic field, the path of an electron in phase space may be either closed (determining a finite motion) or open, which passes continuously through the whole $\vec{p}$-space (and determines an infinite motion).

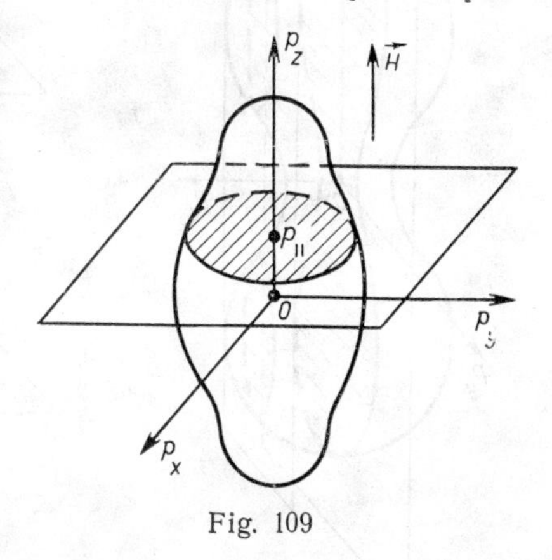

Fig. 109

The particular form of the path of an electron in $\vec{p}$-space depends on the shape of the constant-energy surface $\varepsilon(\vec{p}) = \text{const}$, its orientation relative to the magnetic field, and the magnitude of the projection of the momentum $p_{\parallel}$ onto the direction of the magnetic field. Some various paths along a constant-energy surface having the shape of a dumb-bell are shown in Fig. 110*a* and *b*. The dependence of the path of an electron in $\vec{p}$-space on $p_{\parallel}$ may be explained on an example of a spherical constant-energy surface

$$\frac{p_x^2 + p_y^2 + p_z^2}{2m} = \varepsilon = \text{const}$$

Let the magnetic field $H = H_z$ be directed along the z axis. It is then evident that $p_{\parallel} = p_z = p_{z0} = \text{const}$ and the electron moves along the path described by the equation $p_x^2 + p_y^2 = 2m\varepsilon - p_{z0}^2$, which is exactly the equation of a circle located in the plane $p_z = p_{z0}$. The radius $\sqrt{2m\varepsilon - p_{z0}^2}$ of this circle reduces with an increase of p_{z0}. At $p_{z0} = \sqrt{2m\varepsilon}$, the radius becomes zero and the path of motion of the electron degenerates into a point $p_z = p_{z0}$, $p_x = p_y = 0$.

Let us establish the relationship between the path of an electron in phase space and its path in a crystal lattice ($\vec{r}$-space). As can be seen from equation (2.58), the velocity of electron $\vec{v}_g = \frac{d\vec{r}}{dt}$ in $\vec{r}$-space is perpendicular to its velocity $\frac{d\vec{p}}{dt}$ in

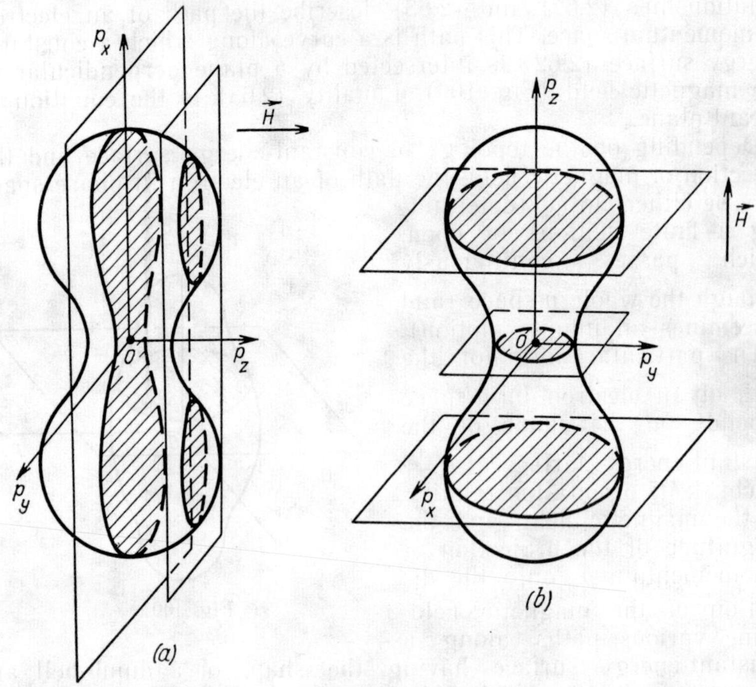

Fig. 110

$\vec{p}$-space. Let equation (2.58) be projected onto a plane perpendicular to the direction of the magnetic field:

$$\frac{d\vec{p}_\perp}{dt} = \frac{e}{c}[\vec{v}_{g\perp} \cdot \vec{H}] = \frac{e}{c}\left[\frac{d\vec{r}_\perp}{dt} \cdot \vec{H}\right] \tag{2.64}$$

where $\vec{p}_\perp$, $\vec{v}_{g\perp}$, and $\vec{r}_\perp$ are components of the quasi-momentum, velocity and radius vector of the electron in the plane of projection. The vectors $\frac{d\vec{r}_\perp}{dt}$, $\vec{H}$, and $\frac{d\vec{p}_\perp}{dt}$ form a right-hand screw triple. Integration of (2.64) for the magnitudes $|\vec{p}_\perp|$ and $|\vec{r}_\perp|$ gives

$$|\vec{p}_\perp| = \frac{|e|H}{c}|\vec{r}_\perp| \tag{2.65}$$

Consequently, the projection of the electron path in $\vec{r}$-space onto the plane perpendicular to $\vec{H}$ coincides in shape with the corresponding path of the electron in $\vec{p}$-space, but differs in size by a factor $\frac{c}{|e|H}$.

Each element of the projection of the path in $\vec{r}$-space is perpendicular to the corresponding element of the path in $\vec{p}$-space, the paths being rotated relative to one another through an angle

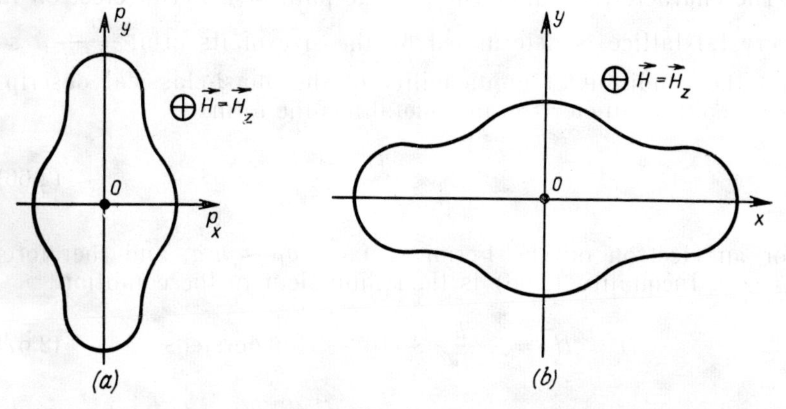

Fig. 111

of 90° around the direction of the magnetic field. Thus, knowing the path of the electron in $\vec{p}$-space (Fig. 111*a*), we can obtain the projection of its path onto the plane perpendicular to the field $\vec{H}$ in $\vec{r}$-space (Fig. 111*b*) by rotating the initial path through 90 degrees clockwise (rotation by the screw rule, the motion of the screw coinciding with the direction $\vec{H}$) around the magnetic field and changing its scale by a factor $\frac{c}{|e|H}$. The direction of motion of the electron along the path after this transformation will evidently remain the same.

In a particular case at $p_{\parallel} = 0$, a closed path in p-space corresponds to a similar closed path in $\vec{r}$-space. In other words, at $p_{\parallel} = 0$, the path of the electron in r-space is completely located in a plane perpendicular to the magnetic field. At $p_{\parallel} \neq 0$, the path closed in $\vec{p}$-space corresponds to the motion along a helix in

$\vec{r}$-space: per each rotation along the path in $\vec{p}$-space during T_H the electron is still additionally shifted in $\vec{r}$-space along the magnetic field.

In a case when the path in $\vec{p}$-space is open in some direction $\vec{n}$, the path of the electron in $\vec{r}$-space is unlimited in the direction perpendicular to $\vec{n}$. Consider an example of the motion of an electron along the surface of a warped cylinder (Fig. 108) in a magnetic field perpendicular to the axis of the cylinder.

The characteristic dimension of the path of a Fermi electron in a crystal lattice is determined by the size of its orbit $\frac{cp_{\perp}}{|e|H}$, so that the condition of applicability of the quasi-classical description (2.59) to an electron in a metal has the form

$$\lambda_B \ll \frac{cp_F}{|e|H} \qquad (2.66)$$

For an electron on the Fermi surface, $p_F \sim \hbar/a$, and therefore, $\lambda_B \sim a$. Inequality (2.66) is then equivalent to the condition

$$H \ll H_a = \frac{c\hbar}{|e|a^2} \sim (10^8 - 10^9) \text{ oersteds} \qquad (2.67)$$

The quasi-classical description of the motion of an electron in a metal placed into a magnetic field is practically always justified, since for magnetic fields of intensities really attainable in laboratory conditions ($H \sim 10^6$ oersted), inequality (2.67) is deliberately fulfilled. Special cases in which the quantum approach to the motion of electron is essential will be considered separately.

Consider the conditions of a finite motion of an electron in a magnetic field. As distinct from a finite motion in an electric field which is practically unfeasible because of the enormous dimensions of the path (see Sec. **2-5** of this chapter), the cyclic motion of an electron in a magnetic field is more or less common phenomenon.

As has been shown earlier, the characteristic size of the orbit of an electron is determined by the dimensions of the area of section of the Fermi surface by a plane perpendicular to the magnetic field. Orbits of finite dimensions, therefore, correspond either to closed Fermi surfaces or to such an orientation of magnetic field $\vec{H}$ relative to an open Fermi surface, at which $\vec{H}$ makes comparatively small angles with the open directions (with $\vec{H}$ being perpendicular to an open direction the length of path is infinite).

For the cyclic motion in a magnetic field, the following inequality must hold

$$l > r_H \sim \frac{cp_\perp}{|e|H} \tag{2.68}$$

where l is the free-path length of the electron.

The meaning of this inequality is clear: an electron must perform at least one cycle of motion along the free-path length. Let $p_\perp$ in inequality (2.68) be replaced with $p_F \sim \frac{\hbar}{a}$ which is approximately of the same order, so that the inequality can be rewritten as

$$l > \frac{c\hbar}{|e|aH}$$

Using the definition of H_a (2.67), we get an equivalent inequality

$$H > \frac{a}{l} \cdot \frac{c\hbar}{|e|a^2} = \frac{a}{l} H_a \sim \frac{a}{l} (10^8 - 10^9) \text{ oersteds} \tag{2.69}$$

This is the condition that a magnetic field must satisfy to provide the cyclic motion of electrons in a metal. At low temperatures and in sufficiently pure and perfect monocrystal specimens of metals the free-path length exceeds tens and hundreds thousands times the interatomic distances. The cyclic motion in a magnetic field can already be observed principally in fields of an order of hundreds oersteds.

Note that the free-path length l of an electron is not the mean distance between successive acts of its scattering. Actually, the motion of an electron in a metal is similar to the motion in a viscous medium during which the electron continuously loses its energy. The magnitude of l is essentially the measure of viscosity of the medium or the measure of the energy lost by the electron along unit length: for the motion along the x axis, $l \sim \left(\frac{\partial \varepsilon}{\partial x}\right)^{-1} \varepsilon$.

The state of energy ε, which is conventionally called an electron and corressponds to the given elementary excitation near a Fermi surface, attenuates exponentially and continuously in time with the characteristic life-time τ (relaxation time), which is determined by the probability of scattering $\omega \sim 1/\tau$. This time via the velocity v_F on the Fermi surface, in turn, determines the free-path length l. Thus, the inequality $\Delta x \ll l$ implies that along the length Δx we can neglect the variation of the energy state of the electron, i.e. neglect the process of attenuation (scattering) and in that sense regard the motion along the path Δx similar to the free motion of the electron in the lattice.

The periodic motion of electrons is of extreme importance in studies of the electron energy spectrum. In such a motion, whole groups of electrons move in synchronism. The contribution of such electrons to some or other physical phenomena is of resonant nature and can be actually observed in experiments.

Let us determine the time of motion T_H of an electron around a closed path in $\vec{p}$-space. This time evidently coincides with the time during which the electron passes along one turn of the spiral in $\vec{r}$-space. At the motion of the electron in free space, the period T_0 of Larmor precession is equal to $\frac{2\pi}{\omega}=\frac{2\pi m_0 c}{|e|H}$. For an electron

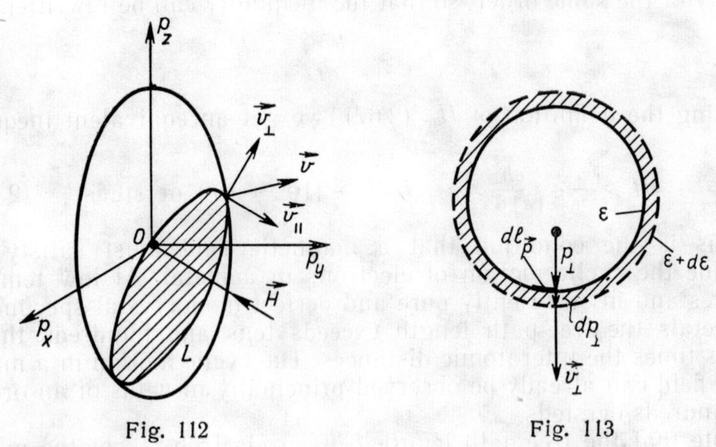

Fig. 112 Fig. 113

in a metal, T_H is determined by the shape of constant-energy surface and the orientation of magnetic field $\vec{H}$.

Consider an arbitrary closed constant-energy surface $\varepsilon(\vec{p})=$ $=$ const. We give the direction of magnetic field $\vec{H}$ and the magnitude of $p_\parallel$, which determines the position of the plane perpendicular to $\vec{H}$, whose intersection with the surface $\varepsilon(\vec{p})=$ const determines the path of the electron in $\vec{p}$-space. Let the quasi-momentum $\vec{p}$ and velocity $\vec{v}_g$ of the electron be resolved into components parallel and perpendicular to the magnetic field ($\vec{p}_\parallel$, $\vec{v}_{g\parallel}$, $\vec{p}_\perp$, and $\vec{v}_{g\perp}$ respectively, see Fig. 112).

The projection of the equation of motion of the electron (2.58) onto the secant plane can be written for the magnitudes of $\vec{v}_{g\perp}$ and $\vec{H}$ as follows:

$$\frac{dl_{\vec{p}}}{dt}=\frac{|e|}{c}v_{g\perp}H \tag{2.70}$$

where $dl_{\vec{p}}$ is an arc element of the path in $\vec{p}$-space. Integration of (2.70) gives the expression for the period of rotation

$$T_H = \frac{c}{|e|H} \oint_L \frac{dl_{\vec{p}}}{v_{g\perp}} \tag{2.71}$$

The integral is taken over the closed contour L of the path in the secant plane. Let $\vec{n}$ be the unit vector of the normal to the path in the plane perpendicular to $\vec{H}$. The velocity $v_{g\perp}$ can then be expressed as

$$v_{g\perp} = \left(\text{grad}_{\vec{p}}\varepsilon\right)_{\perp} = \left(\text{grad}_{\vec{p}}\varepsilon, \vec{n}\right) = \frac{\partial \varepsilon}{\partial p_{\perp}}$$

This means that with the energy increased by $d\varepsilon$, each element of the path $dl_{\vec{p}}$ is shifted along the normal $\vec{n}$ by a distance $dp_{\perp} = \frac{d\varepsilon}{v_{g\perp}}$. The increment of the area of the sections, dS, (hatched portion in Fig. 113) will then be

$$dS = \oint_L dl_{\vec{p}}\, dp_{\perp} \quad \text{or} \quad dS = \oint_L \frac{dl_{\vec{p}}}{v_{g\perp}}\, d\varepsilon$$

But the increment $d\varepsilon$ is independent of the position of the point on the path and can be taken out from the integral sign. The partial derivative $\frac{\partial S}{\partial \varepsilon}$ is then equal to $\oint_L \frac{dl_{\vec{p}}}{v_{g\perp}}$. Hence, for the period T_H we have

$$T_H = \frac{c}{|e|H} \frac{\partial S}{\partial \varepsilon} = \frac{2\pi \left(\frac{1}{2\pi} \cdot \frac{\partial S}{\partial \varepsilon}\right) c}{|e|H} \tag{2.72}$$

By comparing (2.72) with the expression for the period of rotation T_0 of a free electron, we can see that in the cyclic motion of an electron with an arbitrary law of dispersion, the quantity $\frac{1}{2\pi} \cdot \frac{\partial S}{\partial \varepsilon}$ plays the part of the effective mass. This quantity has been termed the cyclotron mass of an electron:

$$m^* = \frac{1}{2\pi} \cdot \frac{\partial S}{\partial \varepsilon} \tag{2.73}$$

The method of introducing the cyclotron mass discussed above can show that this mass is a quantity averaged over the period of Larmor precession of an electron or over the path of precession per one cycle. It follows from (2.73) that m^* is determined by the law of dispersion and the orientation of magnetic field $\vec{H}$. If the law of dispersion is known, we can establish the relationship between the cyclotron mass and the components of the effective-mass tensor (see Sec. **2-4**) for the given direction of $\vec{H}$. Let us illustrate this relationship by two particular examples.

The simplest case is that of a quadratic isotropic dispersion law, $\varepsilon = \frac{p^2}{2m}$. The corresponding tensor of effective mass $m_{ij}^* = \left(\frac{\partial^2 \varepsilon}{\partial p_i \, \partial p_j}\right)^{-1}$ can be written as $m_{ij}^* = m\delta_{ij}$, where δ_{ij} is the Kroneker symbol. Let the magnetic field be directed along the z axis ($H = H_z$). The equations determining the path of the electron in $\vec{p}$-space have the forms:

$$\left.\begin{aligned} \frac{p_x^2 + p_y^2}{2m} &= \varepsilon - \frac{p_{z0}^2}{2m} \\ p_{\parallel} = p_{z0} &= \text{const} \end{aligned}\right\} \tag{2.74}$$

The area $S(\varepsilon, p_{z0})$ of the corresponding section of the constant-energy surface by a plane perpendicular to $\vec{H}$ is

$$S(\varepsilon, p_{z0}) = \pi \left[2m \left(\varepsilon - \frac{p_{z0}^2}{2m} \right) \right] \tag{2.75}$$

Differentiating this expression for energy, we find that the cyclotron mass m^* of electrons (2.73) coincides with m and is independent of the location of the secant plane which is determined by the magnitude of momentum p_{z0}.

Because of the spherical symmetry of the Fermi surface, the cyclotron mass m^* is evidently independent of the direction of $\vec{H}$. Thus, all electrons located on the spherical constant-energy surface have the same cyclotron period and move in synchronism in the magnetic field. As has been mentioned earlier, the area of the orbit in $\vec{p}$-space reduces with an increase of $p_{\parallel} = p_{z0}$ and degenerates into a point at $p_{z0} = \sqrt{2m\varepsilon}$. The velocity of motion of electrons along the orbit therefore reduces with a decrease of its radius. Accordingly, the motion in $\vec{r}$-space along the spiral about the magnetic line of force is the slower, the smaller the radius of the spiral.

Near the reference points (the points where the Fermi surface touches the planes perpendicular to $\vec{H}$) $p_x = p_y = 0$, $p_z = \pm p_{z0} = \pm\sqrt{2m\varepsilon}$, the electrons describe an infinitely small orbit with the same frequency as the electrons located in the plane of the extremal section $p_{z0} = 0$. The former move in $\vec{r}$-space practically with the Fermi velocity in the direction of the magnetic field, while the latter describe closed circles in a plane perpendicular to the magnetic field.

As another example, consider the quadratic anisotropic law of dispersion

$$\varepsilon = \frac{p_x^2}{2m_x} + \frac{p_y^2}{2m_y} + \frac{p_z^2}{2m_z} \tag{2.76}$$

The corresponding constant-energy surface is described by equation (2.25) and has the form of a triaxial ellipsoid with semi-axes $\sqrt{2m_x\varepsilon}$, $\sqrt{2m_y\varepsilon}$, and $\sqrt{2m_z\varepsilon}$. The tensor matrix of the effective mass $m_{ij}^* = \left(\frac{\partial^2 \varepsilon}{\partial p_i \partial p_j}\right)^{-1}$ for the main axes of the ellipsoid is of the form:

$$\{m_{ij}^*\} = \begin{pmatrix} m_x & 0 & 0 \\ 0 & m_y & 0 \\ 0 & 0 & m_z \end{pmatrix}$$

Let the magnetic field be directed, as before, along the z axis. The orbits of electrons will then be determined by the sections $p_{\parallel} = p_{z0} = \text{const}$. As is known, any section of an ellipsoid by a plane is an ellipse. We find the area of the section $S(\varepsilon, p_{z0})$ corresponding to the plane $p_z = p_{z0}$.

The equation of an ellipse in the secant plane is

$$\frac{p_x^2}{2m_x\left(\varepsilon - p_{z0}^2/2m_z\right)} + \frac{p_y^2}{2m_y\left(\varepsilon - p_{z0}^2/2m_z\right)} = 1 \tag{2.77}$$

The area of the section $S(\varepsilon, p_{z0})$ is

$$S(\varepsilon, p_{z0}) = 2\pi \sqrt{m_x m_y}\left(\varepsilon - \frac{p_{z0}^2}{2m_z}\right) \tag{2.78}$$

Hence, according to (2.73), the cyclotron mass $(m^*)_z$ corresponding to any section, with $\vec{H}$ parallel to p_z, can be solely expressed through the components of the effective mass tensor m_x and m_y:

$$(m^*)_z = \sqrt{m_x m_y} \tag{2.79}$$

By directing the magnetic field along the axes x and y, we can find in a similar way the cyclotron masses $(m^*)_x$ and $(m^*)_y$ relating to any sections $p_x = \text{const}$ and $p_y = \text{const}$ respectively

$$(m^*)_x = \sqrt{m_y m_z}, \quad (m^*)_y = \sqrt{m_x m_y} \tag{2.80}$$

It may be finally shown that for an arbitrary direction $\vec{n}_j$ of the magnetic field, the cyclotron mass $(m^*)_{\vec{n}_j}$ can be expressed through the three main components of the tensor m^*_{ii} and the direction cosines of the vector $\vec{n}_j$ relative to the main axes of the ellipsoid:

$$(m^*)_{\vec{n}_j} = \sqrt{\frac{m_x m_y m_z}{m_x \cos^2 \alpha + m_y \cos^2 \beta + m_z \cos^2 \gamma}} \tag{2.81}$$

where α, β, and γ are the angles between $\vec{n}_j$ and axes p_x, p_y, and p_z.

It is thus seen that for the quadratic law of dispersion at which the constant-energy surface is always either an ellipsoid or a sphere, the cyclotron mass m^* is independent of $p_\parallel$ and, in the case of ellipsoid, is determined solely by the orientation of the magnetic field relative to the main axes of the surface. In other words, in the magnetic field of a given direction, all electrons on the surface of an ellipsoid or sphere in $\vec{p}$-space move in synchronism irrespective of the position of the secant plane. For any direction, the ellipsoid, as the sphere, has two reference points (points A and A' in which the surface touches the planes perpendicular to $\vec{H}$) near which the electrons describe infinitely small orbits, but with a frequency coinciding with the general frequency of their rotation in any section (Fig. 114).

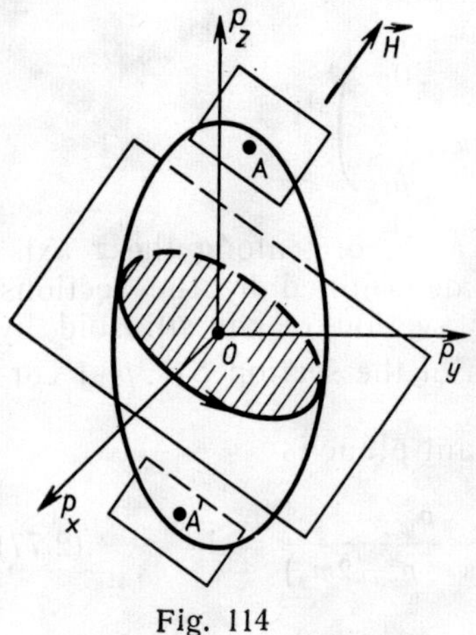

Fig. 114

In a case of an arbitrary law of dispersion, the cyclotron mass m^* is dependent on $p_\parallel$ and is naturally distinct for different electrons on a Fermi surface. For that reason, and as distinct from a free-electron gas, a gas of conduction electrons in a metal cannot rotate in the magnetic field with a frequency common for all the electrons. Different electrons then perform different periodic motions. Those which move over open paths in the plane perpendicular to the magnetic field (in $\vec{p}$-space) perform an infinite motion.

With an arbitrary law of dispersion, it may be observed, however, that individual groups of electrons move synchronously in the magnetic field. Such groups locate near the sections of a constant-energy surface on which the dependence of the cyclotron mass on $p_{\parallel}$ has a local maximum or minimum and $\frac{\partial m^*}{\partial p_{\parallel}}$ turns to zero. Such sections are usually the extremal ones (minimal or maximal) for the given direction of the magnetic field. They may include, for instance, a large section by the plane of symmetry (central section) of a constant-energy surface, having the form of a dumb-bell (Fig. 110*a*) or two similar maximum sections of the same dumb-bell and the minimum section of its "neck" (Fig. 110*b*).

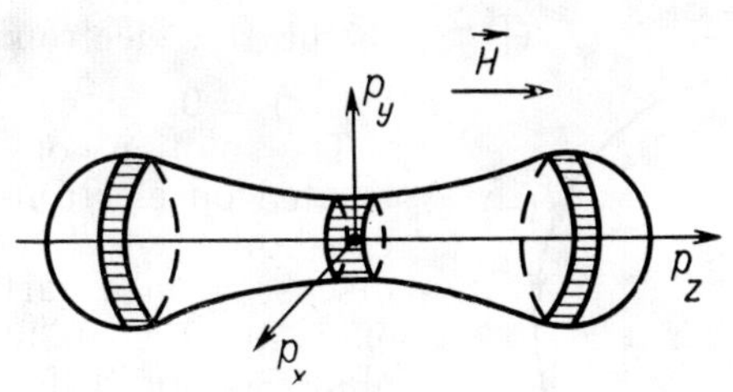

Fig. 115

Indeed, near the section corresponding to $p_{\parallel} = p_{\parallel}^0$, for which $\frac{\partial m^*}{\partial p_{\parallel}} = 0$, the expansion of the cyclotron period $T_H = \frac{2\pi c}{|e| H} m^*$ in powers of the difference $p_{\parallel} - p_{\parallel}^0$ contains no first order terms. Owing to this, there exists a certain vicinity of point $p_{\parallel} = p_{\parallel}^0$, having the width $\Delta p_{\parallel}$, for which the dependence of the period on $p_{\parallel}$ can be neglected.

The electrons located in the "neck" $\sim \Delta p_{\parallel}$ near $p_{\parallel}^0$ form a group of quasi-particles moving practically synchronously in the magnetic field (Fig. 115). Such groups of electrons provide a resonance contribution to various effects. The contribution of the electrons outside the "neck" represents a monotonous (non-resonance) background.

Let us discuss the nature of motion in magnetic fields of quasi-particles located on constant-energy surfaces of the electron and hole type. It should be recalled that the volumes (and therefore, sections S) of the electron-type constant-energy surfaces increase with energy, while those of the hole-type one, on the contrary, decrease. For that reason, $\frac{\partial S}{\partial \varepsilon} > 0$ on an electron-type surface and the effective cyclotron mass is positive, while on a hole-type surface the mass is negative. The different signes of cyclotron masses imply that precession of quasi-particles on electron- and hole-type surfaces occurs in opposite directions.

With the motion in a magnetic field, the Lorentz force $\vec{F}_H = \frac{e}{c}[\vec{v}_g \vec{H}]$ sets up a centripetal acceleration to an electron. The vectors $\vec{v}_g$, $\vec{H}$, and $-\vec{F}_H$ (since $e < 0$ for an electron) form a right-hand screw triple. Consequently, in a magnetic field directed away from the observer, a free electron rotates clockwise (Fig. 116). The same direction of rotation must be observed when the imaging point on Harrison's sphere moves in the magnetic field, since this point corresponds to the free-electron approximation $V_{eff}(\vec{r}) \equiv 0$.

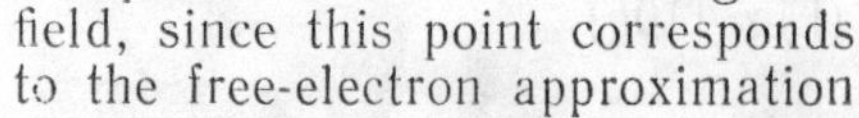

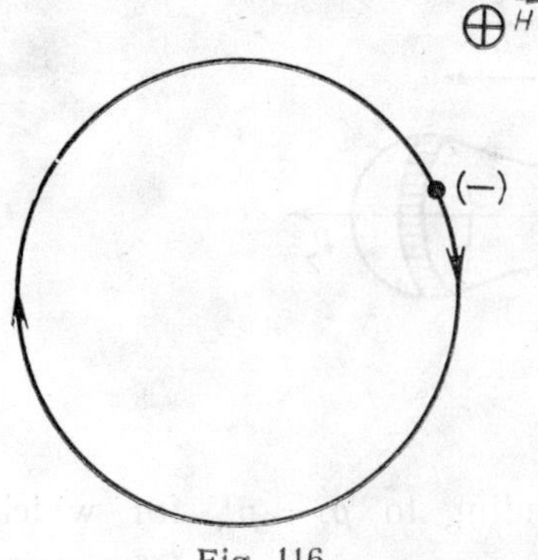

Fig. 116

The motion of quasi-particles located on electron- and hole-type constant-energy surfaces will be discussed on a particular example of a trivalent "metal" with a plane square lattice. The constant-energy surfaces in the first and second zones for such metal have been constructed in Sec. **2-10** (see Figs. 81 and 82).

In a magnetic field perpendicular to the plane of lattice and directed away from the observer, precession of electrons over Harrison's circle (Fig. 79) occurs clockwise. Application of a finite pseudo-potential $V_{eff}(\vec{r})$ results in that the Harrison circle breaks into separate pieces (Fig. 80), but the direction of motion of the electron along each piece cannot be changed.

Let us discuss separately the motion over the pieces of Harrison's circle relating to the second and first zones (Figs. 117 and 118). We start, for instance, from the motion over piece *1* in Fig. 117. When an electron on that piece gets into point A at the boundary of the Brillouin zone, its quasi-momentum along the p_x axis changes stepwise (the electron is reflected from the boundary) and it passes to an equivalent point A'. Further motion over piece *2* occurs continuously up to the next point B' located on the boundary of the Brillouin zone, from which the electron passes to an equivalent point B. A similar motion is observed on pieces *3* and *4* of Harrison's circle.

Note that the electrons located on pieces *1* and *2* remain on them all the time (if no scattering occurs at which the quasi-momentum of an electron changes arbitrarily) and do not pass over onto pieces *3* and *4*. The same is true for the electrons located on pieces *3* and *4*. Since the real motion of an electron in $\vec{r}$-space is continuous, it is reasonable to make this motion continuous also in $\vec{p}$-space by connecting the pieces *1* and *2* or *3* and *4* into continuous curves (i.e. superimposing the equivalent points A and A', B and B', etc.). This namely is the physical essence of translation of pieces of Harrison's circle for a period $2\pi\hbar/a$ along the axes p_x and p_y that has been discussed in Sec. **2-10**.

The idea of mirror reflection of an electron from the boundaries of a Brillouin zone becomes more clear if we regard the pieces of

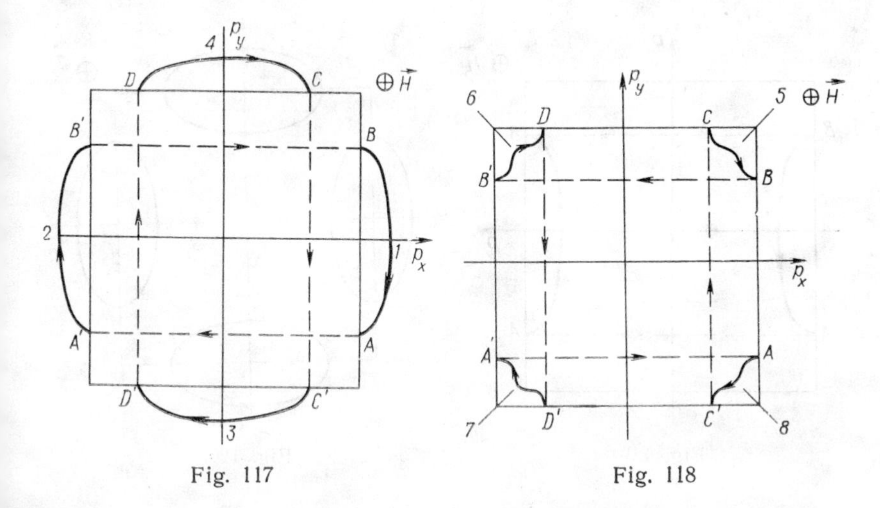

Fig. 117 Fig. 118

circle located inside the first Brillouin zone. The diagram in Fig. 117 will then have an analogous diagram in Fig. 119, in which pieces *1* and *2* are replaced by equivalent pieces *1'* and *2'* after translation into the first zone. It can be seen in that case that the quasi-momentum changes stepwise to an opposite one in point B' and the electron passes over stepwise to point B. A repeated reflection from the boundary of the zone is observed in point A.

If, by means of translation of pieces of curves, we eliminate jumps of electrons from point A to point A' and from point B' to point B, the motion in $\vec{p}$-space will then become equivalent to the motion over closed curves (Fig. 120). The motion over electron-type curves in the second zone that we have discussed occurs clockwise, i.e. in the same direction as the motion of a free electron.

Let us return to the motion of an electron over hole-type pieces of constant-energy curves (Fig. 118). On each such a piece, an electron moves clockwise relative to the centre of Harrison's circle.

Let the electron be present initially, for instance, on piece *5*. In point *B*, it is reflected from the boundary and jumps to an equivalent point *B'* on piece *6*. The whole cycle of motion in $\vec{p}$-space up to the return to piece *5* (a complete rotation in the magnetic field) is shown by arrows in Fig. 118.

By translating the pieces of Harrison's circle over periods $2\pi\hbar/a$ along the axes p_x and p_y, we can eliminate jumps of the electron between the equivalent points and pass to the motion along a closed constant-energy curve having the form of a "rosette" which

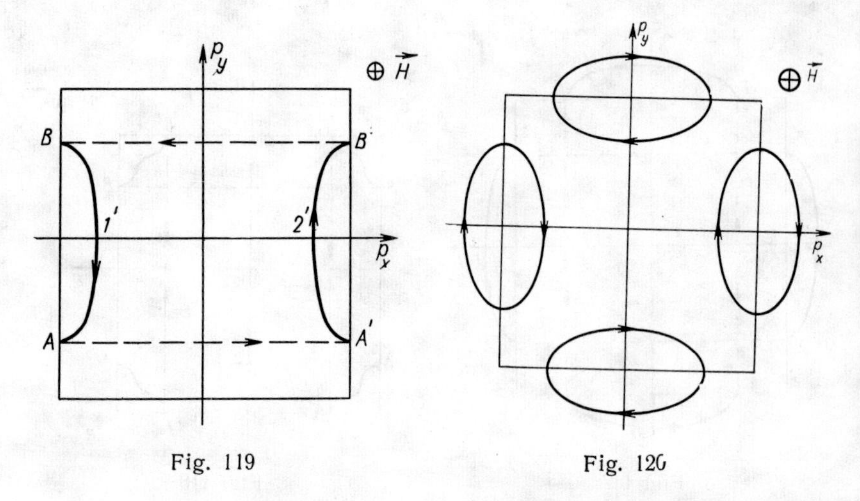

Fig. 119 Fig. 120

surrounds one of the corners of the first Brillouin zone (Fig. 121). The motion over the "rosette" gives an orbit in $\vec{p}$-space which corresponds to the projection of the orbit of real continuous motion of electrons in $\vec{r}$-space onto a plane perpendicular to $\vec{H}$. This indicates once more that the procedure of translating and combining various pieces of Harrison's circle into a continuous closed curve has a deep physical meaning.

As can be seen from Fig. 121, precession over the "rosette" occurs counter-clockwise, i.e. in the direction opposite to the precession of a free electron. Such direction of precession has a positively charged particle ($e > 0$) with a positive effective mass ($m^* > 0$) or a negatively charged particle ($e < 0$) with a negative effective mass ($m^* < 0$). The duality of the concept of a quasi-particle located on a hole-type constant-energy surface is linked with that the nature of motion of a charged particle in electric and magnetic fields is determined not by the signs of its charge e or mass m^* separately, but by the sign of the ratio e/m^*, which does not alter when the concept of $e < 0$, $m^* < 0$ is changed to that of $e > 0$, $m^* > 0$.

Both concepts are entirely equivalent for the description of dynamic effects in external fields. The problem of whether they really differ physically and how it can be revealed will be discussed in Chapter 3 when analysing the contribution of hole-type surfaces to electric conductivity and Hall effect. Here, we have only to note that quasi-particles located on electron- and hole-type constant-energy surfaces really differ from each other in their physical properties, which is revealed in their interactions with external magnetic and electric fields.

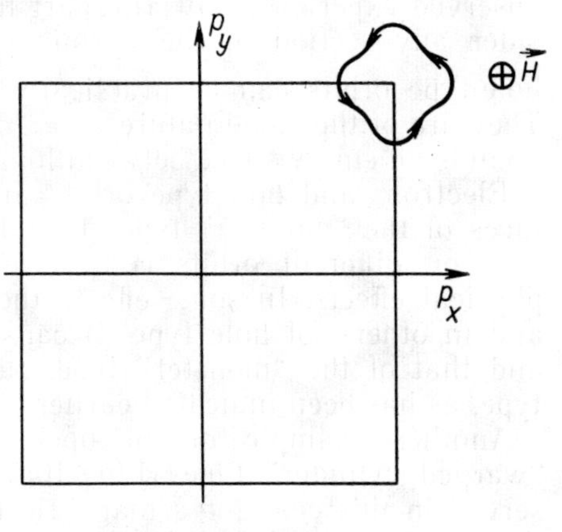

Fig. 121

The motion of quasi-particles located on open constant-energy surfaces in a magnetic field has some specific features which must be discussed separately.

As has been mentioned in Sec. **2-10** of this chapter, open surfaces can be related neither to electron nor hole type. Classification of surfaces into these two types is strictly applicable only for closed surfaces. Quasi-particles on open constant-energy surfaces can move along certain closed orbits of either electron or hole type depending on the orientation of magnetic field and the position of the plane $p_{\parallel} = \text{const}$. In other words, the direction of motion of a quasi-particle in a magnetic field over separate closed orbits located on open surfaces may coincide with the direction of motion of either an electron or a hole. On the other hand, such open surfaces are possible on which orbits of only one type are observed. Let us explain this by means of examples.

Let us take the Fermi surface of such metals as Cu, Ag, or Au and regard it in the scheme of repeating zones. It consists of spheres connected by "necks" in the directions of body diagonals (Fig. 78*a*). If there were no "necks", this Fermi surface would be a closed one and could be related, without doubt, to the electron type. But the presence of "necks" makes this conclusion unfounded. This is confirmed by the existence of orbits of different types in a magnetic field.

Indeed, with the magnetic field orientated along the directions equivalent to [100] or those close to them, electron orbits are observed corresponding, for instance, to the extremum sections of a "belly" of the Fermi surface in each cell of $\vec{p}$-space. With the field orientated along [110], in addition to the same electron orbit around a "belly" and a small electron orbit around a "neck", there appears a hole-type orbit of the "dog's bone" type, which bounds the empty space between four spheres and corresponds to the extremal hole section.

A "dog's bone" type orbit is shown in Fig. 78*a* and *b*. The synchronous motion of a group of electrons along this orbit can be observed experimentally. If, apart from extremal sections, we consider any section of the Fermi surface in Cu, Ag or Au, then hole-type orbits can be practically found for all directions of $\vec{H}$. They are of the same nature as a "dog's bone" orbit: each of them bounds an empty space between four spheres.

Electron- and hole-type orbits are observed for all Fermi surfaces of the "monster" type. The electrons moving in a magnetic field on different orbits contribute differently to some or other physical effects. In some effects the motion is of electron nature, and in others, of hole type. Because of this, such Fermi surfaces and that of the "monster" type, cannot be referred to a definite type, as has been indicated earlier.

Another example of an open Fermi surface is that of the "warped cylinder" type (Fig. 108). Electron-type orbits are observed in this case for a magnetic field directed at an angle Θ to the axis of the cylinder in all sections. They become elongated as the angle Θ approaches 90 degrees. But, for any orientation of the magnetic field, hole-type orbits are non-existent. Thus, a surface of the "warped cylinder" type can be related to the electron-type surfaces.

Note that for open paths the natural concept of effective mass in a magnetic field, connected with the period of rotation on the orbit, is inapplicable. At a passage to an open path, the period becomes infinite. Substantially earlier, when the path has become sufficiently elongated, the cyclic nature of the motion of electrons is disturbed owing to the processes of scattering. With every act of scattering, an electron passes over from one section to another. Because of this the single-valued relation between the motion of the electron and the geometry of the Fermi surface vanishes, though the latter completely determines the nature of motion of electrons during the relaxation time.

To conclude this section, let us dwell upon a new method of introducing the cyclotron mass of electrons in a metal, which is directly connected with the construction of Fermi surfaces according to Harrison. The method is based on calculation of the period of cyclic motion of an electron in a magnetic field along a path located on pieces of Harrison's sphere between Bragg reflection planes. With the zero approximation of the magnitude of effective potential, an electron will move in the magnetic field over a spherical constant-energy surface with the period $T_0 = \frac{2\pi c m_0}{|e| H}$.

If we now pass over to a finite effective potential and account for Bragg reflections, the period T_H of rotation of the electron becomes equal to T_0 multiplied by the ratio of the total angle described by the quasi-momentum of the electron between the planes of energy discontinuity to the total angle 2π (this angle is described by the momentum of a free electron during the same period). It can then naturally be assumed that the effective potential is small and the motion of the electron over an individual piece of the Harrison sphere separated by the boundaries of the Brillouin zones does not practically differ from the motion of a free electron over the same piece of the sphere at $V_{eff}(\vec{r}) \equiv 0$.

As in the first method of introducing the cyclotron mass, the magnitude of m^* is now determined by the period T_H of rotation of the electron on the orbit in $\vec{p}$-space: $m^* = \frac{|e| H}{2\pi c} T_H$. Thus, the ratio of the cyclotron mass to the mass of a free electron, equal to the T_H/T_0 ratio, is correlated with the geometry of the path of the electron on Harrison's sphere. Let this be explained on the example, given earlier, of a trivalent "metal" with a plane square lattice placed into a magnetic field $\vec{H}$ perpendicular to the plane of lattice. For this, we again apply to Figs. 117 and 118.

The path of the electron in the metal considered consists of four pieces *1-4* of Harrison's circle (Fig. 117), which differ from the four pieces *5-8* constituting a hole-type path (Fig. 118). During the period of motion, the quasi-momenta of the electron and hole describe the angles equal to $4\Delta\varphi_{el}$ and $4\Delta\varphi_{hol}$ respectively (Fig. 122). The cyclotron masses of the electron m^*_{el} and hole m^*_{hol} will then be:

$$m^*_{el} = \frac{m_0}{2\pi} 4\,\Delta\varphi_{el}, \quad m^*_{hol} = \frac{m_0}{2\pi} 4\,\Delta\varphi_{hol}$$

In a three-dimensional case, the cyclotron mass is determined by constructing Harrison's sphere and finding all its intersections with the planes of energy discontinuity, as has been done for the construction of Fermi surfaces. Then the projection of these lines onto a plane perpendicular to the magnetic field is found.

The projection of the electron momentum onto this plane moves along a circle and undergoes jumps at reaching the projections of the lines of intersection of Harrison's sphere with Bragg planes. The cyclotron mass corresponding to an orbit is obtained by summing up all the angles between Bragg reflections and multiplying this sum by $\frac{m_0}{2\pi}$.

It can be shown that the cyclotron masses exceed m_0 for some orbits. This is related to the fact that the period of precession of electrons on Harrison's sphere is independent of the position of the plane $p_{\parallel} = \text{const}$ on which the orbit of the electron is located. In other words, the period of precession over the small circle corresponding to a non-central section of Harrison's sphere coincides with the period of precession over the circle

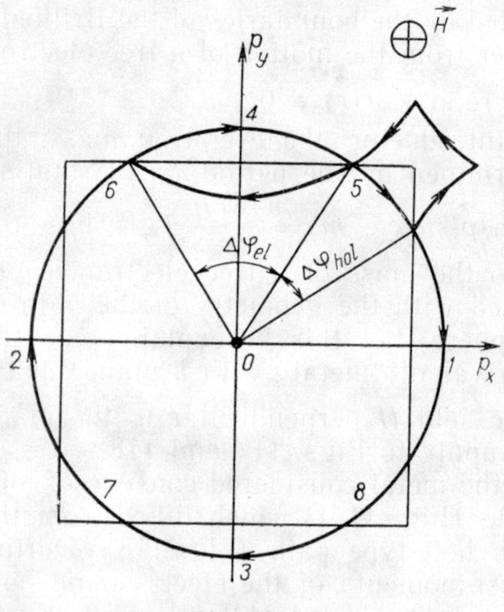

Fig. 122

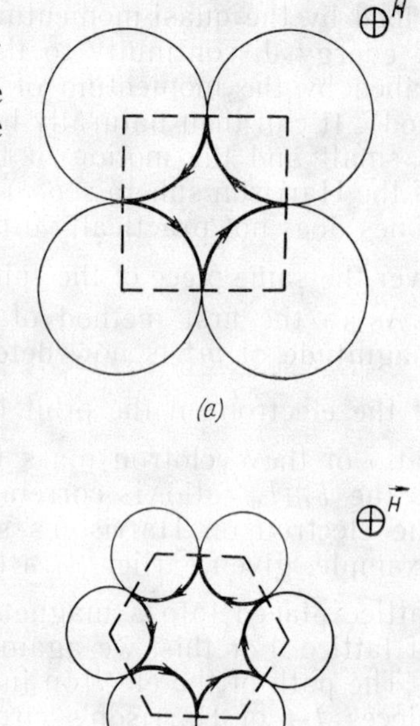

Fig. 123

of the central section. Owing to this the total angle described by the projection of the quasi-momentum of the electron onto the corresponding secant plane perpendicular to the magnetic field H may exceed 2π, and therefore, the cyclotron mass related to this orbit may exceed m_0.

Figure 123*a*, for instance, shows a hole-type orbit consisting of arcs of four contacting circles which correspond to sections of four Harrison's spheres by a plane perpendicular to $\vec{H}$. The quasi-momentum of the hole describes an angle of 2π during motion along this orbit, and therefore, the related cyclotron hole mass is m_0.

Figure 123*b* shows a hole-type orbit consisting of arcs of six contacting circles which are formed by intersection of six neighbouring Harrison's spheres with a plane perpendicular to $\vec{H}$. The corresponding cyclotron mass is evidently $2m_0$.

Using this method, we can determine all masses on the Fermi surface, calculate the number of the electrons dN on the Fermi surface whose cyclotron masses are within the interval from m^* to $m^* + dm^*$, and find the function of distribution of cyclotron masses $\frac{dN}{dm^*} = f(m^*)$ for the given orientation of the field $\vec{H}$.

Figure 124, according to Harrison [26], gives the results of calculation of $\frac{dN}{dm^*}$ for aluminium in the direction of the magnetic field along the crystallographic axis of the type [110]. The values

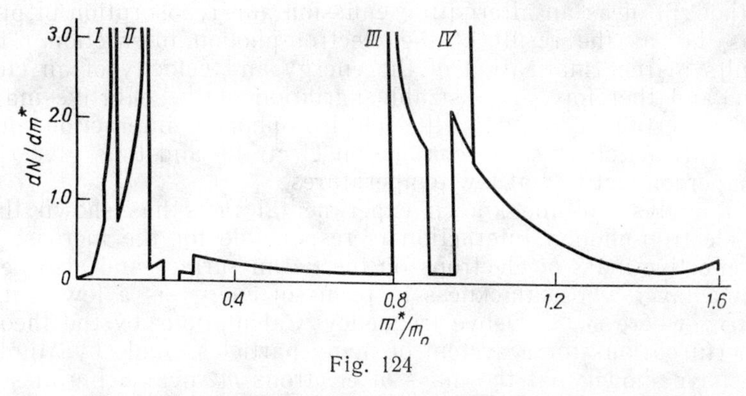

Fig. 124

of m^* obtained fill almost the whole interval from 0 to 1.6 m_0. The curve $\frac{dN}{dm^*} = f(m^*)$ in the figure is normalized so as to equate

$$\int_0^{\frac{m^*_{max}}{m_0}} f(m^*) \frac{dm^*}{m_0}$$

to unity. The singular points on the curve relate to the extremal masses which correspond to the paths of quasi-particles on the hole-type surface in the second zone (peak III at $\frac{m^*}{m_0} = 0.8$) and on the electron-type surfaces in the third zone (peaks I, II, and IV at $\frac{m^*}{m_0} = 0.1$, 0.17, and 1.0 respectively).

The experimental values of cyclotron masses m^* for aluminium found by Moore and Spong [58] exceed the corresponding calculated values 1.3 times. The calculation made by Harrison has shown that such a large discrepancy in the masses cannot be removed by introducing first-order corrections accounting for the non-local nature of pseudo-potential and second-order corrections for the wave function of an electron (i.e. under account of the second-order terms in the expansion of the function over orthogonalized plane waves). These corrections, as a rule, give only a slight variation of the parameters of electrons. As follows from the calculation, the use of any number of orthogonalized plane waves to represent the wave function of an electron cannot remove this discrepancy.

There are two likely sources of error in the masses. One is the use of a self-consistent-field approximation, i.e. a single-electron approximation. But, according to Fletcher and Larson [59], the correction factor is estimated as approximately 1.1. Another source of error is neglecting the motion of the ions. It is known, as electrons move in the lattice, the ions are displaced. The process may be thought of as an alternating emission and reabsorption of phonons, i.e. as the result of the electron-phonon interaction. This results in renormalization of the energy and velocity of an electron, and therefore, in a sizable variation of the effective mass. Under certain conditions the electron-phonon interaction may cause attraction between conduction electrons and thus give rise to superconductivity at low temperatures.

An analysis of the known experimental facts has shown that the electron-phonon interaction is responsible for the increase of the effective mass of electrons on the Fermi surface and also near it in a layer whose thickness in terms of energy is a few tenths of $h\omega_D$, where ω_D is Debye frequency. Calculations by the theory of perturbations for a system of many particles, made by Migdal [60], have shown that the mass of electrons m^* near a Fermi surface is subject to renormalization: $m^{**} = (1 + \lambda) m^*$ (m^{**} being the renormalized mass). The non-renormalized quantity m^* has been termed the band mass. The parameter λ of renormalization is determined by the integral $\lambda = 2 \int_0^\infty \frac{D(\omega)}{\omega} \alpha^2(\omega)\, d\omega$, where $D(\omega)$ is the spectral function of phonons introduced in Chapter 1, and $\alpha(\omega)$ is a phonon-frequency dependent constant of electron-phonon interaction. The calculated values of λ for Na, Al, and Pb are 0.19, 0.50, and 1.6 respectively.

The additional account of the Coulomb repulsion of electrons gives the following expression for a renormalized cyclotron mass in non-superconducting metals: $m^{**} = (1 + \lambda + \mu) m^*$, where λ and μ account respectively for the electron-phonon interaction and Coulomb interaction of electrons. The magnitude of μ is still not known exactly, but it may be expected that it is small (for aluminium, for instance, approximate calculations have given $\mu \cong 0.01$).

It should be emphasized that the electron-phonon interaction does not change the size of the Fermi surface and only manifests itself in corrections to the effective mass. In addition, this effect has no influence on the kinetic properties of metals, for instance, their electric and thermal conductivities.

The renormalization of the effective mass on the Fermi surface results in the variation of the specific heat C_v of the electron gas $(1+\lambda+\mu)$ times. Later, when discussing the various effects in metals, we shall indicate the physical characteristics that vary because of the electron-phonon interaction.

2-13. THE DENSITY OF STATES

The energy distribution of electron states is one of the most important characteristics of the energy spectrum of electrons. This distribution can be described by introducing the concept of the density of states. Let $N(\varepsilon)$ be the number of states with an energy not exceeding ε. The differential $dN(\varepsilon)$ will then give the number of states within an interval of energies from ε to $\varepsilon+d\varepsilon$. The density of states is a function $\nu(\varepsilon)$ equal to the ratio of $dN(\varepsilon)$ to the width $d\varepsilon$ of the interval. This function is evidently linked closely with the law of dispersion of electrons.

Let us derive an expression for the density of states of an electron system for a particular case of the quadratic dispersion law.

Using this simple dispersion law, we can determine the main peculiarities of the function $\nu(\varepsilon)$ in dependence of the number of variations of the system. The dimension of the system will be marked by superscripts III, II, or I at the corresponding functions $N(\varepsilon)$, $\nu(\varepsilon)$, etc. [for instance, $N^{\mathrm{III}}(\varepsilon)$, $\nu^{\mathrm{III}}(\varepsilon)$]. The expression for the density of states will be first derived for a three-dimensional case.

As has been derived earlier, the number of electrons in a unit volume of a three-dimensional phase space (or, in other words, the density of electrons in a three-dimensional phase space) is $\frac{2V}{(2\pi\hbar)^3}$, where V is the volume of crystal. The number of electrons $dN_{\vec{p}}^{\mathrm{III}}$ in an elementary volume $dp_x dp_y dp_z$ of $\vec{p}$-space will then be

$$dN_{\vec{p}}^{\mathrm{III}}=\frac{2V}{(2\pi\hbar)^3}\,dp_x\,dp_y\,dp_z \tag{2.82}$$

With the quadratic dispersion law $\varepsilon=\frac{p^2}{2m}$ it is reasonable to pass from the Cartesian coordinates p_x, p_y, p_z to the spherical p, θ, φ, since the coordinate and constant-energy surfaces in the new variables coincide. The expression for $dN_{\vec{p}}^{\mathrm{III}}$ in the spherical coordinates takes the form:

$$dN_p^{\mathrm{III}}=\frac{2V}{(2\pi\hbar)^3}\,p^2\,dp\sin\theta\,d\theta\,d\varphi \tag{2.83}$$

Integrating by angles, we find the number of electrons in the layer located between p and $p+dp$:

$$dN_p^{\mathrm{III}} = 2\frac{4\pi V}{(2\pi\hbar)^2} p^2\, dp \tag{2.84}$$

Passing from the variable p to $\varepsilon = \frac{p^2}{2m}$, we get for the number of electrons $dN_\varepsilon^{\mathrm{III}}$ in the energy interval from ε to $\varepsilon + d\varepsilon$:

$$dN_\varepsilon^{\mathrm{III}} = \frac{8\sqrt{2}\,\pi V}{(2\pi\hbar)^3} m^{3/2} \varepsilon^{1/2}\, d\varepsilon \tag{2.85}$$

Thus, the density of states $\nu^{\mathrm{III}}(\varepsilon)$ of the three-dimensional electron system with the quadratic dispersion law is:

$$\nu^{\mathrm{III}}(\varepsilon) = \frac{\partial N_\varepsilon^{\mathrm{III}}}{\partial \varepsilon} = \frac{8\sqrt{2}\,\pi V}{(2\pi\hbar)^3} m^{3/2} \varepsilon^{1/2} \tag{2.86}$$

The shape of the function $\nu^{\mathrm{III}}(\varepsilon)$ is shown in Fig. 125.

Expression (2.86) for the density of states has been obtained for the quadratic isotropic dispersion law. A similar expression can be found for an anisotropic quadratic dispersion law which is associated with ellipsoidal constant-energy surfaces. For this, it is sufficient to find the number of electrons $N^{\mathrm{III}}(\varepsilon)$ with an energy not exceeding ε. This number coincides with the number of electrons inside the ellipsoid given by equation (2.25). The semi-axes of the ellipsoid are $\sqrt{2m_x\varepsilon}$, $\sqrt{2m_y\varepsilon}$, and $\sqrt{2m_z\varepsilon}$ and its volume $\Delta_{\vec{p}}$ is:

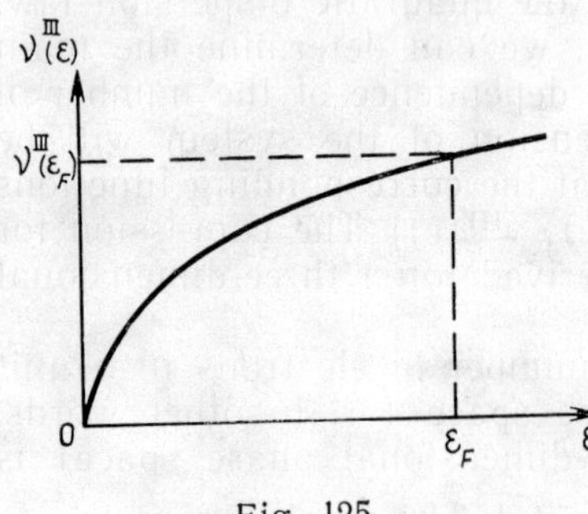

Fig. 125

$$\Delta_{\vec{p}} = \frac{4}{3}\pi 2\sqrt{2}\,(m_x m_y m_z)^{1/2} \varepsilon^{3/2} \tag{2.87}$$

Hence the number of electrons $N^{\mathrm{III}}(\varepsilon)$ is:

$$N^{\mathrm{III}}(\varepsilon) = \frac{16\sqrt{2}\,\pi V}{3\,(2\pi\hbar)^3} (m_x m_y m_z)^{1/2} \varepsilon^{3/2} \tag{2.88}$$

Differentiating by ε, we find the density of states $\nu^{\mathrm{III}}(\varepsilon)$ for the case of ellipsoidal constant-energy surfaces

$$\nu^{\mathrm{III}}(\varepsilon) = \frac{8\sqrt{2}\,\pi V}{(2\pi\hbar)^3} (m_x m_y m_z)^{1/2} \varepsilon^{1/2} \tag{2.89}$$

As will be seen, at the passage from a spherical Fermi surface to an ellipsoidal one the nature of the dependence of ν^{III} on ε has not changed. Let $(m_x m_y m_z)^{1/3}$ be denoted as m_d. This quantity has been termed the effective mass of density of states. In an isotropic case it simply coincides with m. The introduction of the effective mass of density of states makes expressions (2.86) and (2.89) formally identical to one another. The passage from an isotropic case to an anisotropic one thus reduces solely to changing the magnitude of m_d.

As has been mentioned earlier, ellipsoidal constant-energy surfaces are observed in the vicinity of maxima and minima of energy. These surfaces are of the electronic type near a minimum of energy and bound the filled part of p-space. A minimum of energy in an energy band is otherwise called its bottom. Thus, the density of states of electrons, ν^{III}, has the form of (2.89), where ε implies the energy of an electron state $\mathscr{E}$ calculated from the bottom $\mathscr{E}_c$ of the zone upward: $\varepsilon = \mathscr{E} - \mathscr{E}_c$.

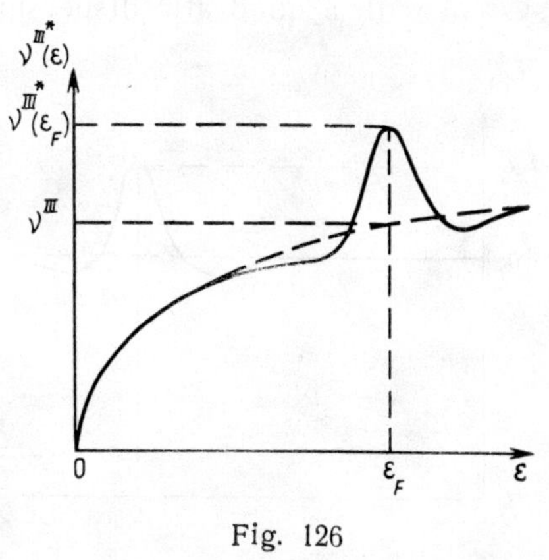

Fig. 126

The constant-energy surfaces are of the hole type near a maximum of energy and bound an empty part of $\vec{p}$-space in a corresponding Brillouin zone. A maximum of energy is otherwise called the top of an energy band. Thus, the density of states of holes also has the form of (2.89), where the energy ε implies the energy of the unfilled electron state $\mathscr{E}$ corresponding to a hole calculated from the top $\mathscr{E}_v$ of the zone downward: $\varepsilon = \mathscr{E}_v - \mathscr{E}$.

Note some properties of the density of states $\nu^{III}(\varepsilon)$ of a three-dimensional system:

(a) the density of states $\nu^{III}(\varepsilon)$ is zero at the boundary of an energy band (in the minimum or maximum of energy in the band) and increases monotonously with energy;

(b) the density of states $\nu^{III}(\varepsilon)$ has no singularities at any values of ε;

(c) the density of states $\nu^{III}(\varepsilon)$ is proportional to the effective mass m_d of electrons (or holes) to the power of 3/2.

Until now, we have discussed the density of states of an electron system obtained without account of the interaction between electrons and also without account of the electron-phonon interaction. As has been shown in the previous section, these two factors result in renormalization of the effective mass of the carriers near a Fermi surface, the electron-phonon interaction providing the greatest contribution. A change (increase) of the effective mass of electrons and holes, in turn, causes an increase of the density of states on the Fermi surface.

Thus, the expression (2.89) obtained for $\nu^{III}(\varepsilon)$ describes a non-renormalized density of states, which is differently called the band density of states. Qualitatively, the dependence of a renormalized quantity $\nu^{III*}(\varepsilon)$ on energy is shown in Fig. 126. The "peak" of density of states near $\varepsilon = \varepsilon_F$ corresponds to a layer of "heavy" electrons or holes formed through the electron-phonon interaction.

Let us derive the density of states for a two-dimensional electron system with a quadratic dispersion law. An analytical expression

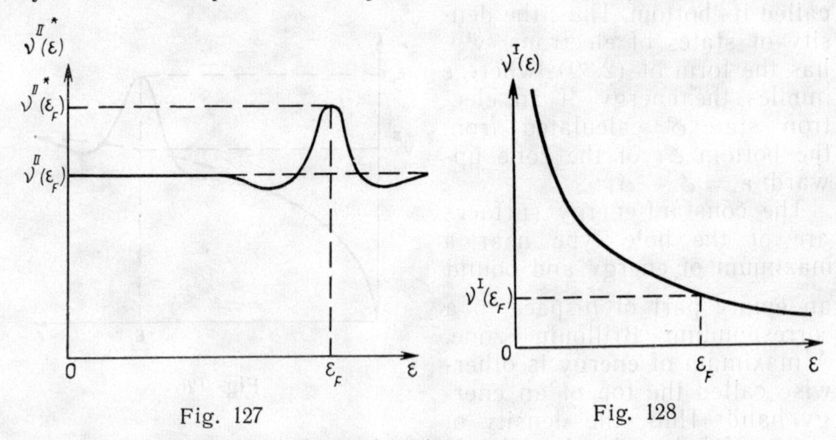

Fig. 127

Fig. 128

for the band density of states $\nu^{II}(\varepsilon)$ can be obtained by the same method as has been used in deriving the expression for $\nu^{III}(\varepsilon)$.

For a two-dimensional crystal of unit area the number of electrons $dN^{II}_{\vec{p}}$ in an element $dp_x dp_y$ of phase space is

$$dN^{II}_{\vec{p}} = \frac{2}{(2\pi\hbar)^2} dp_x\, dp_y \tag{2.90}$$

Passing to cylindrical coordinates and integrating by angle φ, we get the number of electrons dN^{II}_p in the "neck" extending from p to $p + dp$:

$$dN^{II}_p = \frac{4\pi}{(2\pi\hbar)^2} p\, dp \tag{2.91}$$

From the dispersion law $\varepsilon = \frac{p^2}{2m}$ it follows that $pdp = md\,\varepsilon$. The expression for the band density of states $\nu^{II}(\varepsilon)$ can then be written as

$$\nu^{II} = \frac{4\pi}{(2\pi\hbar)^2} m = \frac{m}{\pi\hbar^2} \tag{2.92}$$

The band density of states of a two-dimensional system is constant and proportional to the effective mass of the carriers. The renormalized density of states $\nu^{II*}(\varepsilon)$ has a peak corresponding to the Fermi energy. A qualitative curve of $\nu^{II*}(\varepsilon)$ is shown in Fig. 127.

For a one-dimensional system of unit length, the number of electrons dN_p^I in an element dp of phase space is

$$dN_p^I = \frac{2\,dp}{2\pi\hbar} = \frac{dp}{\pi\hbar} \tag{2.93}$$

Hence the band density of states $\nu^I(\varepsilon)$ is as follows:

$$\nu^I(\varepsilon) = \frac{m^{1/2}}{\sqrt{2}\,\pi\hbar}\varepsilon^{-1/2} \tag{2.94}$$

The function $\nu^I(\varepsilon)$ (Fig. 128) increases with a reduction of ε and has an infinite singularity at $\varepsilon = 0$. This characteristic feature of the density of states of a one-dimensional system $\nu^I(\varepsilon)$ plays a principal part in the distribution of electrons over energy levels in a quantizing magnetic field.

2-14. QUANTIZATION OF THE ENERGY OF AN ELECTRON IN A MAGNETIC FIELD

Because of the cyclic nature of the motion of an electron in a magnetic field $\vec{H}$, its energy related to the motion in a plane perpendicular to $\vec{H}$ becomes discrete, or as it is said, is quantized. This was first indicated by Landau [61] in 1930 when determining the energy levels of an electron in a constant homogeneous magnetic field $\vec{H}$.

Let us recall that the energy ε of a free electron is a continuous quantity and can be expressed through the components of its momentum $\vec{p}$:

$$\varepsilon = \varepsilon(p_x, p_y, p_z) = \frac{p_x^2 + p_y^2 + p_z^2}{2m_0}$$

Let the magnetic field be directed along the z-axis. As has been found by Landau, the motion of an electron in a magnetic field can be resolved into two components: the motion along the field $\vec{H}$ and that in a plane perpendicular to $\vec{H}$. The magnetic field does not change the longitudinal component, i.e. this part of the motion is not quantized. The energy of electron related to it, $\varepsilon_{\parallel}$, as with $H = 0$, is equal to $p_z^2/2m_0$.

The motion in the plane perpendicular to $\vec{H}$ is similar to the motion of a linear harmonic oscillator which oscillates about an equilibrium position $y_0 = \frac{c p_x}{|e| H}$ with a frequency coinciding with the cyclotron frequency $\omega = \frac{|e| H}{m_0 c}$. The energy spectrum $\varepsilon_{\perp}$ of the transverse component of motion coincides with the spectrum of the linear oscillator and consists of discrete levels:

$$\varepsilon_{\perp} = \left(n + \frac{1}{2}\right) \hbar\omega, \text{ where } n = 0, 1, 2, \ldots$$

Thus, the energy of the electron in the magnetic field without account of its spin is as follows:

$$\varepsilon = \varepsilon(n, p_z) = \left(n + \frac{1}{2}\right) \hbar\omega + \frac{p_z^2}{2m_0} \tag{2.95}$$

Quantization of the energy of electrons in a magnetic field is the physical basis of diamagnetism of electron systems.

The motion of a quantum particle differs from that of a classical one in that the energy of cyclic motion is discrete. This distinction can be clearly interpreted by using the quasi-classical concept of the path of motion of an electron in a magnetic field.

Thus, for the classical motion, any values of the energy $\varepsilon_{\perp}$ related to the motion in a plane perpendicular to the field $\vec{H}$ are allowable. The magnitude of $\varepsilon_{\perp}$ is then expressed through the frequency ω of rotation and radius r of the electron orbit: $\varepsilon_{\perp} = \frac{m_0 v_{\perp}^2}{2} = \frac{m_0 \omega^2 r^2}{2}$. The cyclotron frequency $\omega = \frac{|e| H}{m_0 c}$ is independent of the size of the orbit. Thus, each value of the energy $\varepsilon_{\perp}$ has a corresponding definite value of radius r, both values varying continuously.

For a quantum motion, only discrete values of the energy $\varepsilon_{\perp} = \left(n + \frac{1}{2}\right) \hbar\omega$ are allowable.

We shall use, as before, the quasi-classical concept of the path of an electron[1]. The values of radius r of the path can be found

[1]) A quantum particle moves freely in a magnetic field along the z axis and also performs motion within a limited region

$$y_0 - \sqrt{\frac{\hbar}{m_0 \omega}} \leqslant y \leqslant y_0 + \sqrt{\frac{\hbar}{m_0 \omega}}$$

in a plane perpendicular to the magnetic field. The motion in that region corresponds to the classical motion of a charge around a circle with the cyclotron frequency.

by the correspondence principle, used in quantum mechanics to establish the relation between classical and quantum quantities. For this, we correlate the classical and the quantum expressions for the energy $\varepsilon_\perp$:

$$\varepsilon_\perp^{cl} = \frac{m_0\omega^2 r^2}{2} \quad \text{and} \quad \varepsilon_\perp^{qua} = \left(n + \frac{1}{2}\right)\hbar\omega$$

Hence it follows that only discrete values of the radii r_n of orbits are allowed:

$$r_n = \sqrt{\frac{2\hbar}{m_0\omega}\left(n + \frac{1}{2}\right)} = \sqrt{\frac{2c\hbar}{|e|H}\left(n + \frac{1}{2}\right)} \tag{2.96}$$

a definite value of r_n corresponding to each value of the quantum number n.

To pass from one orbit to any other of greater radius, it is required to spend an energy multiple of $\hbar\omega$.

Thus, the energy spectrum of an electron in a magnetic field is determined by the quantum number n and the magnitude of the projection of its momentum p_z onto the direction of the magnetic field which is retained during motion. Such a spectrum can be depicted by a combination of parabolae $\frac{p_z^2}{2m_0}$ shifted relative to one another along the energy axis by $\hbar\omega$ (Fig. 129).

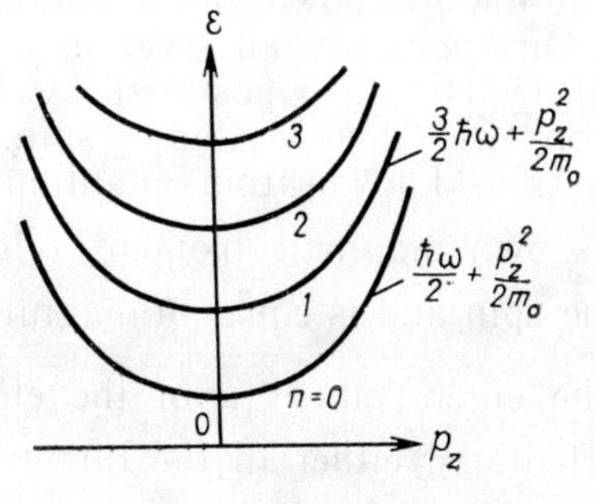

Fig. 129

Each parabola corresponds to its own value of the quantum number n. These parabolae have been termed Landau levels. The numbering of these levels is made by means of the quantum number n.

The energy states on each Landau level, corresponding to different values of p_z, are degenerated in y_0, which actually determines the position of the centre of orbit of an electron along one of the Cartesian axes on a plane perpendicular to the magnetic field. Note that the coordinate of the centre of orbit x_0 along the other Cartesian axis has no definite value simultaneously with y_0, since the quantum operators corresponding to these values do not commutate with one another.

Let us find the degree of degeneration of each energy level (2.95) in a case when the motion of the electron in a plane perpendicular to the magnetic field is restricted within the limits of a finite area S with dimensions L_x and L_y along the axes x and y. With this restriction, the components p_x and p_y of the electron become discrete and multiple of the minimum values of the momenta:

$$\Delta p_x = \frac{2\pi\hbar}{L_x} \text{ and } \Delta p_y = \frac{2\pi\hbar}{L_y} \text{ (see Sec. 2-3)}$$

We assume that the radius of the electron orbit is small compared with L_x and L_y. The minimum distance Δy_0 between neighbouring orbits along the y axis is

$$\Delta y_0 = \frac{c\,\Delta p_x}{|e|H} = \frac{2\pi c\hbar}{|e|HL_x}$$

The number of allowed orbits on the area S is determined by the number of different values of y_0 satisfying the condition $0 < y_0 < L_y$. It is evidently equal to the ratio $\frac{L_y}{\Delta y_0} = \frac{|e|H}{2\pi\hbar_c} L_x L_y$. Then, the number of different states along y_0 for the given n and p_z (degree of degeneration) is $\frac{|e|H}{2\pi\hbar_c} S$; it is proportional to the area S and the first power of the magnetic field.

On each Landau level in a state with the given p_z there are electrons with opposite spins. Thus, each energy state is, in addition, twice degenerated by spin.

As is known from quantum mechanics, an electron possesses its own magnetic moment (eigenmoment) $\vec{\mu}$ which is related to the spin and is equal in magnitude to Bohr magneton $\mu_B = \frac{|e|\hbar}{2m_0c}$. The eigenmoment $\vec{\mu}$ of the electron in a magnetic field can be orientated either in the direction $+\vec{H}$ or in $-\vec{H}$. Account of the energy of the eigenmoment $\vec{\mu}$ of the electron in the magnetic field, which is equal to $-(\vec{\mu}\vec{H})$, results in removal of spin degeneration of Landau levels. Each Landau level then splits into two sub-levels corresponding to two orientations of $\vec{\mu}$. The lower of the two sub-levels is evidently that on which the magnetic moment of the electron coincides in direction with $\vec{H}$.

With an account of the energy-level spin splitting, the expression for the energy of an electron in a magnetic field can be written as

$$\varepsilon = \varepsilon(n, s, p_z) = \left(n + \frac{1}{2}\right)\hbar\omega + s\mu_B H + \frac{p_z^2}{2m_0} \tag{2.97}$$

where s is the spin quantum number, acquiring the values ± 1.

With the account of the spin of an electron, the Landau levels are numbered by means of two quantum numbers n and s. The values $s = +1$ and $s = -1$ are usually written as "+" and "−"

signs at the corresponding number n. With such a notation, the lowest level is 0^-.

Note that for a free electron in the magnetic field the magnitude of spin splitting of levels, equal to $2\mu_B H$, coincides with the distance $\hbar\omega$ between the Landau levels obtained without account of electron spin [1]).

The situation arisen is very peculiar: spin splitting removes degeneration of the Landau levels with the same value of n, but the degeneration of the levels corresponding to the quantum numbers $(n,\ s=+1)$ and $(n+1,\ s=-1)$ is formed. The passage from a system of Landau levels without account of spin to a system of levels accounting for spin splitting can be clearly illustrated on an energy diagram corresponding to the states with $p_z=0$ (Fig. 130). As will be seen from the diagram, the degeneration related to spin is only absent on the 0^- level.

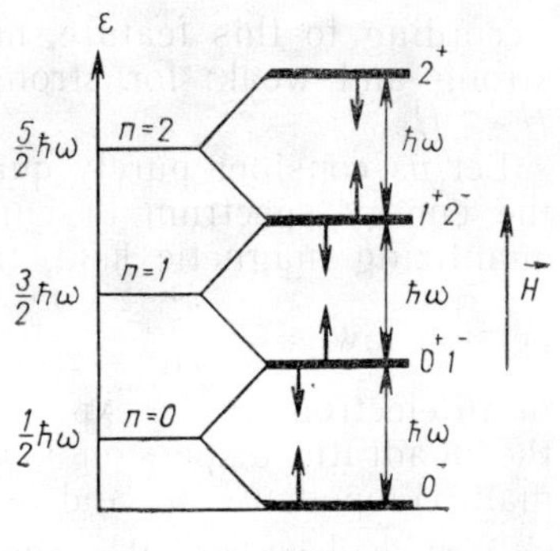

Fig. 130

In expression (2.97) for the energy of an electron, the dependence on quantum numbers n and s can be formally replaced by the dependence on a single number n': $\varepsilon = n'\hbar\omega + \frac{p_z^2}{2m_0}$, where n' can take on a series of values: 0, 1, 2, ...

A circumstance of importance should be emphasized. The energy of an electron in a magnetic field is dependent in a continuous manner only on one projection of the momentum, p_z. Because of this, the system of electrons in the magnetic field possesses many features of a one-dimensional system. This peculiarity has the most substantial effect on the density of states of electrons in the magnetic field.

Quantization of the energy of a free electron in a magnetic field is a consequence of its cyclic motion. For the same reason, the energy of an electron in a metal placed into a magnetic field is also quantized. But quantization of electrons in a metal can only occur when the free-path length l exceeds the size of the orbit of the electron $\frac{cp_\perp}{|e|H}$ in a plane perpendicular to the magnetic field.

---

[1]) Since the formation of Landau levels is connected with the motion of an electron on its orbit, the distance $\hbar\omega$ between the Landau levels is sometimes called orbital splitting.

Because of this the magnitude of a quantizing magnetic field is limited from below by a definite value H_l at which the characteristic size $\frac{cp_F}{|e|H}$ becomes of the same order as l.

Let us recall that estimation of this value of the magnetic field has been done in Sec. **2-12**, Chapter Two:

$$H_l \sim \frac{a}{l} H_a = \frac{a}{l}(10^8\text{-}10^9) \text{ oersteds}$$

According to this feature, magnetic fields are usually classed as strong and weak: for strong fields $H > H_l$, and for weak ones $H < H_l$.

Let us consider purely qualitatively the variations occurring in the energy spectrum of current carriers under the action of a quantizing magnetic field, taking for example the quadratic dispersion law: $\varepsilon = \frac{p_x^2}{2m_x} + \frac{p_y^2}{2m_y} + \frac{p_z^2}{2m_z}$. The frequency of cyclic motion of an electron can be expressed through its cyclotron mass m^*. For the quadratic dispersion law, m^* at $H = H_z$ is related with the main components m_x and m_y of the effective mass tensor: $(m^*)_z = \sqrt{m_x m_y}$ [see (2.80)]. The transition from a quasi-continuous energy spectrum $\varepsilon = \varepsilon(p_x, p_y, p_z)$ to the Landau levels can be made similar to what has been done for free electrons. The energy of an electron in a magnetic field, depending without account of spin on the quantum number n and the projection of momentum p_z, can be written as:

$$\varepsilon = \varepsilon(n, p_z) = \frac{|e|\hbar H}{(m^*)_z c}\left(n + \frac{1}{2}\right) + \frac{p_z^2}{2m_z} \tag{2.98}$$

An energy state with the given n and p_z is twice degenerated by spin, and is also degenerated with the multiplicity $\frac{|e|H}{2\pi\hbar c} L_x L_y$ by the positions of the centre of orbit of the electron in a plane perpendicular to the magnetic field (here L_x and L_y are dimensions of the crystal lattice of the metal in the plane perpendicular to $\vec{H}$).

For quantization of the energy of an electron in a metal, account of the spin cannot be made in the same manner as for a free electron. Analytical computation of the magnitude of spin splitting $(\Delta\varepsilon)^s$ of Landau levels on the basis of the zone structure of a particular metal is a rather complicated problem that has not been solved completely. In particular, it has not still been explained theoretically why the magnitude of spin splitting $(\Delta\varepsilon)^s$ in some metals exceeds orbital splitting $\hbar\omega$.

Spin splitting $(\Delta\varepsilon)^s$ is usually expressed in the units $\mu_B H$:

$$(\Delta\varepsilon)^s = g\mu_B H \tag{2.99}$$

the proportionality factor g being termed g-factor. For a free electron, $g = 2$.

For metals, g-factor may substantially differ from 2 in either side. Note that g-factor may be different for different groups of current carriers even in the same metal. In each particular case, g-factor can also strongly depend on the direction of magnetic field $\vec{H}$.

Thus, g-factor is a new characteristic of current carriers in metals and has to be determined experimentally together with other characteristics, such as cyclotron mass, the size and shape of the Fermi surface, etc.

Upon introduction of the g-factor, the expression for the energy of an electron in a metal placed into a magnetic field $\vec{H}$, with account of spin splitting of levels, can be written in the form similar to (2.97):

$$\varepsilon = \varepsilon(n, s, p_z) = \frac{|e|\hbar H}{(m^*)_z c}\left(n + \frac{1}{2}\right) + \frac{1}{2} sg\mu_B H + \frac{p_z^2}{2m_z} \tag{2.100}$$

where $s = \pm 1$, as for a free electron.

The expression for spin splitting $(\Delta\varepsilon)^s$ can be written in a different way. For this, we introduce a quantity m^s, having the dimension of mass, and relate it to the g-factor as follows:

$$g = 2\frac{m_0}{m^s} \tag{2.101}$$

m^s has been termed the spin mass. Using it, $(\Delta\varepsilon)^s$ can be formally written as for a free electron:

$$(\Delta\varepsilon)^s = 2\frac{|e|\hbar H}{2m^s c} = 2\mu_B^* H \tag{2.102}$$

The spin mass m^s here determines the magnetic momentum μ_B^* which is called the effective magneton.

The use of the spin mass is very convenient for comparing the values of the spin and orbital splitting. With $m^s = m^*$, the spin splitting evidently coincides with the orbital splitting. The system of Landau levels is then similar to the system of levels for a free electron.

Figure 131*a*, *b*, and *c* shows diagrams, similar to that of Fig. 130, for three cases: $m^s > m^*$, $m^s = m^*$, and $m^s < m^*$. With $m^s < m^*$, the spin splitting exceeds the orbital splitting, which results in disturbing the order of sequence of levels. In particular, with the magnitude of spin splitting shown in Fig. 131*c*, the energy of state with $p_z = 0$ on the level n^+ becomes greater than the energy of state on the level $(n+1)^-$.

Let us discuss the quantization of the energy of quasi-particles in the vicinity of a maximum of energy in a band, i.e. the quantization of the energy of holes. Note that the cyclic motion of holes

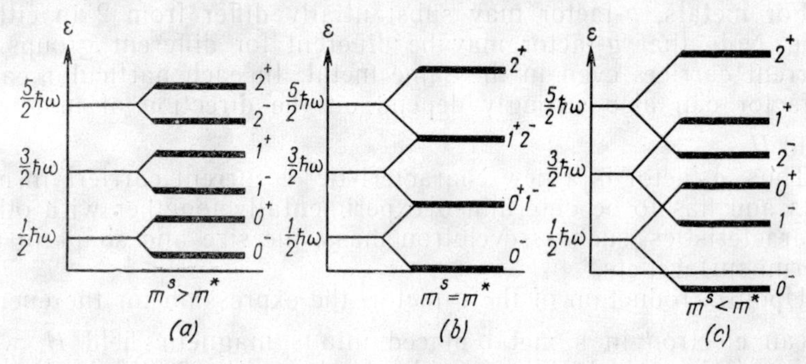

Fig. 131

occurs in a direction opposite to that of electrons, and that the cyclotron mass of holes is negative. As has been indicated, the energy of holes is counted from the top of a band downward. Accordingly, the Landau level of holes with the number n is located above the level with the number $n+1$.

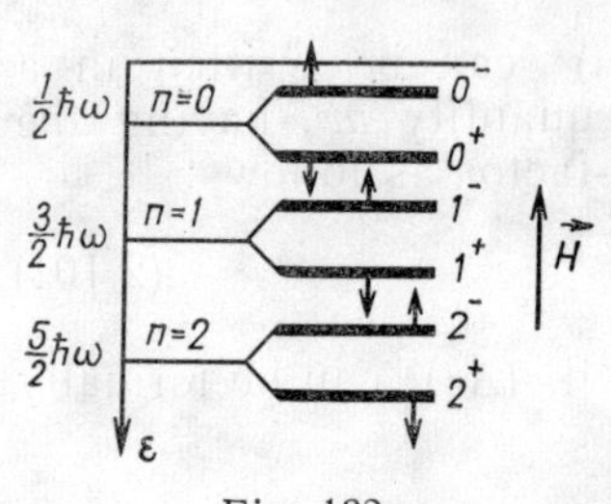

Fig. 132

But the g-factor and spin mass of holes are usually positive quantities [1]). Hence it follows that in a system of hole-type Landau levels the uppermost level under account of spin splitting will be that with 0^-. An energy diagram of hole levels at $p_z = 0$ is shown in Fig. 132. The diagram corresponds to a case when the orbital splitting exceeds the spin splitting. By comparing Figs. 131 and 132, it can be seen that the corresponding energy diagrams for electrons and holes differ from one another only in the direction of counting of energy.

The expression for the energy of a hole in a magnetic field under account of spin splitting of levels is identical with expression (2.100) for the energy of an electron, but ε, $(m^*)_z$ and m^s now denoting the magnitudes of energy, cyclotron mass and effective mass of holes.

---

[1]) In rare cases, g-factor of electrons and holes may be negative. The sign of g-factor can be decided by means of magneto-optical measurements by the nature of optical junctions between the Landau levels from neighbouring bands.

If, in quantization, we regard the energy band as a whole, the following circumstance must be accounted for. The quasi-particle states near the bottom of the band have a positive effective mass and are called electrons, whereas the states near the top have a negative effective mass and are termed holes.

At a departure from the bottom of the band, the effective mass of electrons increases and tends to plus infinity in the middle of the zone. With a departure from the top of the band downward, the effective mass of holes also increases in magnitude and tends to minus infinity in the middle of the band. The passage from the minimum of energy at the bottom of the band, surrounded by closed constant-energy surfaces of electron type, to the maximum at its top, surrounded by closed constant-energy surfaces of hole type, is performed through constant-energy surfaces of open type (corresponding to the energies near the middle of the band).

When electrons approach the open paths, the cyclotron frequencies of motion of electrons and holes in a magnetic field tend to zero, the periods of rotation in the orbits increase to infinity, the motion ceases to be finite, and quantization of the energy is disturbed. Consequently, it is impossible to quantize the energy of all the states in a band, beginning from the bottom and up to the top. With moving farther from the bottom (or top), the Landau levels condense infinitely $\left(\hbar\omega = \frac{|e|\hbar H}{m^* s} \to 0 \text{ at any } H\right)$ and the energy spectrum of current carriers corresponding to the motion in a plane perpendicular to the magnetic field is transformed into a quasi-continuous one.

Thus, the behaviour of quasi-particles on electron- and hole-type constant-energy surfaces in a magnetic field has remained different. This means that the introduction of electrons and holes is connected with a stable physical feature which does not change at quantization of the energy in a magnetic field.

2-15. DISTRIBUTION OF ELECTRONS IN $\vec{p}$-SPACE IN THE PRESENCE OF A QUANTIZING MAGNETIC FIELD

Let us discuss how the distribution of electron states in $\vec{p}$-space varies under the action of a quantizing magnetic field. For simplicity, we shall analyse the case of a one-zone metal with a spherical Fermi surface and neglect spin splitting. At $H = 0$, the allowed states are distributed uniformly inside the Fermi sphere and correspond to the elementary volumes $(2\pi\hbar)^3$. For clarity, these volumes can be denoted by points spaced a distance $2\pi\hbar$ from each other along the axes p_x, p_y, and p_z.

Figure 133 shows the section of the Fermi sphere by the plane $p_z = 0$. The allowed states fill the circle of maximum radius $p_F = \sqrt{2m\varepsilon_F}$ in that plane. In any section $p_z = \text{const}$ the filled states also fill a circle whose radius is $\sqrt{p_F^2 - p_z^2}$. As p_z increases from zero to p_F, this radius decreases and becomes zero in the reference points: $p_z = \pm p_F$.

A uniform filling of the Fermi sphere by the points depicting the allowed states corresponds to a quasi-continuous energy spectrum $\varepsilon = \varepsilon(p_x, p_y, p_z)$, where p_x, p_y, and p_z run through quasi-continuous sets of values from 0 to p_F.

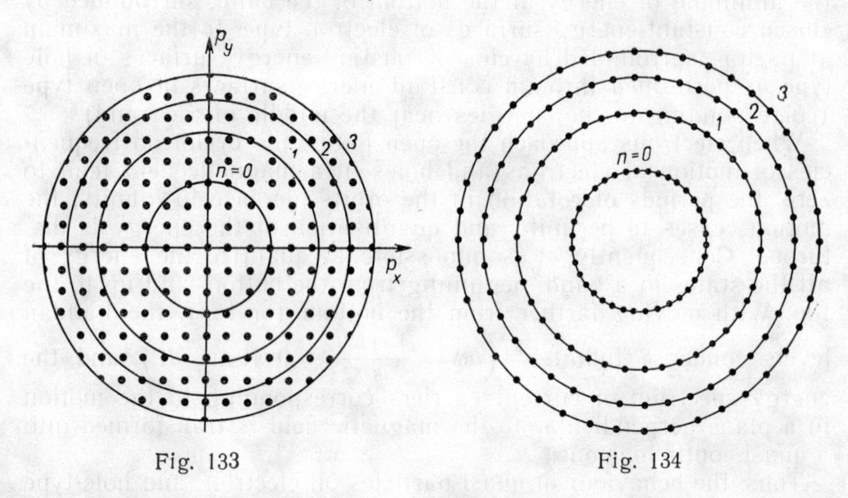

Fig. 133 Fig. 134

Application of a magnetic field does not change the total number of electrons in the metal. But the magnetic field can change the distribution of electrons and, as a consequence, cause redistribution of electrons between bands. In the example considered, only the first band is partly filled, whereas the other bands are empty. The filled states are located around the minimum of energy in the first zone (near its bottom). In that case the number of electrons in the first zone, which coincides with the total number of electrons in the metal, does not change, as well, at application of a magnetic field.

Let the magnetic field be directed along the z axis. The discrete levels $\left(n + \frac{1}{2}\right)\hbar\omega$ then become the allowed values of the energy $\varepsilon_\perp$ related to the motion in a plane perpendicular to the magnetic field. These energy levels determine the discrete allowed orbits of electrons in the planes $p_z = \text{const}$ of $\vec{p}$-space. The radius p_n of the orbit relating to a level with a quantum number n is found by the correspondence principle which can be written as

$$\frac{p_n^2}{2m} = \left(n + \frac{1}{2}\right)\hbar\omega \qquad n = 0, 1, 2, \ldots \tag{2.103}$$

whence

$$p_n = \sqrt{2m\hbar\omega\left(n + \frac{1}{2}\right)} \tag{2.104}$$

All the states which at $H = 0$ were located in the plane $p_z =$ = const between the orbits of radii p_n ($n = 0, 1, 2, \ldots$), become the forbidden ones upon application of magnetic field. In other words, application of the magnetic field results in that all the allowed states in the plane $p_z =$ const are drawn onto the nearest orbit (Fig. 134). It may be definitely assumed that the states from the total area of the central circle of radius p_0 are drawn onto an orbit with $n = 0$, all the states from the area located between the circles of radii p_0 and p_1 are drawn onto an orbit with $n = 1$, etc.

It can be easily seen that the areas located between any two neighbouring orbits coincide with one another and with the area πp_0^2 of the orbit with $n = 0$ and are equal to $2\pi m\hbar\omega$. Because of this, each allowed orbit contains the same number of states $2\frac{2\pi\hbar\omega m}{(2\pi\hbar)^2}$ which is determined by the number of states on the area $2\pi\hbar\omega m$ at $H = 0$ (the factor 2 accounts for the double degeneration of each state by spin). Thus, the degree of degeneration of each allowed orbit of radius p_n in the magnetic field is proportional to the first power of the magnetic field.

The orbits considered are entirely identical to each other in all planes $p_z =$ const. This means that all allowed states in the Fermi surface in the magnetic field are condensed on the surface of coaxial cylinders parallel to the axis p_z (Fig. 135). The radii of cylinders are determined by expression (2.104). The total number of states on each cylinder is proportional to its length in the direction of the p_z axis within the limits of the Fermi sphere. It decreases with the growth of the radius p_n of the cylinder. The number of cylinders inside the Fermi sphere is limited to a number n_{max}, which is found from the following inequality:

$$p_{n_{max}+1} > p_F > p_{n_{max}}$$

whence, noting (2.104), we get

$$2m\hbar\omega\left(n_{max} + 1 + \frac{1}{2}\right) > 2m\varepsilon_F > 2m\hbar\omega\left(n_{max} + \frac{1}{2}\right)$$

or

$$n_{max} + 1 > \frac{\varepsilon_F - \frac{\hbar\omega}{2}}{\hbar\omega} > n_{max} \tag{2.105}$$

It follows from (2.105) that the number of cylinders n_{max} decreases with strengthening of the magnetic field.

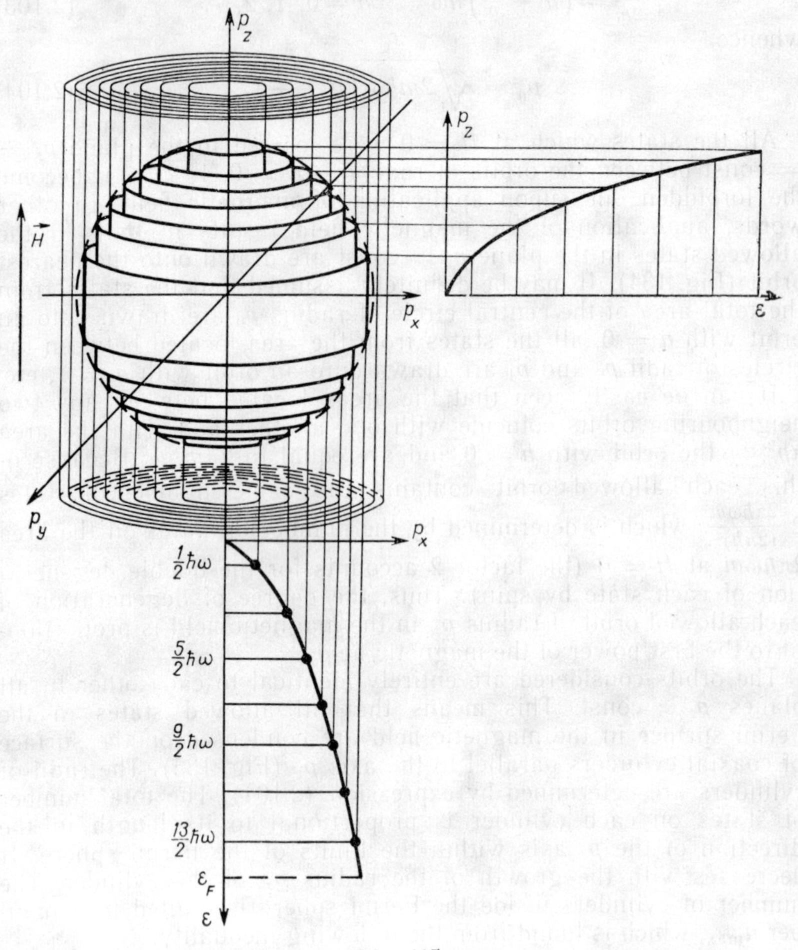

Fig. 135

The lower graph in Fig. 135 shows the dependence of the energy of states on each cylinder in the plane $p_z = 0$, on the radius of the cylinder. These energies represent a discrete series $\left(n+\frac{1}{2}\right)\hbar\omega$. The right-hand curve is the relationship between p_z and the energy of state on each cylinder, counted relative to the magnitude of energy at $p_z = 0$. The length of each cylinder along the p_z axis is determined by the difference between the Fermi energy ε_F and the energy $\left(n+\frac{1}{2}\right)\hbar\omega$ corresponding to that cylinder.

Let us define the total number of states N_n on a cylinder of the number n. The length of the cylinder along p_z is determined as $2\sqrt{p_F^2-p_n^2}$. The number of states along p_z on the cylinder is $\frac{2\sqrt{p_F^2-p_n^2}}{2\pi\hbar}$. The product of this value by the number of states condensed onto the cylinder in the plane $p_z=\text{const}$ is N_n. Thus

$$N_n=2\cdot\frac{2\pi m\hbar\omega}{(2\pi\hbar)^2}\cdot\frac{2\sqrt{p_F^2-p_n^2}}{(2\pi\hbar)}$$

or finally

$$N_n=\frac{8\pi m\hbar\omega\sqrt{2m\left[\varepsilon_F-\left(n+\frac{1}{2}\right)\hbar\omega\right]}}{(2\pi\hbar)^3} \tag{2.106}$$

As can be seen from (2.106), with $\left(n+\frac{1}{2}\right)\hbar\omega=\varepsilon_F$ the number of electrons on the cylinder with the number n turns to zero. At that moment the radius of the cylinder p_n attains p_F and goes outside the Fermi sphere. This passage of the cylinder beyond the Fermi sphere is accompanied with it being freed from the electrons that have been present on it. These electrons are redistributed onto cylinders of a smaller radius which are located inside the Fermi sphere. As the number of the cylinders freed from electrons becomes larger, the degree of degeneration of the cylinders remained inside the Fermi sphere increases. At a certain value of the magnetic field all electrons will be condensed on the last cylinder, having the number $n=0$. This situation can only be considered qualitatively, since $\hbar\omega$ in its order of magnitude attains ε_F at $H=0$ and the quasi-classical approach becomes inapplicable. This problem will be discussed in more detail in the next section.

Let us see how the distribution of electron states in $\vec{p}$-space will be changed at quantization of the energy in a magnetic field in a case of an arbitrary dispersion law. This question can be cleared out by means of the Bohr-Sommerfeld quasi-classical quantization which is applicable under the same conditions as the quasi-classical description of the motion of electrons we have used. According to the quantization principle, an integral of the generalized momentum of an electron taken over the closed contour of its orbit of cyclic motion is equal to $(n+\gamma)2\pi\hbar$, where n is a whole number and γ is a phase addition (with the quadratic dispersion law, $\gamma=1/2$). The generalized momentum of an electron in a magnetic field is of the form $\vec{p}+\frac{e}{c}\vec{A}$, where $\vec{p}$ is the quasi-momentum, and $\vec{A}$ is the vector potential of the magnetic field. The Bohr-Sommerfeld principle of quantization in our case should be written for the projection L of the orbit onto a plane perpendicular to the magnetic field:

$$\oint_L \left(\vec{p}+\frac{e}{c}\vec{A}\right) d\vec{r}=(n+\gamma)\, 2\pi\hbar \tag{2.107}$$

For a magnetic field along the z axis, the vector potential $\vec{A}$ can be taken in the form $\vec{A}=\{-H\cdot y, 0, 0\}$. Substituting $\vec{A}$ into (2.107) gives

$$\oint_L \left(p_x\, dx+p_y\, dy-\frac{e}{c}\, Hy\, dx\right)=(n+\gamma)\, 2\pi\hbar \tag{2.108}$$

In order to transform (2.108), we use the equation of motion

$$\frac{d\vec{p}}{dt}=\frac{e}{c}\left[\frac{d\vec{r}}{dt}\vec{H}\right]$$

whence

$$\frac{dp_x}{dt}=\frac{e}{c}\cdot\frac{dy}{dt}\, H \tag{2.109}$$

It can be found from (2.109) that $p_x=\frac{e}{c}\, Hy$ and $dy=\frac{c}{eH}\, dp_x$. The integral in (2.108) will then be re-written as

$$\frac{c}{eH}\oint_{\mathscr{L}} p_y\, dp_x=(n+\gamma)\, 2\pi\hbar \tag{2.110}$$

where integration is done over the contour $\mathscr{L}$ of the orbit in $\vec{p}$-space. Since $\oint_{\mathscr{L}} p_y dp_x$ equals the area S of the orbit in $\vec{p}$-space (or the area of the section of the Fermi surface by a plane perpendicular to $\vec{H}$), equation (2.110) implies that the areas of electron orbits are quantized in the magnetic field. The discrete area S_n of an orbit depends on the magnetic field H and quantum number n:

$$S_n=\frac{2\pi\hbar\,|e|\,H}{c}\,(n+\gamma) \tag{2.111}$$

The electron states located uniformly in the plane $p_z=\text{const}$ inside the Fermi surface are drawn at application of a magnetic field onto the allowed quantum orbits whose areas are determined by expression (2.111). Since S_n in independent of p_z, the electron states in the volume bounded by the Fermi surface are condensed into tubes whose areas of section by all planes $p_z = \text{const}$ are constant and equal to S_n (Fig. 136).

The degree of degeneration of each tube is proportional to $\hbar\omega$. For a spherical Fermi surface, these tubes are transformed into circular cylinders considered earlier. The shape of a cylindrical tube having the number n is determined by the shape of the section of the Fermi surface by a plane $p_z = \text{const}$, the area of this section being equal in magnitude to S_n. With strengthening of the magnetic field the dimensions of allowed orbits and the distances between them increase. Because of this, the tubes begin to lose electrons and pass beyond the Fermi surface. The picture thus obtained is quite similar to the motion of circular cylinders in the case of quadratic isotropic law of dispersion considered earlier.

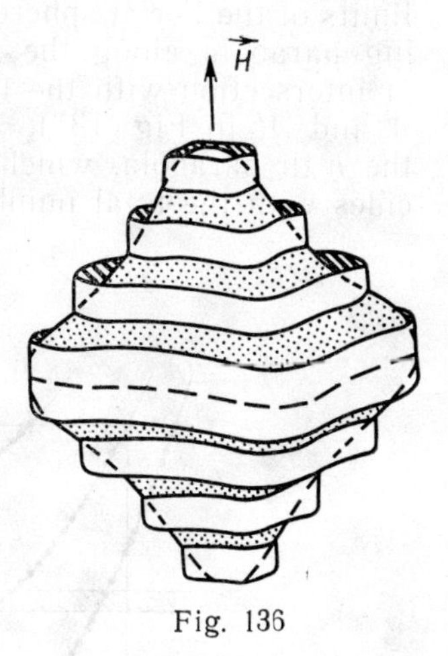

Fig. 136

2-16. THE DENSITY OF STATES IN A MAGNETIC FIELD

Let us come back to the picture of Landau levels in the form of parabolae over p_z shifted relative to each other by $\hbar\omega$ (see Sec. 2-14, Chapter Two, Fig. 129) and consider the set of states with different p_z on each parabola. These states are spaced a distance $2\pi\hbar$ from each other along the p_z axis (Fig. 137). The points at $\varepsilon < \varepsilon_F$ of the parabolae depict filled states. The energy spectrum on each parabola is quasi-continuous.

The parabola with the number n crosses the Fermi level at a definite value of p_z. The greater the number n of a parabola, the smaller p_z corresponds to the energy equal to ε_F. The maximum number $n_{\max}$ of the parabolae located below the Fermi level is evidently determined by the same double inequality (2.105) which determines the maximum number of cylinders inside the Fermi sphere in $\vec{p}$-space (see the previous section).

An energy state with the given value of p_z on a parabola with the number n corresponds to all the states on the discrete orbit with the same number which is located in the plane $p_z = \text{const}$ (see Fig. 134). This means that the state with the given p_z on each parabola has the degeneration multiplicity equal to $2\,\frac{2\pi\hbar\omega m}{(2\pi\hbar)^2}$. Note that the length of the cylinder with the number n within the limits of the Fermi sphere coincides with the size of the corresponding parabola along the p_z axis between two symmetrical points of intersection with the Fermi level (for instance, between points A and A' in Fig. 137). Therefore, the total number of states on the n-th parabola, which is located below the Fermi level, coincides with the total number of the states N_n on the n-th cylinder

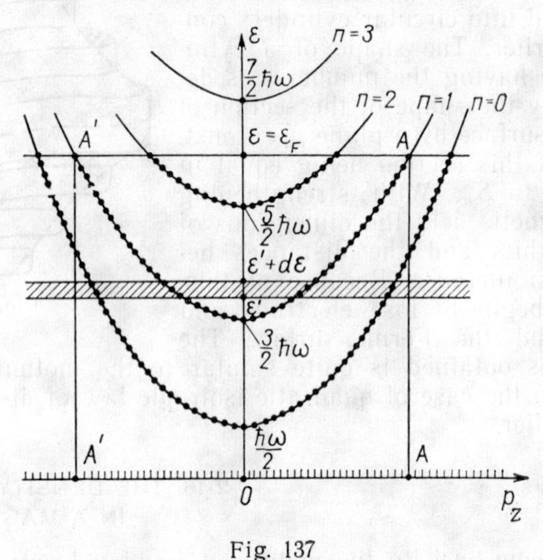

Fig. 137

within the Fermi sphere [see expression (2.106)]. This number of states reduces as the parabola moves upward with strengthening of the magnetic field, and the filled portion of the parabola between the points of intersection with the Fermi level becomes shorter.

If each Landau level is considered separately, then the energy of electrons on it is only dependent on p_z, as in the model of a one-dimensional electron gas. In order to describe the density of states on each parabola, we can use formula (2.94) for a one-dimentional system (see Sec. **2-13**, Chapter Two). We then only have to take into account that the formulas were applicable, the energy of electrons on a Landau level with the number n must be made not from zero, but from the value $\left(n+\frac{1}{2}\right)\hbar\omega$ corresponding to the energy of state with $p_z = 0$ on the n-th parabola. This means that the argument in formula (2.94) will now be the difference $\varepsilon - \left(n+\frac{1}{2}\right)\hbar\omega$.

Thus, taking into account the degeneration multiplicity of each state, equal to $2\,\frac{2\pi\hbar\omega m}{(2\pi\hbar)^2}$, the density of states $\nu_n(\varepsilon)$ on a parabola with the number n can be written on the basis of (2.94) as

$$\nu_n(\varepsilon) = \frac{4\pi m\hbar\omega}{(2\pi\hbar)^2} \cdot \frac{m^{1/2}}{\sqrt{2}\,\pi\hbar} \left[\varepsilon - \left(n + \frac{1}{2}\right)\hbar\omega\right]^{-1/2} \tag{2.112}$$

Then, in order to obtain the total density of states $\nu_H(\varepsilon)$ of the electron system in the magnetic field for the given value of energy ε, it is sufficient to sum up the expressions for $\nu_n(\varepsilon)$ over all parabolae for which the argument $\varepsilon - \left(n + \frac{1}{2}\right)\hbar\omega$ is positive. For instance, for the value of energy $\varepsilon = \varepsilon'$ in Fig. 137, only the states located on two parabolae with $n = 0$ and $n = 1$ contribute to the total number of states $dN(\varepsilon')$ within the energy interval from ε' to $\varepsilon' + d\varepsilon$. The total density of states $\nu_H(\varepsilon')$ can therefore be obtained by summing up only $\nu_0(\varepsilon')$ and $\nu_1(\varepsilon')$.

Consequently, the total density of states $\nu_H(\varepsilon)$ of the electron system with the quadratic isotropic dispersion law placed into magnetic field H can be written as

$$\nu_H\varepsilon = \frac{\hbar\omega m^{3/2}}{\sqrt{2}\,\pi^2\hbar^3} \sum_{n=0}^{n=n_\varepsilon} \left[\varepsilon - \left(n + \frac{1}{2}\right)\hbar\omega\right]^{-1/2} \tag{2.113}$$

The maximum value of the number n_ε of the parabolae over which the summation is made is determined from the double inequality:

$$n_\varepsilon < \frac{\varepsilon - \frac{\hbar\omega}{2}}{\hbar\omega} < n_\varepsilon + 1 \tag{2.114}$$

The density of states $\nu_H(\varepsilon)$ on the n-th parabola has an infinite singularity at $p_z = 0$, or else, at $\varepsilon = \left(n + \frac{1}{2}\right)\hbar\omega$. The functions $\nu_n(\varepsilon)$ for $n = 0, 1, 2$, etc. are shown in Fig. 138. The total density of states $\nu_H(\varepsilon)$ is given in Fig. 139. The dotted curve in the last figure illustrates the dependence of the band density of states $\nu^{\mathrm{III}}(\varepsilon)$ on energy at $H = 0$ (see Fig. 125) [1].

Since the total number of electrons in a magnetic field cannot change, the areas below the curves $\nu_H(\varepsilon)$ and $\nu^{\mathrm{III}}(\varepsilon)$ within the limits from 0 to ε_F must be equal [it is to be recalled that the total number of electrons $N = \int_0^{\varepsilon_F} \nu(\varepsilon)\, d\varepsilon$]. With a weakening of

[1]) In the analysis of density of states in a magnetic field the effect of renormalization of the effective mass is disregarded for simplicity.

the magnetic field H the density of Landau levels increases. It is evident that in the limit at $H \to 0$ we must again pass from the discrete Landau levels to the quasi-continuous spectrum of the three-dimensional system $\varepsilon = \varepsilon(p_x, p_y, p_z)$. The density of states of such a system is described by formula (2.86).

Let us show that the limit transition at $H \to 0$ in the expression for the density of states $\nu_H(\varepsilon)$ (2.113) actually gives expression (2.86) for $\nu^{III}(\varepsilon)$. At low values of H, the Landau levels come very

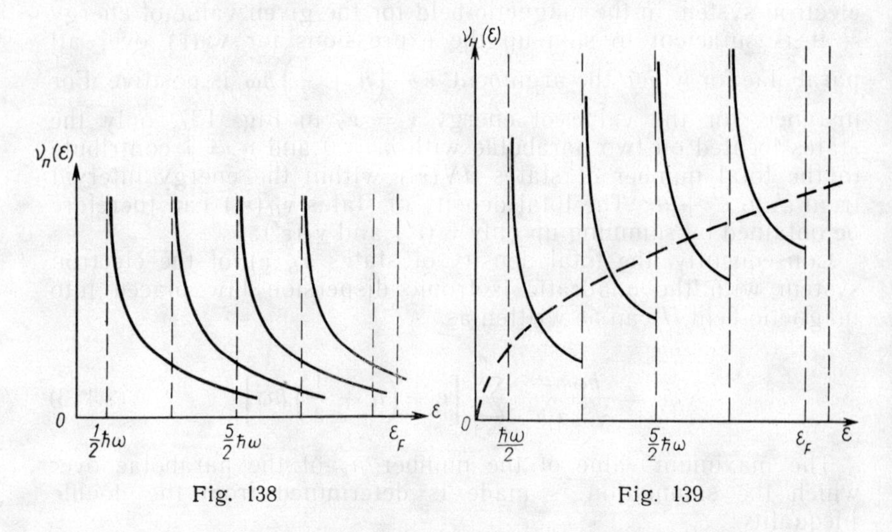

Fig. 138 Fig. 139

near to each other, so that in formula (2.113) we can pass from summation over n to integration over a continuous variable u:

$$\sum_{n=0}^{n=n_\varepsilon} \left[\varepsilon - \left(n + \frac{1}{2}\right)\hbar\omega\right]^{-1/2} \to \int_0^{n_\varepsilon} \left[\varepsilon - \left(u + \frac{1}{2}\right)\hbar\omega\right]^{-1/2} du =$$

$$= -\frac{2}{\hbar\omega}\left[\varepsilon - \left(u + \frac{1}{2}\right)\hbar\omega\right]^{1/2}\Bigg|_0^{n_\varepsilon}$$

With $u = n_\varepsilon$, the difference $\varepsilon - \left(u + \frac{1}{2}\right)\hbar\omega$ does not exceed $\hbar\omega$ (see Fig. 137). We now write this difference as $\alpha\hbar\omega$, where $0 < \alpha < 1$. The limit transition from $\nu_H(\varepsilon)$ at $H \to 0$ has the form as follows

$$\lim_{H\to 0} \nu_H(\varepsilon) = \lim_{H\to 0} \frac{2m^{3/2}}{\sqrt{2}\,\pi^2\hbar^2}\left\{\left(\varepsilon - \frac{\hbar\omega}{2}\right)^{1/2} - (\alpha\hbar\omega)^{1/2}\right\} =$$

$$= \frac{\sqrt{2}\, m^{3/2}}{\pi^2\hbar^3}\,\varepsilon^{1/2} = \frac{8\sqrt{2}\,\pi}{(2\pi\hbar)^3}\, m^{3/2}\varepsilon^{1/2}$$

The expression obtained coincides with formula (2.86) for $\nu^{III}(\varepsilon)$. Let us see how the density of states at the Fermi level will change at an increase of the magnetic field.

As can be seen from Fig. 139, the infinite singularities of the density of states correspond to the values of energy equal to $\left(n+\frac{1}{2}\right)\hbar\omega$ at $n = 0, 1, 2, \ldots$. With an increase of the magnetic field the distances $\hbar\omega$ between the singularities grow and the infinite maxima ("peaks") on the curve $\nu_H(\varepsilon)$ pass in succession through the Fermi level. Each time the bottom of a next Landau parabola coincides with the Fermi level, an infinite singularity of the density of states appears on the latter, which in turn causes a singularity of all the thermodynamic and kinetic characteristics of the electron system which depend on the number of electrons on the Fermi level.

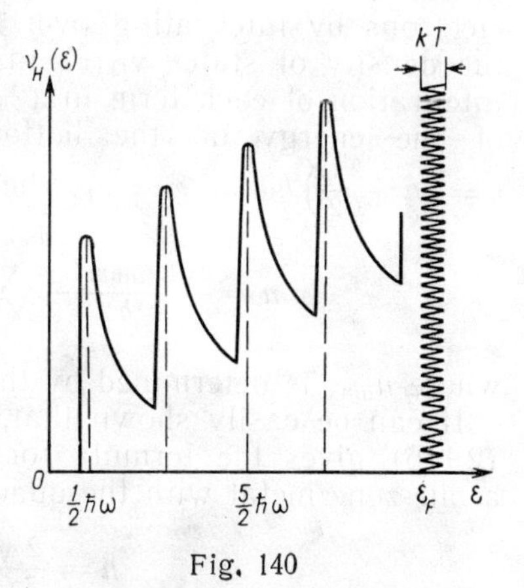

Fig. 140

The periodic repetition of these singularities at an increase of H is physically the cause of the oscillating effect of the magnetic field on such parameters as electric and thermal conductivity, magnetic susceptibility, specific heat, etc. In real cases, owing to that the lifetime τ of quasi-particles is finite, each Landau level is blurred by a value $\Delta\varepsilon$, which can be estimated from Heisenberg's uncertainty relationship: $\Delta\varepsilon \gtrsim \hbar/\varepsilon$. In addition, at a finite temperature T the boundary of Fermi distribution is also blurred by a value $\sim kT$ (Fig. 140). Because of these reasons the singularities of the density of states at a Fermi level become more or less pronounced peaks.

In order to observe oscillatory dependence of physical quantities in a magnetic field, it is evidently required that the widening of Landau levels $\sim\hbar/\tau$ and blurring of the Fermi level $\sim kT$ were substantially smaller than the energy state $\hbar\omega$ between the singularities. For this, the following inequalities must be satisfied:

$$\hbar\omega \gg kT, \qquad \hbar\omega \gg \frac{\hbar}{\tau} \text{ or } \omega\tau \gg 1 \tag{2.115}$$

Oscillations of the density of states at the Fermi level in a magnetic field cause oscillations of the position of the Fermi level itself. When the bottom of a next Landau level reaches the Fermi level, the latter tends to move upward: the Fermi sphere expands.

After the given Landau level has emerged beyond the Fermi level and lost electrons, the next level is still sufficiently deep and the Fermi sphere contracts.

This motion of the Fermi level occurs since the total number of electrons in the metal at any value of the magnetic field (H) is a constant value. From this condition, it is possible to find the dependence of the Fermi energy ε_F of electrons on magnetic field. Let it be illustrated by calculating ε_F for certain values of H on an example of a one-zone metal with the quadratic isotropic dispersion law. For this, we determine the concentration n of electrons by integrating over energy the expression (2.113) for the density of states $\nu_H(\varepsilon)$ in the magnetic field. Note that the integration of each term in (2.113) must be done from the value of the energy on the bottom of a corresponding parabola $\varepsilon = \left(n + \frac{1}{2}\right)\hbar\omega$ to $\varepsilon = \varepsilon_F$. The integration gives

$$n = \frac{\sqrt{2}\,\hbar\omega m^{3/2}}{\pi^2 \hbar^3} \sum_{n=0}^{n=n_{\max}} \left[\varepsilon_F(H) - \left(n + \frac{1}{2}\right)\hbar\omega\right]^{1/2} \tag{2.116}$$

where $n_{\max}$ is determined by the double inequality (2.105).

It can be easily shown that, in the limit at $H \to 0$, expression (2.116) gives the formula for the concentration of electrons in a one-zone metal with the quadratic isotropic dispersion law:

$$n = \frac{2\sqrt{2}}{3\pi^2\hbar^3} m^{3/2} (\varepsilon_F^0)^{3/2} \tag{2.117}$$

where ε_F^0 is the Fermi energy at $H = 0$.

At any H, the sum (2.116) contains a finite number $(n_{\max} + 1)$ of terms. This expression determines the dependence of ε_F on the magnetic field. It can be shown that at $\hbar\omega \ll \varepsilon_F^0$ the relative variation of Fermi energy $\frac{\varepsilon_F(H) - \varepsilon_F^0}{\varepsilon_F^0}$ in the magnetic field is an effect proportional to the second power of a small parameter $\left(\frac{\hbar\omega}{\varepsilon_F^0}\right)^{1/2}$. For that reason, in magnetic fields for which $\hbar\omega \ll \varepsilon_F^0$, the Fermi energy can be assumed practically constant and equal to ε_F^0. But with $\hbar\omega \sim \varepsilon_F^0$ the periodic variation of the Fermi energy in the magnetic field becomes a rather noticeable effect which cannot be disregarded in calculations of the quantities dependent on ε_F.

With an increase of the magnetic field H, the number of the Landau levels located below the Fermi level reduces. The Landau levels cross in succession the Fermi level and lose electrons. At the exit of a next Landau level, the electrons on that level pass to lower levels. The last level which principally can be freed from electrons is that with $n = 1$. After its exit, only a single Landau level with $n = 0$ remains under the Fermi level. All electrons in the metal condense on that level.

The region of magnetic fields at which electrons condense onto the last level is called the ultra-quantum region. In that region the quasi-classical approximation is no more applicable, in particular, the concept of the path of an electron loses sense. The description of motion of quasi-particles in the ultra-quantum region requires the apparatus of quantum mechanics.

Let H_n be the value of the magnetic field at which the n-th Landau level crosses the Fermi level, and let ω_n be the cyclotron frequency at $H = H_n$. We shall calculate ε_F at the exit from several first levels, i.e. at $H = H_1$, H_2, etc.

At $H = H_1$ the bottom of a parabola with $n = 1$ coincides with the Fermi level. Consequently, at this value of the field, $\varepsilon_F(H_1) = \left(1 + \frac{1}{2}\right)\hbar\omega_1$ (see Fig. 137).

In a magnetic field exceeding H_1 by a small magnitude δ, only one parabola with $n = 0$ is under the Fermi level, and accordingly the sum entering the expression (2.116) has only one term $\sqrt{\varepsilon_F(H) - \frac{\hbar\omega}{2}}$ left. In fields smaller than H_1 by δ the same sum contains two terms

$$\sqrt{\varepsilon_F(H) - \frac{3}{2}\hbar\omega} + \sqrt{\varepsilon_F(H) - \frac{\hbar\omega}{2}}$$

In the limit at $\delta \to 0$, H tends to H_1 from the right in the former case and from left in the latter. With $\delta = 0$ $(H = H_1)$ the sum of the two terms equals the single one:

$$\sqrt{\varepsilon_F(H_1) - \frac{3}{2}\hbar\omega_1} + \sqrt{\varepsilon_F(H_1) - \frac{1}{2}\hbar\omega_1} = \sqrt{\varepsilon_F(H_1) - \frac{1}{2}\hbar\omega_1}$$

Hence it follows that

$$n(H_1) = \frac{\sqrt{2}\, m^{3/2}\hbar\omega_1}{\pi^2\hbar^3}\sqrt{\varepsilon_F(H_1) - \frac{\hbar\omega_1}{2}}$$

or

$$\varepsilon_F(H_1) = \frac{3}{2}\hbar\omega_1, \quad n(H_1) = \frac{4m^{3/2}\,[\varepsilon_F(H_1)]^{3/2}}{3\sqrt{3}\,\pi^2\hbar^3}$$

Comparing the expression for $n(H_1)$ with expression (2.117) for $n(0)$ we find that $\varepsilon_F(H_1) = \left(\frac{3}{2}\right)^{1/3}\varepsilon_F^0 \approx 1.146\varepsilon_F^0$.

Similarly, it can be shown that

$$\varepsilon_F(H_2) = \frac{\frac{5}{2}\left(\frac{2}{3}\right)^{2/3}}{(\sqrt{1}+\sqrt{2})^{2/3}}\,\varepsilon_F^0 \approx 1.058\varepsilon_F^0$$

The general formula for $\varepsilon_F(H_n)$ is

$$\varepsilon_F(H_n) = \left(\frac{2}{3}\right)^{2/3} \frac{\left(n+\frac{1}{2}\right)}{(\sqrt{1}+\sqrt{2}+\cdots+\sqrt{n})^{2/3}}\,\varepsilon_F^0 \qquad (2.118)$$

Consider expression (2.116) for the concentration of electrons $n(H)$ in the ultra-quantum region of magnetic fields $H > H_1$:

$$n = \frac{\sqrt{2}\ \hbar\omega m^{3/2}}{\pi^2\hbar^3}\sqrt{\varepsilon_F(H) - \frac{\hbar\omega}{2}} \qquad (2.119)$$

The Fermi energy in that region of fields increases rapidly with H:

$$\varepsilon_F(H) = \frac{\hbar\omega}{2} + \frac{\pi^4\hbar^6 n^2}{2m^3}\left(\frac{1}{\hbar\omega}\right)^2 \qquad (2.120)$$

and tends to an asymptotic expression $\varepsilon_F \to \hbar\omega/2$. The dimension along p_z of the last parabola under the Fermi level then decreases continuously, similar to $1/\hbar\omega$, at the degree of degeneration of the state with each p_z on that Landau level increases proportional with $\hbar\omega$.

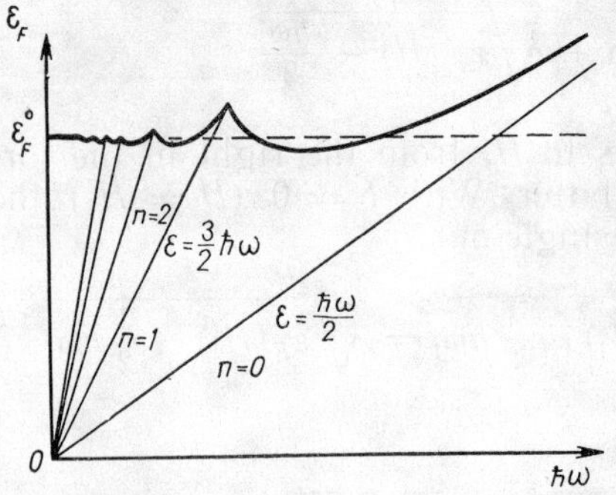

Fig. 141

The dependence of Fermi energy ε_F on magnetic field in a one-zone metal with the quadratic isotropic dispersion law is shown in Fig. 141, from which we can see that the variation of ε_F in a magnetic field should be essentially accounted for in the region of fields $H \gtrsim (H_4 - H_5)$.

In weaker fields, ε_F can be considered, with a satisfactory degree of accuracy, independent of H and equal to ε_F^0. Strictly speaking, this approximation is valid with $H \ll H_1$ or with sufficiently large values ($n \gg 1$) of the quantum numbers of the Landau levels crossing the Fermi level.

Let us use the approximation $\varepsilon_F(H) = \varepsilon_F^0 = \text{const}$ to determine the period of oscillations of the density of states at a Fermi level.

The singularity corresponding to the level with number n appears at such value of the magnetic field H_n for which

$$\left(n+\frac{1}{2}\right)\frac{|e|\hbar H_n}{m^* c}=\varepsilon_F \tag{2.121}$$

whence

$$\frac{1}{H_n}=\left(n+\frac{1}{2}\right)\frac{|e|\hbar}{m^* c\varepsilon_F} \tag{2.122}$$

Similarly, for the singularity corresponding to the level with number $n+1$:

$$\frac{1}{H_{n+1}}=\left(n+1+\frac{1}{2}\right)\frac{|e|\hbar}{m^* c\varepsilon_F} \tag{2.123}$$

Subtracting (2.122) from (2.123) we get:

$$\frac{1}{H_{n+1}}-\frac{1}{H_n}=\frac{|e|\hbar}{m^* c\varepsilon_F} \tag{2.124}$$

It can be seen from (2.124) that the interval between two consecutive singularities of the density of states in a reciprocal magnetic field is independent of the number n and is a constant under the assumption adopted. This means that singularities are formed periodically in such a field. The quantity $\frac{1}{H_{n+1}}-\frac{1}{H_n}$ is called the period of oscillations in reciprocal field and is denoted as $\Delta\left(\frac{1}{H}\right)$. The period of oscillations is

$$\Delta\left(\frac{1}{H}\right)=\frac{|e|\hbar}{m^* c\varepsilon_F} \tag{2.125}$$

Knowing it, we can find the product $m^*\varepsilon_F$ which determines the area of an extremal section S_{extr} of Fermi sphere: $S_{extr}=\pi p_F^2=2\pi m^*\varepsilon_F$. Expression (2.125) can be written for an extremal section as follows:

$$S_{extr}=\frac{2\pi|e|\hbar}{c}\cdot\frac{1}{\Delta\left(\frac{1}{H}\right)} \tag{2.126}$$

Formula (2.126), obtained for the particular case of a Fermi sphere, is valid for any law of dispersion if S_{extr} implies the area of an extremal section of the Fermi surface by a plane perpendicular to the direction of the magnetic field. In the general form it is called the Lifshits-Onsager formula and follows from the formula for quantization of areas S_n of orbits in a magnetic field (2.111).

Indeed, the singularity of the density of states on a Fermi level is observed when a next tube goes beyond the Fermi surface and its section by a plane $p_z=$ const attains an extremal value S_{extr}. Denoting the magnitude of the magnetic field at the moment of singularity by H_n, formula (2.111) can be written in the form

$$S_{extr} = \frac{2\pi\hbar\,|e|\,H_n}{c}(n+\gamma) \tag{2.127}$$

In order to pass from (2.127) to (2.126), we have to assume that the extremal section of the Fermi surface S_{extr} is independent of the number of singularity. This is equivalent of assuming that the Fermi level is constant, which is only true at $\hbar\omega \ll \varepsilon_F$. Thus, the region of applicability of formula (2.126) is also limited by the values of fields $H \ll H_1$ or the values of larger quantum numbers $n \gg 1$. With a reduction of the number n, the interval $\frac{1}{H_{n+1}} - \frac{1}{H_n}$ ceases to be constant and the positions of singularities are shifted towards greater fields relative to the values $H = H_n$ calculated at $\varepsilon_F(H) = \varepsilon_F^0 = \text{const}$. Indeed, $H_n = \frac{m^* c}{|e|\hbar} \cdot \frac{1}{\left(n+\frac{1}{2}\right)} \varepsilon_F(H_n)$, where $\varepsilon_F(H_n)$ is related to ε_F^0 and n through expression (2.118). At all values $H = H_n$, $\varepsilon_F(H_n) > \varepsilon_F^0$.

It is not difficult to calculate the position of singularities $1/H_n$ in a reciprocal magnetic field for the first numbers of Landau levels $n = 1, 2, 3, 4$. In the units $\frac{|e|\hbar}{m^* c\varepsilon_F^0}$, the values of $1/H_n$ are

$$\frac{1}{H_n} = \frac{|e|\hbar}{m^* c\varepsilon_F^0}\left(n+\frac{1}{2}\right)\beta_n \tag{2.128}$$

The factor β_n accounting for oscillations of a Fermi level is equal to 0.872, 0.945, 0.966, and 0.977 for $n = 1, 2, 3$, and 4 respectively.

Thus, the position of the first singularity $1/H_1$ in the reciprocal magnetic field is shifted by approximately 13 per cent towards smaller $1/H$ compared with the theoretical value $1/H_1$ at $\varepsilon_F(H_1) = \varepsilon_F^0$. The distance between the two first singularities $\frac{1}{H_2} - \frac{1}{H_1}$ exceeds approximately 1.05 times the distance $\Delta\left(\frac{1}{H}\right)$ from expression (2.125).

Let us find the expression for the density of states of electrons in a metal with the account of spin splitting of the Landau levels. In that case each parabola in Fig. 137 is splitted into two parabolae shifted relative to the initial one upward and downward by $\frac{1}{2} g\mu_B H$. The density of states on each of the new parabolae is described, as before, by formula (2.94) for a one-dimensional electron system, with taking into account the reference point of energies on the parabola.

Thus, the density of states $\nu_{n,s}(\varepsilon)$ on the Landau level corresponding to the quantum numbers n and s, by analogy with formula (2.112), can be written as:

$$\nu_{n,s}(\varepsilon)=\frac{2\pi\hbar\omega m}{(2\pi\hbar)^2}\cdot\frac{m^{1/2}}{\sqrt{2}\,\pi\hbar}\left[\varepsilon-\left(n+\frac{1}{2}\right)\hbar\omega-\frac{1}{2}sg\mu_BH\right]^{1/2} \quad (2.129)$$

Let us interprete geometrically the distribution of electrons in $\vec{p}$-space in the presence of a magnetic field and with an account of spin splitting of levels to verify this analogy. For this, we consider the structure of expression (2.100) for the energy in a magnetic field accounting the spin splitting. Let us recall that at $H\to 0$ the energy levels (2.100) form a quasi-continuous energy spectrum $\varepsilon=\varepsilon(p_x, p_y, p_z)$, each state of which is twice degenerated by spin.

The first term in expression (2.100) $(n+1/2)\hbar\omega$, as the first term in (2.98), describes discrete levels of the energy of orbital motion. This component has a classical analogue in the form of the motion of an electron on an orbit in a plane perpendicular to the magnetic field $\vec{H}$, because of which it can be interpreted by means of the correspondence principle.

According to this principle, the allowed quasi-classical orbits in a plane $p_z=$ const of $\vec{p}$-space are circles whose radii p_n are determined by expression (2.104). Without account of spin splitting, each state on the orbit of radius p_n in the plane $p_z=$ const has an energy $\varepsilon=\left(n+\frac{1}{2}\right)\hbar\omega+\frac{p_z^2}{2m_z}$ for both orientations of the spin. The combination of all states with the given n has a corresponding Landau cylinder of radius p_n.

Under account of spin splitting, the energy of each state with the given n and p_z becomes by $\frac{1}{2}g\mu_BH$ greater for the one orientation of spin and by the same magnitude lower, for the other. The energy addition $\frac{1}{2}g\mu_BH$ related to the magnetic eigenmomentum of an electron, has no classical analogue, and therefore, cannot be interpreted by the correspondence principle. But the concept of Landau cylinder that we have introduced earlier, can be retained with an account of spin splitting, if we introduce a new variable having the dimension of energy $\in_s=\varepsilon-\frac{1}{2}sg\mu_BH$ and by means of it write expression (2.100) in the form coinciding with (2.98)

$$\in_s=\left(n+\frac{1}{2}\right)\hbar\omega+\frac{p_z^2}{2m_z} \quad (2.130)$$

As before, we assume by the correspondence principle that at application of a magnetic field the allowed states in $\vec{p}$-space are drawn onto the discrete Landau cylinders with radii p_n determined by formula (2.104). As earlier, the number of states with the given projection of spin which are condensed onto the cylinder in a plane $p_z = \text{const}$ is determined by the area between neighbouring orbits and is equal to $\frac{2\pi\hbar\omega m}{(2\pi\hbar)^2}$. This value is independent of p_z and characterizes the degree of degeneration of the states on the cylinder. Then, for each value of the quantum number n, one and the same Landau cylinder of radius p_n and one and the same value of the variable $\mathcal{E}_s$ at the given value of p_z correspond to the states with either orientation of the spin.

Thus, relative to the variable p_z, each cylinder is, as before, twice degenerated by spin and formally is absolutely identical with the Landau cylinder introduced without account of spin splitting. But the actual values of the energy ε of the electron states on the cylinder corresponding to the quantum numbers $s = \pm 1$ differ from $\mathcal{E}_s$ by $\pm \frac{1}{2} g\mu_B H$. Therefore, the geometrically identical Landau cylinders differ from one another in the magnitude of the energy of the electron states with the same n and p_z. This means that each value of s can be correlated with its own system of cylinders.

The distinction between the two systems of cylinders consists in the magnitude of energy determining the boundary of filled states on the cylinders and the number of cylinders on which electrons are present. For the system corresponding to $s = \pm 1$, the boundary energy is $\mathcal{E}^F_{s=+1} = \varepsilon_F - \frac{1}{2} g\mu_B H$, whereas for the system corresponding to $s = -1$, it is $\mathcal{E}^F_{s=-1} = \varepsilon_F + \frac{1}{2} g\mu_B H$.

The filled states on the cylinders of the first system ($s = +1$) are confined within the sphere of a smaller radius $p^F_{s=+1} = \sqrt{2m\left(\varepsilon_F - \frac{1}{2} g\mu_B H\right)}$ and the cylinders inside the sphere are shorter along p_z. The corresponding radius of sphere for the second system is $p^F_{s=-1} = \sqrt{2m\left(\varepsilon_F + \frac{1}{2} g\mu_F H\right)}$ and the cylinders are longer along p_z. The maximum number $n_{\max}(s)$ of cylinders of a system relating to the given value of s is determined by the double inequality:

$$n_{\max}(s) < \frac{\varepsilon_F - \frac{\hbar\omega}{2} - \frac{1}{2} sg\mu_B H}{\hbar\omega} < n_{\max}(s) + 1 \qquad (2.131)$$

Thus, account of spin degeneration has resulted in formation of two systems of Landau cylinders corresponding to two projections of the spin of an electron onto the direction of the magnetic field. The two systems of cylinders geometrically coincide, but each of them is inserted into its own "Fermi sphere" whose radius depends on the orientation of spin and the magnitude of magnetic field H. In this connection, the exit of the n-th cylinder of one system beyond its own "Fermi sphere" does not occur simultaneously with the exit of the n-th cylinder of the other system.

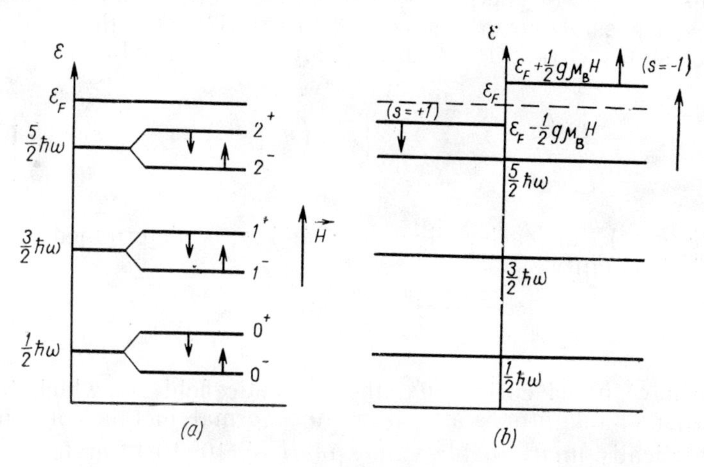

Fig. 142

The spin splitting of Landau levels can be given a quite analogous interpretation by means of an energy diagram corresponding to $p_z = 0$ (see Fig. 131a, b, and c). With an increase of the magnetic field the levels with different n and s cross the boundary of Fermi distribution $\varepsilon = \varepsilon_F$ (Fig. 142a) in a certain sequence depending on the ratio between spin and orbital splitting, the parabola passing out from the Fermi level being each time liberated from electrons which flow over onto lower levels. But instead of it we can construct a single diagram of levels $(n + {}^1/_2)\hbar\omega$, and to introduce for each value $s = \pm 1$ its own boundary of Fermi distribution $\varepsilon^{F}_{s=\pm 1} = \varepsilon_F \mp \frac{1}{2} g\mu_B H$ depending on the magnetic field.

Singularities of the density of states appear when the levels cross both boundaries.

This method of interpretation is sometimes more convenient for the analysis of oscillational dependences of various characteristics of a metal on the magnetic field and will be employed later in the book.

At a passage to the ultra-quantum limit, under the Fermi level there remains one Landau 0^- level on which all electrons condense[1]).

The motion of the 0^- level over the scale of energies is determined by the ratio between the spin and orbital splitting. For instance, at $m^s = m^*$ the energy of that level is zero at any magnitude of the magnetic field.

This analysis of energy levels of an electron in a magnetic field makes it possible to generalize formula (2.113) for the total density of states $\nu_H(\varepsilon)$ in the case of spin splitting of Landau levels:

$$\nu_H(\varepsilon) = \frac{\hbar\omega m^{3/2}}{2\sqrt{2}\,\pi^2\hbar^3} \sum_{s=\pm 1} \sum_{n=0}^{n=n_\varepsilon(s)} \left[\varepsilon - \left(n + \frac{1}{2}\right)\hbar\omega - \frac{1}{2} s g \mu_B H\right]^{-1/2} \tag{2.132}$$

where $n_\varepsilon(s)$ for each value of ε and s is determined by the double inequality

$$n_\varepsilon(s) < \frac{\varepsilon - \frac{\hbar\omega}{2} - \frac{1}{2} s g \mu_B H}{\hbar\omega} < n_\varepsilon(s) + 1 \tag{2.133}$$

We have noted earlier that the magnetic fields in which $\hbar\omega$ in its order of magnitude attains ε_F^0 for normal metals correspond to practically unattainable values of $H \sim (10^8\text{-}10^9)$ oersted.

But this does not mean that the ultra-quantum limit in a magnetic field is inaccessible for experimental studies and that it is practically meaningless to discuss the phenomena in this region of fields. The point is that the estimation obtained for magnetic fields H_a is an average one and actually relates to the majority of normal metals, in which the size of the Fermi surface is of the same magnitude as the characteristic size of the Brillouin zone.

But, as can be shown by constructing a Fermi surface by Harrison's method, in many metals there also exist, apart from large electron- and hole-type surfaces, little surfaces whose size is small compared with the characteristic size $\hbar/a$ (see, for instance,

---

[1]) When a Landau level corresponding to the given direction of spin is freed from electrons, they pass onto lower levels; this passage being possible either onto the levels with the same direction of spin or onto those with the opposite direction.

But in the second case, the spin of the electron must change its direction at the passage. This process can be forbidden with a sufficiently large magnitude of spin splitting. Then two Landau levels 0^+ and 0^- remain in the ultra-quantum limit below the Fermi level, since electrons cannot pass from level 0^+ to level 0^-. The last among the passing levels is then that with 1^-, from which electrons flow over onto level 0^-.

Fig. 97—the electron surface of lead in the fourth energy band). Quasi-particles located on these surfaces represent small groups of carriers whose effective mass is, as a rule, several orders of magnitude smaller than m_0. The corresponding Fermi energy ε_F is also rather small compared with the value (5-10) eV characteristic of large groups of carriers.

Small groups of carriers are present in bivalent metals, such as Be, Mg, Zn, Cd, trivalent metals—Al, Ga, In, Tl, and also in semimetals As, Sb, and Bi, in which the concentration of electrons is 10^{-4}-15^{-5} per atom. Various alloys of these metals with each other also possess these properties. Finally, the substances with a degenerated electron system, similar to a one-zone metal, include high-alloyed semiconductors in which the boundary of Fermi distribution is determined by the degree of alloying and may be made rather low.

The Fermi energy of small groups of carriers in the substances indicated lies within the interval from a few MeV to 1 eV and their cyclotron masses equal $(10^{-1}$-$10^{-3})m_0$. In accordance with these values, the ultra-quantum limit is attained in magnetic fields of strength from a few kilo-oersteds to approximately 10^6 oersted, which are now easily obtained experimentally. Thus, for certain groups of carriers, it is practically possible to observe the passage of the last Landau levels, and also the process of condensation of carriers onto the single level 0^-.

2-17. A TWO-ZONE METAL IN A MAGNETIC FIELD

Our discussion will be stated from the simplest case of an isotropic quadratic spectrum of electrons and holes. This approximation is sufficiently accurate for studying the main peculiarities of the behaviour of a two-zone metal in a magnetic field. For definiteness, we shall consider a bivalent metal in which there is an overlap between the first and the second energy bands. The concentration of electrons n in the second band is equal to the concentration p of holes in the first. Let us recall that the existence of free carriers in a bivalent metal is only connected with overlap of the bands, i.e. with the circumstance that the minimum of energy in the second band is lower than the maximum of energy in the first. This makes a partial filling of the second band energetically reasonable, while the first band remains unfilled.

Let the origin of coordinates be placed for convenience into the point of minimum energy in the second band (conduction zone). The law of dispersion of electrons can be written in the form

$$\mathscr{E} - \mathscr{E}_c = \frac{\vec{p}^2}{2m_{el}^*} \tag{2.134}$$

where $\mathscr{E}_c$ is the magnitude of energy in the minimum and m^*_{el} is the isotropic effective mass of electrons.

Let the maximum of energy in the first band (valence band) be displaced relative to the origin of coordinates by a vector $\vec{G}$. The law of dispersion for holes is of the form

$$\mathscr{E}_v - \mathscr{E} = \frac{(\vec{p} - \vec{G})^2}{2m^*_{hol}} \tag{2.135}$$

where $\mathscr{E}_v$ is the maximum magnitude of energy in the valence band and m^*_{hol} is the absolute value of the isotropic effective mass of holes.

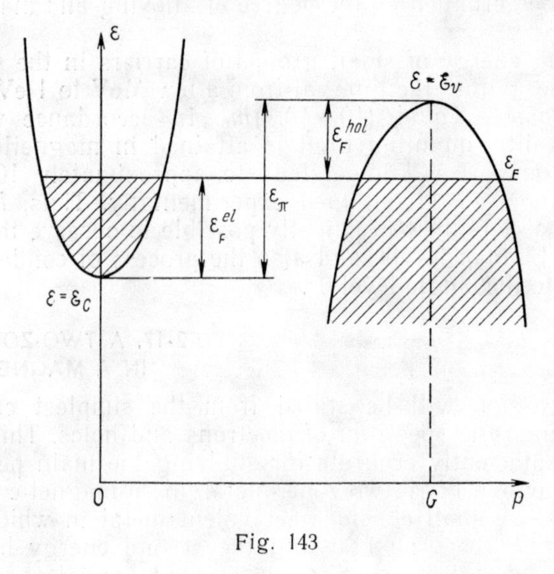

Fig. 143

The distribution of electrons between the zones occurs as with a liquid in communicating vessels. The Fermi level determining the boundary of filling must be the same in both bands. A diagram showing symbolically the energy spectrum of a two-zone metal is given in Fig. 143. The shaded areas are the regions of filled states below Fermi level. The sum of ε_F^{el} and ε_F^{hol} is the energy of overlap ε_{ov} of the bands. The value of Fermi energy in each band is determined from the condition of equality of the concentrations of electrons n and holes p.

The concentration of current carriers in the absence of magnetic field can be calculated by formula (2.117). The concentrations n

and p may be written as

$$n = \frac{2\sqrt{2}}{3\pi^2\hbar^3}\left(m^*_{el}\varepsilon^{el}_F\right)^{3/2} \tag{2.136}$$

$$p = \frac{2\sqrt{2}}{3\pi^2\hbar^3}\left(m^*_{hol}\varepsilon^{hol}_F\right)^{3/2}$$

Equating $n = p$, we get

$$\frac{\varepsilon^{el}_F}{\varepsilon^{hol}_F} = \frac{m^*_{hol}}{m^*_{el}} \tag{2.137}$$

or, expressing ε^{el}_F and ε^{hol}_F through ε_π:

$$\varepsilon^{el}_F = \frac{m^*_{hol}}{m^*_{el} + m^*_{hol}}\varepsilon_\pi, \qquad \varepsilon^{hol}_F = \frac{m^*_{el}}{m^*_{el} + m^*_{hol}}\varepsilon_\pi \tag{2.138}$$

Let the metal be placed into a magnetic field. Systems of Landau levels for electrons and holes are formed as a result of quantization of the energy of carriers. The corresponding energy diagram of the levels is shown schematically in Fig. 144. With an increase of the magnetic field the electron- and hole-type Landau levels pass across the Fermi level, causing oscillations of the density of states.

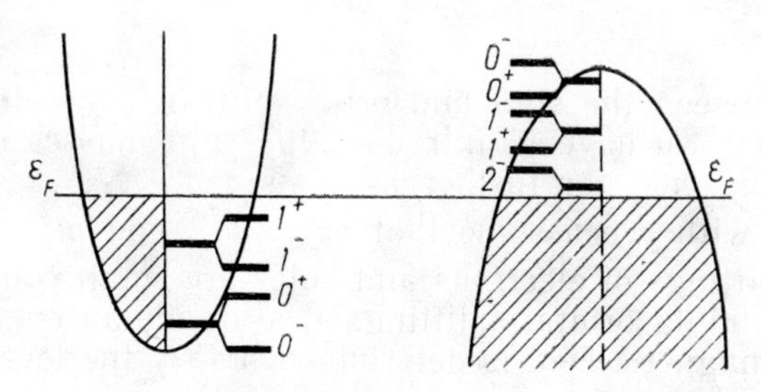

Fig 144

The mechanism that causes singularities of the density of states on the Fermi level for each group of carriers is completely identical with the similar mechanism in a one-zone metal which has been described in the previous section. Because of this in a two-zone metal placed into magnetic field H all thermodynamic and kinetic characteristics depending on the number of carriers of the Fermi level also oscillate. But this process in the two-zone metal has a new qualitative singularity related to the presence of two groups of carriers of different types. The point is that the concentrations of electrons n and holes p vary under the action of the magnetic field.

This singularity is not found in a one-zone metal, since the total number of filled electron states in all the bands does not change in a magnetic field. Quantization of the energy of carriers in the magnetic field in the presence of two overlapping bands can cause redistribution of the filled electron states between the bands, retaining, naturally, the equilibrium between them, which is expressed by the equality $n = p$.

There are two different causes of variation of the concentration of electrons and holes in a two-zone metal under the action of a magnetic field. They will be considered separately.

1. The first is the variation of the energy of overlap of the bands under the action of the magnetic field. It is directly determined by

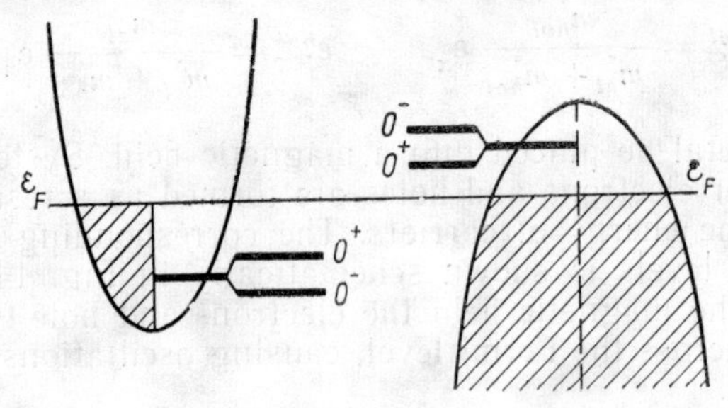

Fig. 145

the ratio between the spin and orbit splittings for electrons and holes, so that we have to introduce the spin masses of electrons m^s_{el} and holes m^s_{hol} for further analysis.

To begin with, we assume that $m^s_{el} > m^*_{el}$ and $m^s_{hol} > m^*_{hol}$, i.e. the spin splittings of electrons and holes are small compared with the corresponding orbital splittings. The minimum energy of electrons in a magnetic field is determined not by the level of energy $\mathscr{E} = \mathscr{E}_c$ corresponding to the bottom of the conduction band at $H = 0$, but by the bottom of the Landau 0^- level.

Using expressions (2.100) and (2.102), it can be shown that this energy $\Delta\varepsilon_{el}$, calculated from the level $\mathscr{E} = \mathscr{E}_c$, is

$$\Delta\varepsilon_{el} = \frac{|e|\,\hbar H}{2c} A_{el} \tag{2.139}$$

where

$$A_{el} = \left(\frac{1}{m^*_{el}} - \frac{1}{m^s_{el}}\right)$$

Since there are no electrons below the 0^- level, the bottom of the conduction band in the magnetic field coincides with the Landau 0^- level, rather than with the level $\mathscr{E} = \mathscr{E}_c$. As can be seen from Fig. 145, with $m^s_{el} > m^*_{el}$, the edge of the conduction zone moves upward along the energy scale with an increase of the magnetic field. Since variations in the system are determined by displacements of various levels relative to each other, let us consider the motion of the edge of the valence zone for holes.

The maximum energy in the valence zone is equal to the energy on the top of the parabola 0^- ($n = 0$, $s = -1$). This energy $\Delta\varepsilon_{hol}$, calculated from the level $\mathscr{E} = \mathscr{E}_v$ corresponding to the edge of the valence zone at $H = 0$, is

$$\Delta\varepsilon_{hol} = \frac{|e|\hbar H}{2c} A_{hol} \tag{2.140}$$

where

$$A_{hol} = \left(\frac{1}{m^*_{hol}} - \frac{1}{m^s_{hol}}\right)$$

There are no holes in the magnetic field above the Landau 0^- level. Thus, the edge of the valence zone in the magnetic field coincides with the 0^- level. With $m^s_{hol} > m^*_{hol}$, the 0^- level moves downward along the energy scale with an increase of the magnetic field.

The energy of overlap of the zones, ε_{ov}, equal to the energy distance between the edge of the valence zone and the bottom of the conduction zone in the magnetic field, can be written as

$$\varepsilon_{ov}(H) = \varepsilon^0_{ov} - \Delta\varepsilon_{ov} \tag{2.141}$$

Here ε^0_{ov} is the energy of overlap at $H = 0$, and $\Delta\varepsilon_{ov}$ is the variation of the energy of overlap in the magnetic field which is equal to

$$\Delta\varepsilon_{ov} = \Delta\varepsilon_{el} + \Delta\varepsilon_{hol} = \frac{|e|\hbar H}{2c} B \tag{2.142}$$

where

$$B = A_{el} + A_{hol} = \left(\frac{1}{m^*_{el}} - \frac{1}{m^s_{el}} + \frac{1}{m^*_{hol}} - \frac{1}{m^s_{hol}}\right)$$

When B is positive, the overlap of the zones in the magnetic field decreases, the conduction zone moving upward along the energy scale relative to the valence zone. With an unvariable filling of the zones, this process would result in that the boundary of filling (Fermi level) in the conduction zone exceeded the boundary of filling in the valence zone. In fact, the boundary of filling in both zones remains at the same level, and the motion of the zones causes the excess electrons to flow over from the conduction zone into the valence zone (Fig. 146).

A decrease of overlap of the bands in the magnetic field should result in that the energy of overlap $\varepsilon_{ov}(H)$ at a definite value $H = H_k$ becomes zero. The edge of the 0^- parabola for electrons then coincides with the edge of the 0^- parabola for holes

(Fig. 147). The value $H = H_k$ is found from the condition $\varepsilon_{ov}(H) = 0$ or $\Delta\varepsilon_{ov} = \varepsilon^0_{ov}$ and is equal to

$$H_k = \frac{2c\varepsilon^0_{ov}}{|e|\hbar B} \tag{2.143}$$

With a further increase of the magnetic field, the bottom of the 0^- parabola for electrons turns to be higher on the energy scale than the edge of the 0^- parabola for holes; an interval of forbidden energies ε_g, termed the energy gap (Fig. 147), is then formed in the energy spectrum. Thus, at $H = H_k$, the system is transformed from the metallic state into the semiconductor state

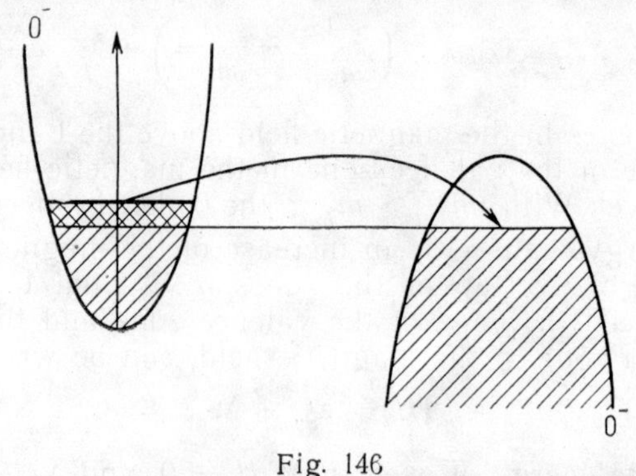

Fig. 146

in which a work equal to ε_g is to be done in order to generate electrons and holes.

Note that the monotonous reduction of overlap of the bands in a magnetic field and continuous transition to the state with an energy gap, as has been described above, can evidently take place only when there is no qualitative change of the energy spectrum of electrons of the system, at which the concept of boundaries of the bands loses sense. This qualitative change of the spectrum near a metal-semiconductor junction is connected with the formation of a new state of the substance, which has been called the stationary phase of exciton insulator. An exciton insulator is a system in which electrons and holes form bound states, i.e. electron-hole pairs, or excitons. Because of neutrality of the electron-hole pairs the system of excitons cannot conduct electric current and therefore is an insulator.

Let us briefly discuss the properties of an exciton insulator and the conditions under which it can be formed.

The formation of excitons in a semiconductor or a two-zone metal can be explained by Coulomb attraction between negatively charged electrons and holes which behave as positively charged particles. The forces of Coulomb attraction in a system of two particles—an electron and a hole—can result in the formation of a bound state which is similar to the bound state of an electron and a proton in the atom of hydrogen. Such bound states can freely move over the crystal and behave as quasi-particles.

The energy of bonding ε_0 of an electron and a hole in an exciton and the effective radius r_0 of exciton can be estimated by the well-known Bohr's formulae, noting that Coulomb interaction of two

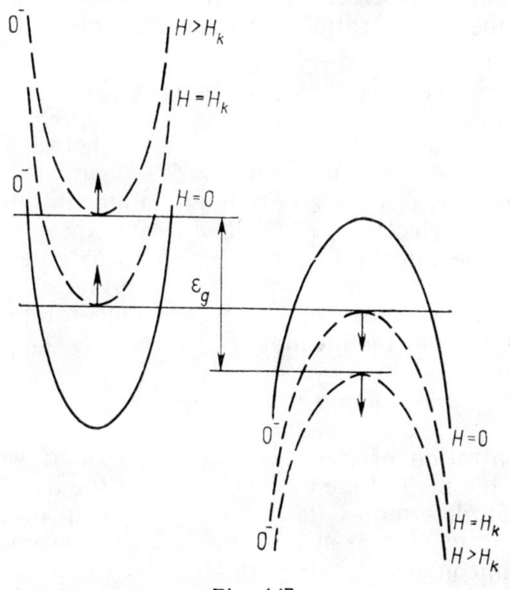

Fig 147

charges in a crystal is inversely proportional to the dielectric constant $\varkappa$ of the medium:

$$\varepsilon_0 = \frac{1}{2} \cdot \frac{m^* e^4}{\varkappa^2 \hbar^2}, \quad r_0 = \frac{\varkappa \hbar^2}{m^* e^2} \tag{2.144}$$

Here $m^* = \left(1/m^*_{el} + 1/m^*_{hol}\right)^{-1}$ is the reduced effective mass of an electron or hole. Typical values of ε_0 and r_0 for the known metals and semiconductors are (10^{-2}-10^{-1}) eV and (10^{-6}-10^{-7}) cm. Thus, the energy of bonding an exciton is two or three orders lower than the characteristic energies in the metal (for instance, Fermi energies) and their radii are many times the interatomic distances in the crystal. It may be visualized that the orbits of an electron or hole at their rotation around the common centre of masses in an exciton envelop tens or hundreds of thousands of atomic cells.

For that reason, expression (2.144) contains an averaged macroscopic value of $\varkappa$.

The formation of excitons in metals with equal numbers of electrons and holes ($n = p$) is prevented by the effect of screening of Coulomb interaction by free carriers, which consists in that the electric potential of an individual electron or ion in the metal becomes short-acting [see expression (2.2) in Sec. 2-3].

The effective radius r_D of Coulomb interaction, which is termed the Debye radius of screening, is determined by the quantities $\varkappa$ and m^* and the concentration of free carriers n:

$$r_D = \sqrt{\frac{\varkappa \hbar^2}{4m^* e^2} n^{-1/3}} = \frac{1}{2} \sqrt{r_0 n^{-1/3}}$$

Since at distances $r > r_D$ the attraction between an electron and hole becomes weak, no formation of bound states occurs at $r_0 > r_D$. Therefore, to form an exciton insulator in a metal with the equal number of electrons and holes $n = p$, the concentration of carriers must be limited to n_A at which r_D becomes of the same order as r_0. Thus, the condition for the formation of an exciton insulator in a two-zone metal at the temperature of absolute zero is in the form of the inequality

$$n = p < n_A = \left(\frac{m^* e^2}{4\varkappa \hbar^2}\right)^3 \qquad (2.145)$$

The concentration of electrons and holes in a two-zone metal is related to the magnitude of overlap ε_{ov} of the bands. Thus, condition (2.145) determines the minimum overlap of the bands at which the metallic state is still stable against electron-hole pairing.

For the semiconductor state with an energy gap ε_g, the minimum energy to be spent in order to form an exciton is $\varepsilon_g - \varepsilon_0$. At $\varepsilon_g - \varepsilon_0 > 0$, the bound state of an electron or hole is unstationary. Such states, formed, for example, by means of irradiation of a semiconductor with a light of the quantum energy $\hbar\nu$ exceeding the difference $\varepsilon_g - \varepsilon_0$, have a certain mean lifetime τ_e and vanish through recombination of electrons and holes. But with reduction of the energy gap, beginning from $\varepsilon_g = \varepsilon_0$, formation of excitons becomes energetically reasonable, i.e. excitons in a stationary state must appear in a system. In that case the semiconductor is transformed into an exciton insulator.

Thus, the condition for the formation of an exciton insulator in a semiconductor with an energy gap ε_g at the temperature of absolute zero is

$$\varepsilon_g < \varepsilon_0 \qquad (2.146)$$

An increase of temperature results in destruction of a part of electron-hole pairs in the exciton insulator and the formation of free electrons and holes. The concentration of free (non-paired) carriers $n = p$ is then proportional to the factor $e^{-\frac{\varepsilon_0}{kT}}$. At the critical temperature $T = T_{cr}$, which coincides in the order of magnitude with ε_0/k, a phase transition from the state of exciton insulator to that of the initial metal or semiconductor occurs.

Because of this, with condition (2.145) fulfilled for a metal, or condition (2.146), for a semiconductor, an exciton insulator can only be formed at temperatures T satisfying the inequality

$$T < T_{cr} \sim \frac{\varepsilon_0}{k} \tag{2.147}$$

The conditions of formation of an exciton insulator depend substantially on magnetic field. The point is that the energy of bonding of the exciton in a strong quantizing magnetic field increases compared with ε_0, while its effective size in a plane perpendicular to the magnetic field decreases compared with r_0.

At $H = 0$, the effective dimensions of an exciton are the same in all direction: the exciton represents a spherically symmetrical formation of an electron and a hole. With an increase of the magnetic field the exciton contracts in a transverse direction: it becomes an ellipsoid of rotation around a magnetic line of force with the anisotropy increasing with H. In the limit of very strong magnetic fields, excitons acquire the form of thin "cigars" elongated along the direction of $\vec{H}$. The centres of masses of the electron and hole in such a "cigar" are located on the same magnetic line of force.

Hazegawa [65] has shown theoretically that in strong magnetic fields for which the inequality $\hbar\omega^* \gg \varepsilon_0 \left[\omega^* = \frac{|e| H}{m^* c}\right.$, m^* being the reduced mass of an electron or hole, see (2.144)], holds the energy of bonding $\varepsilon(H)$ and effective radius $r(H)$ of an exciton can be written as

$$\varepsilon(H) \approx \varepsilon_0 \left[\ln \frac{\hbar\omega^*}{\varepsilon_0}\right]^2$$
$$r(H) \approx \frac{r_0}{\left[\ln \frac{\hbar\omega^*}{\varepsilon_0}\right]^2} \tag{2.148}$$

For substances of a small cyclotron mass m^* and a sufficiently high dielectric constant $\varkappa$ the factor $\hbar\omega^*/\varepsilon_0$ becomes rather high in really attainable magnetic fields. This makes it possible to increase noticeably the energy of bonding of excitons in the magnetic field.

In addition, a strong magnetic field reduces the harmful effect which the anisotropy of the initial energy spectrum and the impurities inevitably present in real crystals have on the formation of the exciton insulator phase. For these reasons, the formation of the exciton insulator phase can only be observed experimentally in strong magnetic fields. Transition into the state of exciton insulator under such conditions was first established in 1970 [66].

Thus, in the region of sufficiently low temperatures, a continuous transition from the metallic state into semiconductor state under the action of a magnetic field H attaining a value $H \sim H_{cr}$, is unfeasible. A monotonous decrease of the overlap of the bands and of the concentration of carriers in a magnetic field occur only to a certain value H' ($H' < H_{cr}$) at which condition (2.145) becomes valid and bound states of electrons and holes are formed in the system. The concentration of free carriers in the point $H = H'$

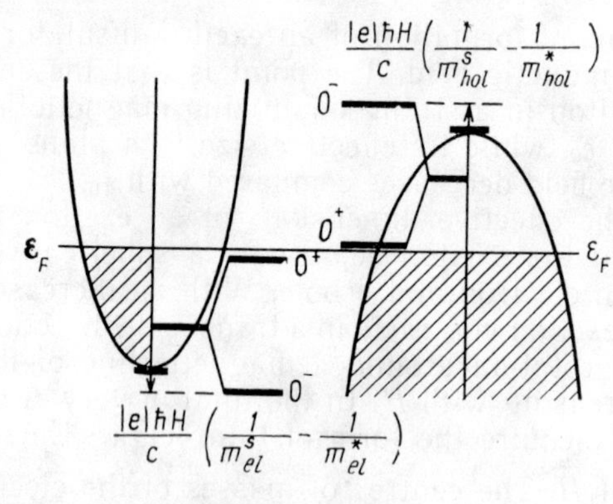

Fig. 148

drops stepwise down to an exponentially small value proportional to $e^{-\frac{\varepsilon(H)}{kT}}$, and the energy spectrum of the metal consisting of Landau 0^- levels of electrons and holes is recombined into the energy spectrum of an exciton insulator.

An exciton insulator can exist in a wide interval of fields from $H' < H_{cr}$ to $H'' > H_{cr}$, H'' being the value of the magnetic field at which the inequality $\varepsilon_g(H) < \varepsilon(H)$ is not true. Electron-hole pairs are destroyed in that point and the spectrum of exciton insulator is recombined into a semi-conductor spectrum consisting of Landau 0^- levels for electrons and holes separated by an energy gap $\varepsilon_g(H'')$.

Let us now analyse the case when the spin splitting exceeds the orbital splitting in both bands: $m^s_{el} < m^*_{el}$, $m^s_{hol} < m^*_{hol}$. The position of 0^- levels for electrons and holes for that case is shown in Fig. 148. As can be seen, the edge of the conduction band (0^- level) has lowered by $-\frac{|e|\hbar H}{2c}\left(\frac{1}{m^*_{el}} - \frac{1}{m^s_{el}}\right)$ relative to the level

$\mathscr{E} = \mathscr{E}_c$, while the band edge of the valence zone has lifted by $-\frac{|e|\hbar H}{2c}\left(\frac{1}{m^*_{hol}} - \frac{1}{m^s_{hol}}\right)$ relative to the level $\mathscr{E} = \mathscr{E}_v$. The overlap of the bands has therefore increased by the sum of these values.

An increase of overlap of the bands is equivalent to the conduction zone moving in the magnetic field along the energy scale onto the valence zone. The concentration of electrons n and the equal concentration of holes p increase owing to the flow of the excess electrons from the valence zone into the conduction zone (Fig. 149). The growth of concentration occurs quicker than it

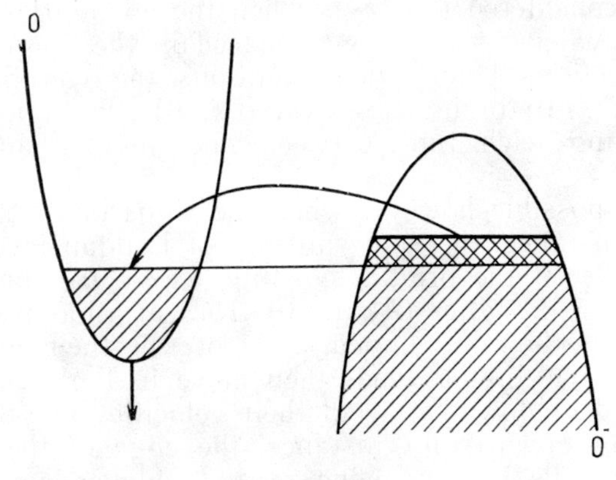

Fig. 149

would be determined solely by an increase of overlap of the bands, since the degree of degeneration of Landau levels increases simultaneously. This will be discussed later in the analysis of the second cause of variation of concentration of current carriers in magnetic fields.

Thus, the direction of motion of the band edge of the conduction zone in the magnetic field is determined by the sign of A_{el} in expression (2.139): at $A_{el} > 0$, the edge of the band is displaced upward, and at $A_{el} < 0$, downward. The magnitude of the velocity of displacement is

$$\left|\frac{\partial \Delta\varepsilon_e}{\partial H}\right| = \frac{|e|\hbar}{2c}|A_{el}| \tag{2.149}$$

Similarly, the band edge of the valence zone moves downward along the energy scale in the magnetic field if A_{hol} is positive in expression (2.140), and upward if A_{hol} is negative. The magnitude of the velocity of displacement of the edge of the valence zone is

$$\left|\frac{\partial \Delta\varepsilon_{hol}}{\partial H}\right| = \frac{|e|\hbar}{2c}|A_{hol}| \tag{2.150}$$

Relative motion of the bands in the magnetic field, which can cause the energy of overlap ε_{ov} to increase or decrease, is determined by the magnitude and sign of the sum $\Delta\varepsilon_{el} + \Delta\varepsilon_{hol}$, which in turn depends on the magnitude and sign of B [see (2.142)].

As has been shown, a decrease of overlap of the bands in the magnetic field corresponds to the case $B > 0$. On the contrary, at $B < 0$ the overlapping increases. The magnitude of the velocity of variation of overlapping in both cases is

$$\left| \frac{\partial \varepsilon_{ov}}{\partial H} \right| = \frac{|e|\hbar}{2c} |B| \tag{2.151}$$

We have considered the cases when the boundaries of the conduction and valence zones are displaced in the magnetic field in opposite directions. Under such conditions, the overlap either decreases (at $B > 0$) or increases (at $B < 0$), the signs of A_{el} and A_{hol} coinciding each time between one another and with the sign of B.

Cases are possible however when the signs of A_{el} and A_{hol} are unlike. For instance, the spin splitting of Landau levels for electrons may exceed the orbital splitting ($A_{el} < 0$), and for holes, be smaller than the latter ($A_{hol} > 0$). The variation of overlapping of the bands is then determined, as before, by the magnitude of B.

The boundaries of the zones then move in the same direction, either upward or downward, but their velocities may be different. When moving upward, for instance, the edge of the conduction zone may lag behind the edge of the valence zone ($A_{el} > 0$, $A_{hol} < 0$; $A_{el} < |A_{hol}|$, and $B < 0$), and therefore, overlap of the zones will increase ($B < 0$). If, on the contrary, the edge of the conduction zone moves swifter than the edge of the valence zone ($A_{el} > 0$; $A_{hol} < 0$; $A_{el} > |A_{hol}|$, and $B > 0$), overlap of the zones will decrease and at a definite value of the magnetic field H_{cr} the transition into a semiconductor state will occur [1]). A similar analysis may be made for the motion of the zone boundaries downward.

Thus, the ratio between the spin and orbital masses of electrons and holes (B) determines mutual position of the bands along the energy scale at each value of magnetic field. This position remains invariable if only $B = 0$, the energy of band overlapping being independent of magnetic field. With $B = 0$, three different cases are possible:

(a) the bands move upward with the same velocity (A_{el} and A_{hol} are other than zero but have unlike signs, with $A_{el} = |A_{hol}|$);

---

[1]) With observance of the conditions discussed above, a stationary phase of exciton insulator can be formed in a certain interval of fields near $H = H_{cr}$.

(b) the bands move downward with the same velocity (A_{el} and A_{hol} are other than zero but have unlike signs, with $|A_{el}| = A_{hol}$);

(c) the bands are immovable: $A_{el} = 0$, $A_{hol} = 0$.

Let the last case be discussed in more detail. If A_{el} and A_{hol} become zeros, this implies that the spin splittings of Landau levels for electrons and holes are equal to the corresponding orbital splittings. For $A_{el} = 0$, the Landau 0^- level for electrons in any magnetic field coicides with the edge of the conduction zone at $H = 0$. It is then said that the 0^- level is "frozen in" into the edge of the conduction zone. Similarly, at $A_{hol} = 0$, the Landau 0^- level for holes is "frozen in" into the edge of the valence zone (Fig. 150). This situation is of interest, because it makes it possible to exclude

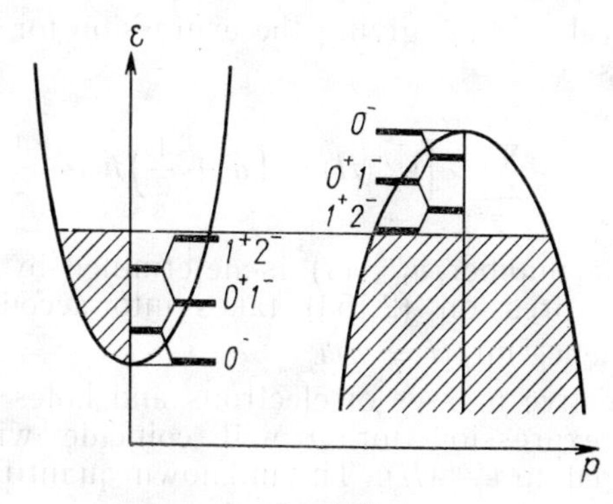

Fig. 150

the influence of the motion of zone boundaries onto the variation of concentration of carriers in the magnetic field and to elucidate the role of other factors, apart from the motion of zone boundaries.

II. The second cause of variation of the concentration of carriers in the magnetic field is a continuous growth of the degree of degeneration of Landau levels for electrons and holes. In the absence of motion of the zone boundaries this results in an increase of the concentrations n and p.

In order to see into the physical aspect of this problem, we simplify it by excluding all additional factors that can vary in the magnetic field. First of all, we exclude the influence of motion of the zone boundaries by assuming $A_{el} \equiv A_{hol} \equiv 0$. Further, we assume for simplicity that the electron and the hole zones are similar, i.e. $m^*_{el} = m^*_{hol} = m$. In the absence of magnetic field, the Fermi level in such a system is located exactly half-way between the edge of the conduction zone and the edge of the valence zone, so that $\varepsilon_F^{el} = \varepsilon_F^{hol} = \varepsilon_F^0 = \varepsilon_{ov}/2$ [see expressions (2.137) and (2.138)]. The concentrations of electrons and holes at $H = 0$ are

$$n = p = \frac{2\sqrt{2}}{3\pi^2\hbar^3}\left(m\varepsilon_F^0\right)^{3/2} = \frac{1}{3\pi^2\hbar^3}\left(m\varepsilon_{ov}\right)^{3/2} \tag{2.152}$$

The concentration of carriers in a magnetic field can be obtained by integrating over energy the expression (2.132) for the density of states $\nu_H(\varepsilon)$, in which spin splitting is taken into account. Integration of each term in the sum is done from the edge of the corresponding Landau level

$$\varepsilon = \varepsilon_{n,\,s} = \left(n + \frac{1}{2}\right)\hbar\omega + \frac{1}{2}sg\mu_B H$$

to the Fermi level $\varepsilon = \varepsilon_F$, giving the expression for the concentration of electrons

$$n = \frac{\hbar\omega m^{3/2}}{\sqrt{2}\,\pi^2\hbar^3} \sum_{s=\pm 1} \sum_{n=0}^{n=n_{\max}(s)} \left[\varepsilon_F^{el}(H) - \left(n + \frac{1}{2}\right)\hbar\omega - \frac{1}{2}s\hbar\omega\right]^{1/2} \tag{2.153}$$

where the whole number $n_{\max}(s)$ is determined by double inequality (2.131). Expression (2.153) takes into account that $m_{el}^* = m_{el}^s = m$, whence $g\mu_B H = \hbar\omega$.

Since the effective masses of electrons and holes are equal, the corresponding expression for p will coincide with (2.153) if $\varepsilon_F^{el}(H)$ is changed to $\varepsilon_F^{hol}(H)$. The unknown quantities $\varepsilon_F^{el}(H)$ and $\varepsilon_F^{hol}(H)$ for each value of the magnetic field are found from two equations: $n = p$ and $\varepsilon_F^{el}(H) + \varepsilon_F^{hol}(H) = \varepsilon_{ov}$. It follows from these equations that the Fermi level, as at $H = 0$, remains exactly halfway between the edge of the conduction zone and the edge of the valence zone (Fig. 151*a*) and

$$\varepsilon_F^{el}(H) = \varepsilon_F^{hol}(H) = \varepsilon_F^0 = \frac{\varepsilon_{ov}}{2}$$

In order to simplify the analysis of expression (2.153) for the concentration of carriers, we pass to the ultra-quantum limit of magnetic fields. Then only one term, corresponding to $s = -1$, $n_{\max}(-1) = 0$, will remain in the double sum:

$$n = p = \frac{\hbar\omega \cdot m^{3/2}}{\sqrt{2}\,\pi^2\hbar^3}\left(\varepsilon_F^0\right)^{1/2} \tag{2.154}$$

It may be seen from (2.154) that though the Fermi energy is constant, the concentration of carriers increases proportionally to the growth of the degree of degeneration of Landau 0^- levels for electrons and holes.

Let us show that an increase of concentration occurs, as before, owing to the flow of electrons from the valence zone into the conduction zone. As will be seen from Fig. 151a, all filled electron states of the second zone (conduction zone) in the ultra-quantum limit are located under the Fermi level on a parabola which is the last Landau 0^- level. The electron system is perfectly one-dimensional, the energy of each state being dependent only on quasi-momentum p_z. The hole states are also located on the Landau level and occupy the part of the parabola above the Fermi level in the first zone. The first (valence) zone under the Fermi level is filled with electrons.

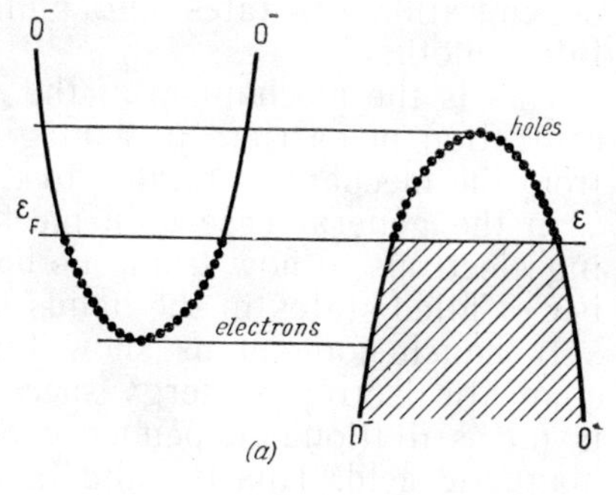

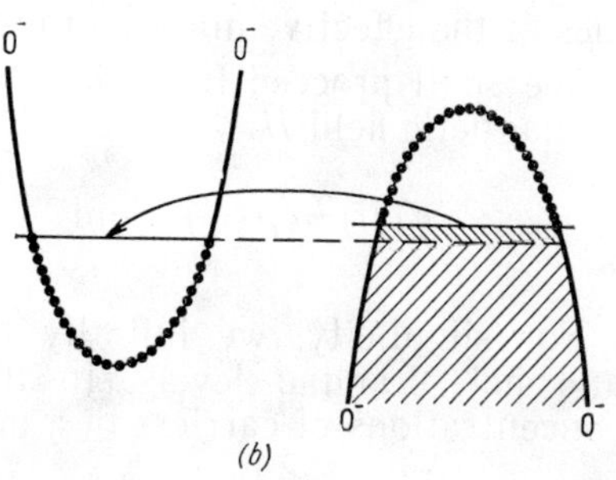

Fig. 151

The increasing degeneration of each state in the magnetic field at the 0^- level implies that the capacity of the conduction zone increases continuously at the constant geometric dimensions of the parabola 0^- in Fig. 151a (the dimensions of the parabola 0^- are specially fixed in the case considered). If the number of electrons in the conduction zone were not replenished, the boundary of filling of that zone, i.e. the height of the Fermi level, would come lower, since, with an increase of the capacity of each state along p_z, a constant number of electrons would fill ever smaller part of the electron parabola.

But the Fermi level in one zone cannot be lower than in the other. Its lowering in the conduction zone would result in that part of the electrons in the valence zone near the Fermi level were in energetically unprofitable position (Fig. 151b). These electrons would immediately flow into the conduction zone, which would equalize the levels in both zones.

Actually, no lowering of the Fermi level occurs in the conduction zone: at an increase of the capacity of that zone in the magnetic field, electrons flow continuously from the valence zone, corresponding exactly to the increase of the capacity of the conduction zone and thus compensates the variation of the boundary of its filling. It is then natural that the number of electrons under the

Fermi level in the valence zone decreases as a result of diminishing of the mean density of states, but the number of vacant places (holes) increases in full correspondence with this and always remains equal to the number of electrons in the conduction zone. The magnetic field acts as a pump which controls the degeneration of states and pumps the electrons from one band into the other.

This is the mechanism of the second cause of variation of concentration of carriers in a magnetic field. It differs in its nature from the mechanism related to the motion of zone boundaries.

In the general case both mechanisms act simultaneously causing electrons to flow from one band into the other. The total number of filled states in the bands in either case does not change.

In conclusion, let us show that in a two-zone metal with the quadratic isotropic energy spectrum of electrons and holes, there is no oscillational dependence of the concentrations n and p on magnetic field. This is valid in the general case for different values of the effective masses of carriers: $m^*_{el} \neq m^*_{hol}$.

We shall proceed from that, for a two-zone metal at any value of magnetic field H:

$$n(H) = p(H) \quad \text{and} \quad \varepsilon_F^{el}(H) + \varepsilon_F^{hol}(H) = \varepsilon_{ov}(H)$$

For simplicity, we initially neglect spin splitting of electron and hole Landau levels. In that case the expressions for the concentrations of carriers in a magnetic field will be

$$n = \frac{\sqrt{2}\,\left(\hbar\omega_{el}\left(m^*_{el}\right)\right)^{3/2}}{\pi^2\hbar^3} \sum_{n=0}^{n=(n_{\max})_{el}} \left[\varepsilon_F^{el}(H) - \left(n + \frac{1}{2}\right)(\hbar\omega_{el})\right]^{1/2}$$

$$p = \frac{\sqrt{2}\,(\hbar\omega)_{hol}\left(m^*_{hol}\right)^{3/2}}{\pi^2\hbar^3} \sum_{n=0}^{n=(n_{\max})_{hol}} \left[\varepsilon_F^{hol}(H) - \left(n + \frac{1}{2}\right)(\hbar\omega)_{hol}\right]^{1/2} \tag{2.155}$$

Here the quantities $(n_{\max})_{el}$ and $(n_{\max})_{hol}$ are determined by inequalities (2.105) written for electrons or holes respectively.

As is known, the effective mass, which enters the expression for the quadratic isotropic dispersion law, coincides with the isotropic cyclotron mass and the mass of density of states. Then we have

$$(\hbar\omega)_{el}\, m^*_{el} = \left(\hbar\omega_{hol}\right) m^*_{hol} = \frac{|e|\,\hbar \cdot H}{c}$$

Equating expressions (2.155) for $n(H)$ and $p(H)$ to one another and cancelling the common multipliers, we get

$$\sum_{n=0}^{n=(n_{\max})_{el}} \left[m^*_{el}\varepsilon^{el}_F(H) - \left(n+\frac{1}{2}\right)\frac{|e|\hbar H}{c}\right]^{1/2} =$$

$$= \sum_{n=0}^{n=(n_{\max})_{hol}} \left[m^*_{hol}\varepsilon^{hol}_F(H) - \left(n+\frac{1}{2}\right)\frac{|e|\hbar H}{c}\right]^{1/2} \quad (2.156)$$

where, according to inequalities (2.105), $(n_{\max})_{el}$ and $(n_{\max})_{hol}$ are determined by the ratios $\frac{\varepsilon^{el}_F(H)}{(\hbar\omega)_{el}}$ and $\frac{\varepsilon^{hol}_F(H)}{(\hbar\omega)_{hol}}$ respectively. Hence it follows that

$$m^*_{el}\varepsilon^{el}_F(H) = m^*_{hol}\varepsilon^{hol}_F(H) \quad (2.157)$$

and therefore, the relationship of the form (2.137) and (2.138) are valid at any value of the magnetic field:

$$\frac{\varepsilon^{el}_F(H)}{\varepsilon^{hol}_F(H)} = \frac{m^*_{hol}}{m^*_{el}} \quad (2.158)$$

$$\varepsilon^{el}_F(H) = \frac{m^*_{hol}}{m^*_{el} + m^*_{hol}}\,\varepsilon_{ov}(H)$$

$$\varepsilon^{hol}_F(H) = \frac{m^*_{el}}{m^*_{el} + m^*_{hol}}\,\varepsilon_{ov}(H) \quad (2.159)$$

It can easily be verified that these expressions are also true with the account of spin splitting, provided that

$$\frac{m^*_{el}}{m^s_{el}} = \frac{m^*_{hol}}{m^s_{hol}} \quad (2.160)$$

The dependence of the energy of overlap on magnetic field is related to monotonous motion of the zone boundaries and is determined by formulae (2.141) and (1.142). Thus, expressions (2.159) show that $\varepsilon^{el}_F(H)$ and $\varepsilon^{hol}_F(H)$ are monotonous functions of magnetic field.

The absence of oscillations of the Fermi energy of electrons and holes and, as a cosequence, the absence of oscillations of concentrations n and p, can be related to that, conditions (2.158) and (2.160) being satisfied, the Landau levels for electrons and holes move in opposite directions in the magnetic field and cross simultaneously the Fermi level.

Under these conditions, the diagrams of Landau levels for electrons and holes (see, for instance, Fig. 144) are geometrically similar to one another and differ only in scale by a factor $\varepsilon_F^{el}/\varepsilon_F^{hol}$. But the levels for electrons and holes with the same number n travel the distances in the magnetic field that are connected by the same ratio.

The magnitude of reciprocal magnetic field $1/H_n$ at which the n-th Landau level intersects the Fermi level, is determined by expression (2.122). By equality (2.157), the values of $1/H_n$ for electrons and holes coincide:

$$\left(\frac{1}{H_n}\right)_{el} = \frac{|e|\hbar}{m_{el}^* \varepsilon_F^{el}(H_n)c} = \left(\frac{1}{H_n}\right)_{hol} = \frac{|e|\hbar}{m_{hol}^* \varepsilon_F^{hol}(H_n)c} \tag{2.161}$$

As the next Landau levels for electrons and holes approach simultaneously the Fermi level from above and below, the latter neither lowers nor moves upward, as was the case with the one-zone metal.

In a two-zone metal with an anisotropic spectrum, crossing of the Fermi level by the electron and hole Landau levels is no more simultaneous. Such a metal may serve to demonstrate that the dependence of the Fermi energy of electrons and holes on magnetic field is of non-monotonous nature.

Consider a case of an anisotropic quadratic spectrum of electrons and holes whose general form coincides with expression (2.76). The spectrum of each type of carriers is characterized by the three main components of the effective mass tensor: $(m_x^*)_{el}$, $(m_y^*)_{el}$, $(m_z^*)_{el}$ and $(m_{x'}^*)_{hol}$, $(m_{y'}^*)_{hol}$, $(m_{z'}^*)_{hol}$. The main axes p_x, p_y, and p_z of the electron ellipsoid may be not coincident with the main axes $p_{x'}$, $p_{y'}$, $p_{z'}$ of the hole ellipsoid. The concentrations of carriers n and p are now determined by the masses of the densities of states of electrons $(m_d^*)_{el}$ and holes $(m_d^*)_{hol}$:

$$(m_d^*)_{el} = [(m_x^*)_{el}(m_y^*)_{el}(m_z^*)_{el}]^{1/3}$$

$$(m_d^*)_{hol} = [(m_{x'}^*)_{hol}(m_{y'}^*)_{hol}(m_{z'}^*)_{hol}]^{1/3} \tag{2.162}$$

Using formula (2.88) to determine the concentrations of electrons n or holes p at $H=0$, and also the equality $n=p$, we can find the ratio of $\varepsilon_F^{el}(0)$ to $\varepsilon_F^{hol}(0)$ at $H=0$:

$$\frac{\varepsilon_F^{el}(0)}{\varepsilon_F^{hol}(0)} = \frac{(m_d^*)_{hol}}{(m_d^*)_{el}} \tag{2.163}$$

But in the general case the cyclotron masses of electrons and holes for an arbitrary direction of magnetic field $\vec{H}$ do not coincide with the corresponding masses of the density of states. For that reason, with the anisotropic spectrum of current carriers, the equality of the ratios

$$\frac{\varepsilon_F^{el}(H)}{\varepsilon_F^{hol}(H)} = \frac{(\hbar\omega)_{el}}{(\hbar\omega)_{hol}}$$

which would generalize (2.157) for that case, is non-existent. This can be strictly shown by means of the expressions for n and p which are equivalent to formulae (2.155):

$$n = \frac{\sqrt{2}\,(\hbar\omega)_{el}\left(m_d^*\right)_{el}^{3/2}}{\pi^2\hbar^3} \sum_{n=0}^{n=(n_{\max})_{el}} \left[\varepsilon_F^{el}(H) - \left(n+\frac{1}{2}\right)(\hbar\omega)_{el}\right]^{1/2}$$

$$p \frac{\sqrt{2}\,(\hbar\omega)_{hol}\left(m_d^*\right)_{hol}^{3/2}}{\pi^2\hbar^3} \sum_{n=0}^{n=(n_{\max})_{hol}} \left[\varepsilon_F^{hol}(H) - \left(n+\frac{1}{2}\right)(\hbar\omega)_{hol}\right]^{1/2} \qquad (2.164)$$

Thus, in the general case with an anisotropic spectrum of current carriers, the quantum levels of electrons and holes pass across the Fermi level not in synchronism with each other. The ultra-quantum limits for electrons and holes can be attained at different values of magnetic fields.

To simplify calculations, we shall discuss a two-zone metal in which only electrons have the anisotropic nature of the spectrum. This is sufficient to describe the effects related to anisotropy. Let the following designations be introduced:

$$(m_{x'}^*)_{hol} = (m_{y'}^*)_{hol} = (m_{z'}^*)_{hol} = M$$
$$(m_x^*)_{el} = m_1; \quad (m_y^*)_{el} = m_2; \quad (m_z^*)_{el} = m_3$$
$$(m_d^*)_{el} = m_d \qquad (2.165)$$

Then, obviously, $(m_d^*)_{hol} = M$.

For real substances, the inequality $m_d < M$ is usually valid. We assume that it is satisfied. In addition, we assume a strong anisotropy of the electron ellipsoid, i.e. $m_2 \ll m_1$ and $m_3 \ll m_1$. The discussion will be made for the magnetic field $\vec{H}$ orientated along the main axis of the first electron ellipsoid. The cyclotron mass m^* of electrons then is $\sqrt{m_2 m_3}$ and obeys the inequalities

$$m^* \ll m_d, \quad m^* \ll M \qquad (2.166)$$

The cyclotron mass of holes coincides with M at any orientation of the field.

It is convenient to start the discussion for fixed zone boundaries, assuming the spin masses of electrons m^s and holes M^s coinciding with the corresponding cyclotron masses. The result obtained can be generalized for the case of an arbitrary motion of zone boundaries.

Thus, we initially assume that the band overlap ε_{ov} is fixed ($A_{el} = A_{hol} = 0$). Re-write relationship (2.163) in the new symbols:

$$\frac{\varepsilon_F^{el}}{\varepsilon_F^{hol}} = \frac{M}{m_d} \tag{2.167}$$

In our case, $m_d < M$ and $\varepsilon_F^{hol} < \varepsilon_F^{el}$.

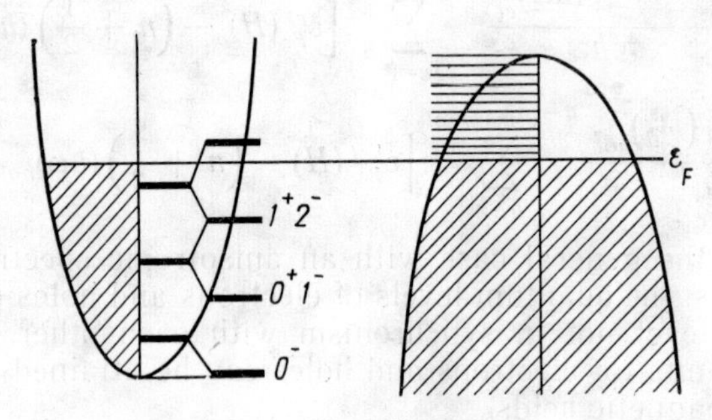

Fig. 152

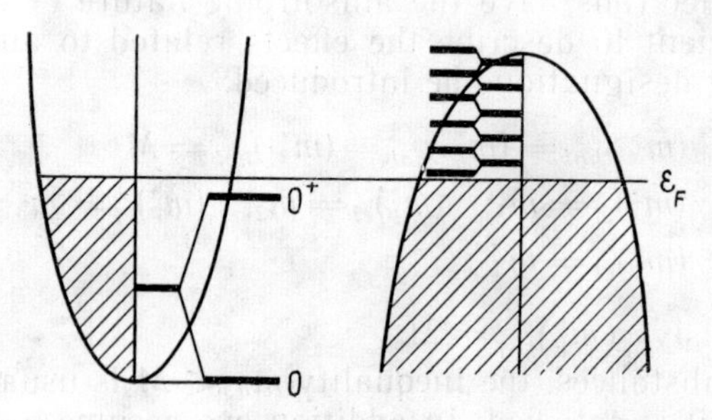

Fig. 153

The diagram of the band spectrum at $H = 0$ coincides with that shown in Fig. 143. The condition $m^* \ll M$ implies that $(\hbar\omega)_{el} \gg$ $\gg (\hbar\omega)_{hol}$, so that, in magnetic fields for which $(\hbar\omega)_{el} < \varepsilon_F^{el}$ at $H = 0$, quantization of energy of holes can be neglected and their energy spectrum regarded as quasi-continuous (because of extremely small distances between Landau levels for holes). In other words, the numbers n_{hol} of Landau levels for holes crossing the Fermi level exceed substantially the numbers of Landau levels for electrons, $n_{hol} \gg n_{el}$. A diagram of Landau levels in such fields is shown in Fig. 152.

Under these conditions, the position of the Fermi level in the magnetic field depends mainly only on the motion of electron Landau levels. When a next Landau level for electrons approaches the Fermi level, it moves upward and the Fermi ellipsoid expands. After the Landau level has been freed from electrons, the Fermi ellipsoid contracts and the Fermi level approaches the next quantum level. This results in oscillations of Fermi energy of electrons ε_F^{el} and, because of the equality $\varepsilon_F^{el}+\varepsilon_F^{hol}=\varepsilon_{ov}=\text{const}$, also of Fermi energy of holes.

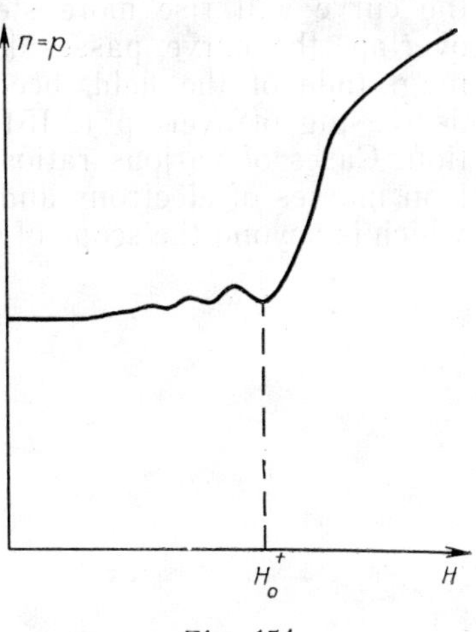

Fig. 154

When the Fermi level moves upward, ε_F^{hol} decreases. The concentration of holes p and that of electrons n, which is equal to p, then diminish (more exactly, n and p pass through a local minimum): the electrons from the Landau level which has been freed flow over into the valence zone. Then, when the quantum level is completely freed of electrons and crosses the Fermi boundary, the Fermi level begins to lower rapidly toward a next quantum level onto which the electrons flow back from the valence zone. Thus, oscillations of n and p are related to the periodic flow of electrons from one zone to the other at which their total number is not changed.

Consider the transition to the ultra-quantum limit for electrons, which is accompanied by passing out of the last quantum level 0^+. The corresponding diagram of Landau levels is given in Fig. 153.

After passing out of the 0^+ level, the Fermi ellipsoid for electrons begins to contract quickly, and the Fermi level lowers to the last 0^- level which has been left. This causes the Fermi energy of electrons to decrease and that of holes to increase. The electrons from the valence zone begin to flow over onto the electron 0^- level, causing an increase of the concentrations n and p.

While at $H=0$ the ratio between ε_F^{el} and ε_F^{hol} was determined by the ratio M/m_d (2.167), now $\varepsilon_F^{el}\to 0$ as $[1/(\hbar\omega)_{el}]^2$ and ε_F^{hol} approaches the value of fixed overlap ε_{ov} of the bands. The concentration of carriers, which initially increases sharply after the passage of the electron 0^+ level because of the sharp lowering of the Fermi level, is further determined only by increasing degeneration of the 0^- level and becomes proportional to the magnetic field H.

A qualitative curve of the dependence of concentrations $n = p$ on magnetic field at the transition to the ultra-quantum limit for electrons is shown in Fig. 154, where H_{0+} denotes the magnetic field at which the level 0^+ passes out.

If the motion of zone boundaries is taken into account, the slope of the curve in Fig. 154 may slightly change on each of its sections. For instance, if the band overlap increases in the magnetic field, the curve will rise more steeply. On the contrary, at a smaller overlap, the curve passes lower and, beginning from a certain magnitude of the field, becomes dropping. At a large speed of decreasing of overlap (2.151), the curve will have no rising section. Cases of various ratios between the spin masses and cyclotron masses of electrons and holes require more detailed analysis which is beyond the scope of the book.

CHAPTER THREE

ELECTRIC CONDUCTIVITY OF METALS

3-1. PHENOMENOLOGICAL DESCRIPTION OF ELECTRIC CONDUCTIVITY

In the absence of temperature gradients and diffusion of current carriers the relationship between the electric current density $\vec{j}$ in a conductor and the electric field strength $\vec{E}$ in sufficiently weak fields is established by Ohm's law: $\vec{j} = \sigma\vec{E}$, where the proportionality factor σ is termed the electric conductivity.

In an isotropic metal, the direction of current $\vec{j}$ coincides with that of the external field $\vec{E}$ and σ is here a scalar quantity. Anisotropy of a metal makes the vectors $\vec{E}$ and $\vec{j}$ non-collinear (in an anisotropic case, collinearity of $\vec{E}$ and $\vec{j}$ is observed with $\vec{E}$ directed along the main axes of symmetry of the crystal). Transformation of vector $\vec{E}$ into vector $\vec{j}$ is made by means of a second-rank symmetric tensor: the tensor of electric conductivity σ_{ij} which has six independent components.

The expression for Ohm's law [1])

$$j_i = \sigma_{ik} E_k \tag{3.1}$$

can be inverted

$$E_i = \rho_{ik} j_k \tag{3.2}$$

where the components of electric field $\vec{E}$ can be expressed through the component of current density $\vec{j}$ by means of resistivity tensor ρ_{ik}.

The resistivity tenzor ρ_{ik} is an inverse of the conductivity tensor σ_{ik}. The components σ_{ik} and ρ_{ik} are related as follows:

$$\sigma_{ik}\rho_{kj} = \delta_{ij}$$

where δ_{ij} are Kroneker delta-symbols. The resistivity tensor ρ_{ij} is also symmetrical: $\rho_{ij} = \rho_{ji}$. Its components can be found

---

[1]) Summation is to be made over repeating indices.

by measuring the electric field along the axes 1, 2, and 3 when a current flows along one of these axes. For instance, if the current flows only along the axis 1 ($j = j_1$), then

$$E_1 = \rho_{11} j_1, \quad E_2 = \rho_{21} j_1, \quad E_3 = \rho_{31} j_1 \tag{3.3}$$

Measurements of fields E_1, E_2 and E_3 give the components of the tensor, ρ_{11}, ρ_{21}, ρ_{31}. If the current is now directed along the axis 2 ($j = j_2$), we get three new expressions:

$$E_1 = \rho_{12} j_2, \quad E_2 = \rho_{22} j_2, \quad E_3 = \rho_{32} j_2 \tag{3.4}$$

which are used to determine ρ_{12}, ρ_{22}, ρ_{32}, etc.

A cubic crystal possesses no anisotropy of electric conductivity. For it the non-diagonal components of the tensor ρ_{ij} are equal to zero, and the diagonal ones are equal to each other. In such a crystal, $\rho_{ij} = \rho_0 \delta_{ij}$. With uniaxial crystals (those with prevailing direction 3), $\rho_{11} = \rho_{22} \neq \rho_{33}$.

3-2. ELECTRIC CONDUCTIVITY AND ENERGY SPECTRUM OF ELECTRONS IN A METAL

According to the concepts developed in the previous chapters, a metal may be regarded as a potential box containing various groups of current carriers and a phonon gas. The effective mass and nature of motion of the i-th group of carriers are uniquely determined by the shape of the corresponding constant-energy surface and the dispersion law: $\varepsilon_F^i = \varepsilon^i (\vec{p})$. Constant-energy surfaces in $\vec{p}$-space may have very diverse shape and may be regarded as rigid structures on which the gas of current carriers is concentrated.

Each group of carriers is characterized by its concentration $n_i = \frac{2\Delta_i}{(2\pi\hbar)^3}$, where Δ_i is the volume bounded by the set of constant-energy surfaces in the i-th band. Note that each energy band (each group of carriers) retains its individuality even when the zones are overlapped.

The current carriers representing an ideal gas of quasi-excitations are diffracted by phonons in their motion in an electric field $\vec{E}$. The processes of electron scattering on phonons will be discussed in the next section from the standpoint of their contribution to electric conductivity. Here, it should only be noted that each group of current carriers is characterized by its free-path length l_i. The mean free time (or relaxation time) τ_i is connected with l_i by the relationship $\tau_i = l_i / v_F^i$, where v_F^i is the velocity of electrons on Fermi surface in the i-th band.

Consider a group of current carriers in an external electric field $\vec{E}$ assumning its constant-energy surface differing not strongly from an ellipsoid. Let the electric field $\vec{E}$ be directed along one of the main axes of this surface, for instance, along p_x (Fig. 155).

The electrons inside the volume bounded by the constant-energy surface cannot be accelerated by the electric field, since all the states are filled. Only the electrons located near the Fermi surface can be accelerated. Under the action of the force $-|e|\vec{E}$, the momenta of these electrons vary in accordance with the equation of

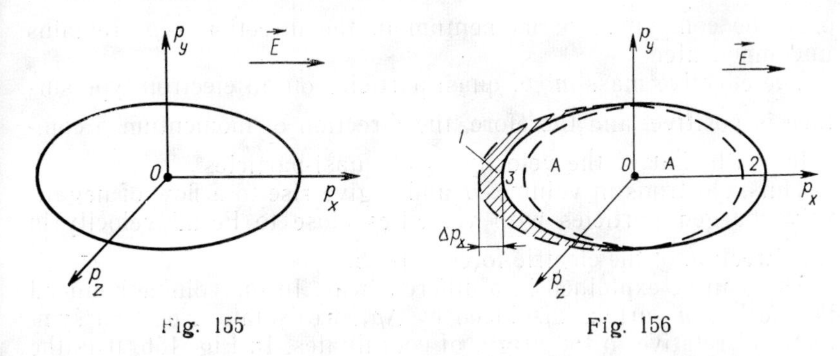

Fig. 155 Fig. 156

motion (2.20), the variation Δp_x of the momentum during the relaxation time τ_x being equal to

$$\Delta p_x = |e| E_x \tau_x \tag{3.5}$$

Since all the electrons on the surface in the external electric field $\vec{E}$ have acquired an additional mean momentum Δp_x, the constant-energy surface is displaced as a rigid structure by Δp_x in the direction opposite to the electric field vector $\vec{E}$ (Fig. 156).

There was no electric current before application of the field E, and the distribution of electrons was symmetrical relative to the origin of coordinates in the momentum space. This means that for each electron with a momentum $+\vec{p}$ we could always find an electron with $-\vec{p}$.

As a result of displacement of the Fermi surface, the central symmetry of distribution of electrons is disturbed. In volume *1* (Fig. 156), there appear electrons whose total momentum has a non-compensated component in the direction $-\vec{p}_x$. Simultaneously, in volume *2*, there disappear the electrons whose motion has compensated the flow of electrons present in volume *3*.

Thus, the displacement of the Fermi surface gives rise to the appearance of non-compensated electrons in volumes *1* and *3*. Since the displacement Δp_x is small compared with the Fermi momentum p_F, the volumes *1* and *3* are practically equal to one another.

For electrons in volumes *1* and *3*, the projections of momenta along the axes $\vec{p}_y$ and $\vec{p}_z$ are mutually compensated. In other words, for an electron with the momentum components $+p_y$, $+p_z$ along these axes we can always find an electron with the corresponding components $-p_y$, $-p_z$. For electrons in volumes *1* and *3*, only the component of momentum in the direction $-\vec{p}_x$ remains uncompensated.

The effective mass m^* of quasi-particles on an electron-type surface is positive, and therefore, the direction of momentum $\vec{p}$ coincides with that of the velocity $\vec{v}_g$ of quasi-particles.

Thus, electrons in volumes *1* and *3* give rise to a flow of negatively charged particles with velocities close to Fermi velocity in the direction of the electric force $-|e|\vec{E}$.

This can be explained in a different way. In the volume bounded by the Fermi surface displaced by Δp_x, we isolate a portion symmetrical relative to the origin of coordinates. In Fig. 156, it is the internal volume *A-A* bounded from the left and right by dotted lines. Electrons in this volume do not contribute to electric conductivity because of the symmetry of their distribution over momenta.

The remaining portion of the volume (volumes *1* and *3*) located asymmetrically relative to the origin of coordinates, forms a non-compensated flow of electrons (electric current) in the direction $-\vec{p}_x$. At an increase of the electric field E, the displacement of the surface increases, the non-compensated volumes *1* and *3* become larger, and therefore, the number of particles contributing to electric conductivity also increases.

Consider now a hole-type Fermi surface bounding unfilled states (Fig. 157). Let the electric field be again directed along the $\vec{p}_x$ axis. The variation of the momentum of quasi-particles on a hole-type surface can also be described by the equation of motion (2.20). Then, under the action of electric field, as in the earlier case, the surface will be displaced in the direction of the force $-|e|\vec{E}$ (Fig. 158), the magnitude of displacement Δp_x being again determined by expression (3.5) where τ_x should now be understood as the relaxation time of holes.

Displacement of the Fermi surface disturbs the central symmetry of distribution of particles over momenta. There appear particles in volume *1* (Fig. 158) whose total momentum has a non-compensated component in the direction $+\vec{p}_x$. In volume *2*, particles vanish whose motion has compensated the flow of particles contained in volume *3*. Thus, as in the earlier case, displacement of the surface gives rise to the appearance of particles with noncompensated components of momenta in the direction $+\vec{p}_x$ in volumes *1* and *3*. But the effective mass m^* of particles on a hole-type surface is negative, so that the direction of velocity $\vec{v}_g$ of a particle is opposite to that of its momentum $\vec{p}$.

Thus, negatively charged particles in volumes *1* and *3* move with velocities near the Fermi velocity in the direction $-\vec{p}_x$, i.e.

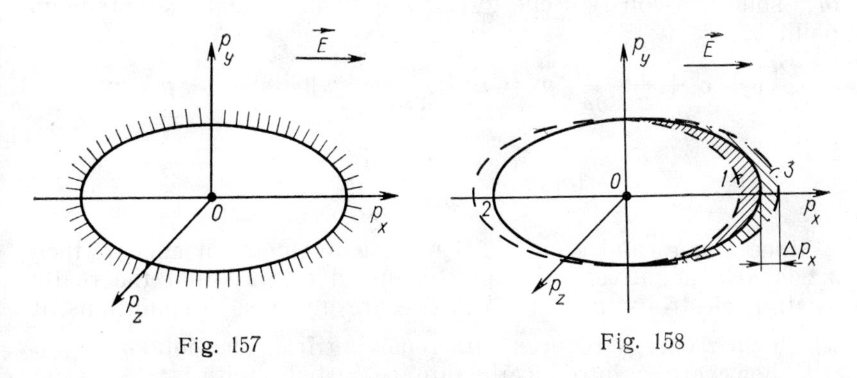

Fig. 157 Fig. 158

form an electric current of the same direction as the electrons do. Hence it follows that in metals having constant-energy surfaces of both hole and electron type the total electric current is composed of the currents formed by the carriers of both types. Electrons and holes have an additive contribution to electric conductivity. Let us see whether the phenomenon of electric conductivity makes it possible to reveal differences between the two concepts of quasi-particles on hole-type constant-energy surface that we have introduced in Sec. **2-12**, Chapter Two.

As has been just shown, regarding a hole as a negatively charged particle ($e < 0$) with a negative effective mass ($m^* < 0$) gives an electric current coinciding in direction with that formed by electrons ($e < 0$, $m^* > 0$).

Measuring the electric conductivity of two-zone metals with equal concentrations of electrons n and holes p corresponds to that the current of electrons and holes add together. If these currents were subtracted from one another, then it would turn out in particular that a two-zone metal in which electrons and holes have the same kinetic properties possesses no condictivity. But this contradicts all the known experimental data.

Thus, only such presentation of a hole is correct at which the contribution of holes to electric conductivity is additive to that of electrons.

From this standpoint, let a hole be regarded as a quasi particle with a positive charge $(e > 0)$ and positive effective mass $(m^* > 0)$.

We again recall that the negative effective mass of quasi-particles located on constant-energy surfaces of hole type is a consequence of expansion of their energy into a series over momenta near the points $\vec{p} = \vec{p}_0$ in which the maximum values of the energy ε are observed in the band. For instance, for the simplest case of a spherical constant-energy surface we have near the maximum point

$$\varepsilon = \varepsilon_0 + \frac{1}{2} \cdot \frac{\partial^2 \varepsilon}{\partial p'^2} p'^2 = \varepsilon_0 + \frac{p'^2}{2m^*}, \quad \text{where} \quad \vec{p}' = \vec{p} - \vec{p}_0$$

and

$$(m^*)^{-1} = \frac{\partial^2 \varepsilon}{\partial p'^2} \quad \text{at } \vec{p}' = 0.$$

Since $\varepsilon < \varepsilon_0$ and $\varepsilon - \varepsilon_0 < 0$ near a maximum of energy, then m^* is also negative. This only means in essence that the really existing electrons in a metal lattice are under such conditions at which their energy reduces with removing from the point $\vec{p}_0$.

If, however, we introduce imaginary particles with $m^* > 0$ (and $e > 0$) to describe such electrons, then we have to change the direction of energy counting and the value $\varepsilon - \varepsilon_0$ must be positive. This in turn means that with these new particles $(m^* > 0, e > 0)$ their energy in point $\vec{p}_0$ is at the minimum and increases with moving away from this point. Since filling of $\vec{p}$-space with quasi-particles always begins in the points where the energy of particles is at the minimum, then we have now to assume that the region of $\vec{p}$-space that has been empty, is now filled with imaginary quasi-particles with $m^* > 0$ and $e > 0$.

A hole-type constant-energy surface then becomes entirely similar in its properties to an electron-type surface to an accuracy of the sign of charge of quasi-particles. In other words, quasi-particles with $m^* > 0$ and $e > 0$ fill a constant-energy surface of the hole type, while quasi-particles with $m^* > 0$, $e < 0$ (electrons) fill a surface of the electron type.

Thus, the transition from one concept of a hole to the other consists not only in the change of sign of charge and mass, but also in a different nature of filling of $\vec{p}$-space with quasi-particles: particles of negative mass are located beyond the corresponding constant-energy surface which in this case bounds an empty part of the space, whereas particles of positive mass are located within the constant-energy surface.

It can be easily shown that holes, if regarded as particles with $m^* > 0$, $e > 0$, also give a contribution to the electric conductivity which is additive to that of electrons ($m^* > 0$, $e < 0$). Let a surface similar to that shown in Fig. 155 be filled with quasi-particles with $m^* > 0$, $e > 0$. As before, the electric field $\vec{E}$, is directed along the $+\vec{p}_x$ axis. The force acting on particles is then equal to $+|e|\vec{E}$ and coincides in direction with the electric field.

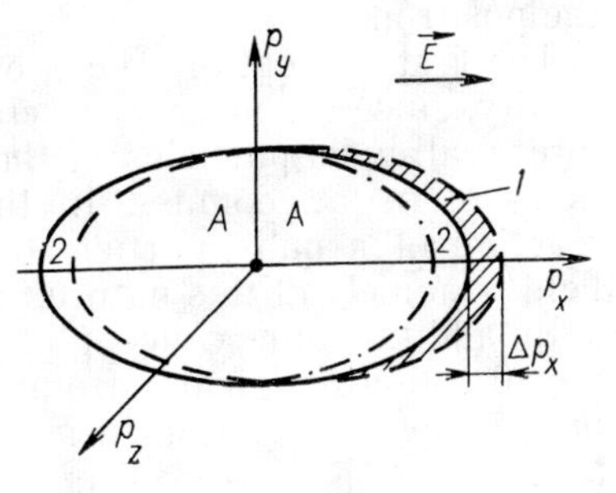

Fig. 159

Under the action of the electric field the constant-energy surface will be displaced by a distance $\Delta p_x = |e| E_x \tau_x$ in the direction $+\vec{p}_x$ (Fig. 159). Particles with $m^* > 0$, $e > 0$, having a non-compensated component of momenta along the $+\vec{p}_x$ axis, will then appear in volume *1*. Simultaneously, particles whose motion was compensating the current of particles in volume *3* will vanish in volume *2*.

Thus, the resultant current is determined by the particles in volumes *1* and *3*. We now note that at $m^* > 0$ the directions of momentum and velocity of a particle are coincident. The current of particles therefore corresponds to the transition of a positive electric charge in the direction of the electric field $\vec{E}$, which is equivalent to the transition of a negative charge in the opposite direction, i.e. to the current of electrons.

The conclusion we have come to is of deep sense. The sign of the effective mass of a particle, irrespective of the sign of its charge, is only determined by the sign of the curvature of the constant-energy surface relative to the volume filled with charge carriers. If a closed surface is empty then it has a negative curvature relative to the external filled volume and the particles on that surface possess a negative effective mass. If however a closed surface is filled, then its curvature is positive relative to the internal volume and the particles filling it, of a charge of any sign, have a positive effective mass.

As regards the contribution of particles to various physical phenomena, it is entirely equivalent whether we consider a particle on a filled surface with $m^* > 0$ or on an empty one, but with $m^* < 0$, provided we have changed the sign of charge at the transition from $m^* > 0$ to $m^* < 0$. In that sense, negatively charged quasi-particles on electron-type (filled) surfaces (i.e. simple electrons with $m^* > 0$, $e < 0$) can also be replaced by quasi-particles with $m^* < 0$, $e > 0$, but located on a hole-type surface, i.e. filling the space around electron-type surface that has been empty before.

Considering this analysis, the transition from the concept of a hole with $m^* < 0, e < 0$ to that of a hole with $m^* > 0, e > 0$ is the transition from a real hole to a certain equivalent analogue of the positron.

The first concept of a hole, evidently, corresponds to the physical reality, since the maxima of energy really exist in bands, and therefore, real quasi-particles in the vicinity of maxima behave as particles of negative mass. In this region of momentum space, the reaction of a quasi-particle on the action of an external electric field is namely of this nature.

In addition, the concept of a hole as a particle with $m^* < 0$, $e < 0$ follows logically from the general concept of construction of the energy spectrum in metals. The concept of a hole as a particle with $m^* > 0$, $e > 0$ is, however, a conditional one and holds true only for closed constant-energy surfaces surrounding the points of maxima of energy.

Namely for that reason the second concept of a hole has been firmly implanted in the physics of semiconductors, which deals only with small closed constant-energy surfaces.

Note that the use of particles of positive effective mass is much more convenient in practice. Indeed, regarding a particle assumed to have a negative effective mass, we have always to operate with a constant-energy surface surrounding an empty space, since these two concepts are inseparable. A particle of a positive effective mass is however similar qualitatively to a particle in free space moving under the action of external fields. In some cases this particle may therefore be separated from the corresponding constant-energy surface and regarded as an isolated particle in the field of external forces.

For that reason, in order to visualize various phenomena, we shall later regard a hole as a particle with $m^* > 0$ and $e > 0$.

Let us derive the formula of electric conductivity. For this, we write down expression for the vector of density of electric current $\vec{j}$.

As has been shown earlier, both with electron-type and hole-type constant-energy surfaces, an electric current is formed by Fermi quasi-particles present in volumes *1* and *3* (see Figs. 156 and 158).

Let us recall that for simplicity we consider constant-energy surfaces close to ellipsoidal, with the electric field $\vec{E}$ directed along one of the main axes $\vec{p}_x$. Because of this the current density vector $\vec{j}$ also coincides in its direction with the $\vec{p}_x$ axis. Vector $\vec{j}$ can be written as the following sum:

$$\vec{j} = -|e| \sum (v_F)_{\vec{p}_x} \tag{3.6}$$

where $(v_F)_{\vec{p}_x}$ is the projection of Fermi velocity of a quasi-particle on the $\vec{p}_x$ axis, and the summation is done for all particles present in volumes *1* and *3* of $\vec{p}$-space in Figs. 156 and 158. We have obtained an analytical expression for current density $\vec{j}$ and electric

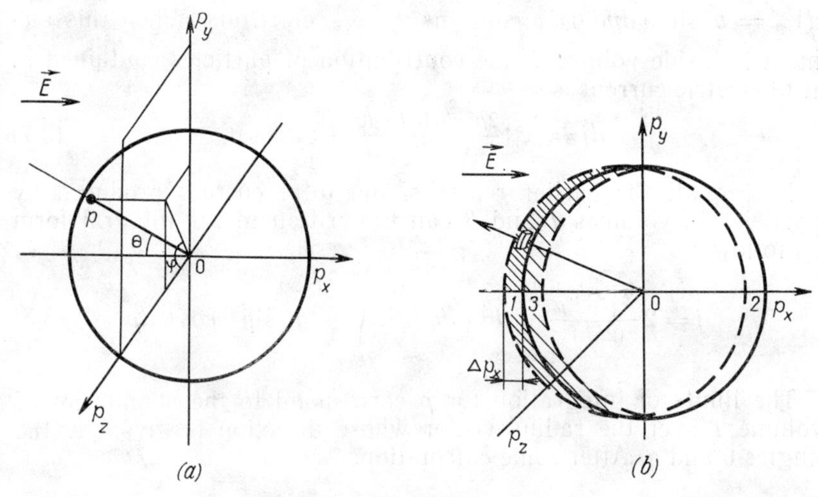

Fig. 160

conductivity σ for a particular case of a spherical constant-energy surface. This simplest case, to which the scalar quantity σ corresponds, makes it possible to establish the parameters on which the electric conductivity of a metal depends.

For definiteness, let us consider a spherical constant-energy surface of the electron type in an external electric field $\vec{E}$. We introduce a spherical system of coordinates whose polar axis is directed opposite to vector $\vec{E}$ (Fig. 160*a*).

Let quasi-particles on the surface be characterized by the isotropic relaxation time τ. Under the action of the electric field $\vec{E}$ the surface is displaced a distance $\Delta p = |e|E\tau$ in the direction of the polar axis giving rise to an electric current which is determined by quasi-particles in volumes *1* and *3* (Fig. 160*b*).

Since $\Delta p \ll p_F$, the velocity of any particle in volumes *1* and *3* is directed along the radius vector $\vec{p}$ and practically coincides in magnitude with the Fermi velocity v_F.

The expression for current density (3.6) includes the projection of velocity of particles onto the direction of the vector $-\vec{E}$, equal to $v_F \cos\theta$ in the given case.

At $\Delta p \ll p_F$, the volumes *1* and *3* are equal to one another to an accuracy of the values of the second order of Δp, so that it is sufficient to calculate the current formed by particles in volume *1* and then to double the value obtained.

According to formula (2.57), an element of spherical volume $dV_p = p^2 \sin\theta\, dp\, d\theta\, d\varphi$ contains $\frac{2\, dV_p}{(2\pi\hbar)^3}$ electrons. When this element is inside volume *1*, the contribution of particles contained in it to electric current is

$$dj = |e| \frac{2p^2 \sin\theta\, dp\, d\theta\, d\varphi}{(2\pi\hbar)^3} v_F \cos\theta \tag{3.7}$$

Noting all the earlier remarks, the total current produced by particles in volumes *1* and *3* can be written in the integral form as follows:

$$j = 2\frac{2|e|v_F}{(2\pi\hbar)^2} \int_0^{\pi/2} d\theta \int_0^{2\pi} d\varphi \int_{p_F}^{p_F+\Delta p\cos\theta} p^2 \sin\theta \cos\theta\, dp \tag{3.8}$$

The limits of integration for p correspond to the boundaries of volume *1* over the radius vector whose direction is given by the angles θ and φ. After some calculations, we have

$$j = \frac{8\pi |e| v_F p_F^2 \Delta p}{3(2\pi\hbar)^3} \tag{3.9}$$

The quantity $4\pi p_F^2$ equals the area S_F of the constant-energy surface of the given group of current carriers. Replacing Δp by $|e|E\frac{l}{v_F}$, expression (3.9) is transformed as follows:

$$j = \frac{2}{3} \cdot \frac{|e|^2 S_F l}{(2\pi\hbar)^3} E \tag{3.10}$$

Comparison with Ohm's law $j = \sigma E$ gives the electric conductivity σ related to the given group of current carriers as follows:

$$\sigma = \frac{2}{3} \cdot \frac{|e| S_F l}{(2\pi\hbar)^3} \tag{3.11}$$

Expression (3.11) for electric conductivity has been first derived by Lifshits [80], [13] and is called Lifshits' formula. It is valid for any arbitrary law of dispersion at which the dependence of the energy of electrons on momentum is of the general form $\varepsilon = f(p^2)$, where $f(p^2)$ is a continuous function of its argument. The constant-energy surfaces corresponding to this dispersion law are concentric spheres with the centre at point $\vec{p} = 0$.

The expression $\varepsilon = \frac{p^2}{2m}$ for the quadratic isotropic law of dispersion is a particular case of the law $\varepsilon = f(p^2)$. Another particular case, for instance, is a linear law of dispersion $\varepsilon = \pm \alpha \sqrt{p^2}$, $\alpha = \text{const}$, for which the point $\vec{p} = 0$ is conical.

It is of interest to compare Lifshits' formula for the quadratic isotropic dispersion law with the formula of electric conductivity obtained in the Drude-Lorentz theory on the basis of the model of free electrons.

Let us recall that according to the Drude-Lorentz theory a metal is regarded as a potential box filled with the gas of free electrons that obey Boltzmann's statistics. In an electric field $\vec{E}$, each electron acquires an additional velocity $\Delta v = \frac{|e| E\tau}{m_0}$ during the relaxation time τ in a direction opposite to $\vec{E}$. The quanity Δv has been termed the drift velocity, since the whole ensemble of electrons drifts with this velocity under the action of the electric field. The density of the electric current produced by drift of electrons is

$$j = |e| n \Delta v(E) \tag{3.12}$$

where n is the concentration of electrons. Substituting $\Delta v(E)$ into (3.12), we can re-write the expression for j as follows:

$$j = \frac{|e|^2 n\tau}{m_0} E \tag{3.13}$$

whence we get the Drude-Lorentz formula for electric conductivity:

$$\sigma = \frac{|e|^2 n\tau}{m_0} \tag{3.14}$$

The Lifshits formula will now be transformed by using the expression for the quadratic isotropic dispersion law $\varepsilon = \frac{p^2}{2m^*}$. To do this, we express the area S_F of the Fermi surface through the volume Δ_F it bounds in $\vec{p}$-space and the magnitude of Fermi momentum p_F: $S_F = 3\frac{\Delta_F}{p_F}$. $\left(\text{Indeed, } S_F = 4\pi p_F^2,\ \Delta_F = \frac{4}{3}\pi p_F^2\right)$. Substituting this quantity into (3.11) and noting that the relationship $p_F = m^* v_F$ holds true for the quadratic isotropic dispersion law, we obtain

$$\sigma = \frac{|e|^2 2\Delta_F}{m^* (2\pi\hbar)^3} \cdot \frac{1}{v_F} = \frac{|e|^2 \tau}{m^*} \cdot \frac{2\Delta_F}{(2\pi\hbar)^3} \tag{3.15}$$

According to (2.57), $\frac{2\Delta_F}{(2\pi\hbar)^3}$ is the concentration n of electrons in the given group, including both the electrons located on the Fermi surface and those in the volume bounded by that surface. Then the Lifshits formula (3.11) for the quadratic isotropic dispersion law can be written as

$$\sigma = \frac{|e|^2 n\tau}{m^*} \tag{3.16}$$

which coincides in the form with the Drude-Lorentz formula (3.14).

The process of conduction in this particular case may be thought of as if all the electrons in $\vec{p}$-space were accelerated in the electric field, and not only those located on the Fermi surface. For the quadratic dispersion law, the product of the small number of electrons located in volumes *1* and *3* near the Fermi surface (i.e. those really contributing to electric conductivity) by the Fermi velocity is equal to the product of the total number of the electrons present in the volume of the sphere by the small addition of drift velocity $\Delta v(E)$.

In other words, in the case considered the small number of fast Fermi electrons produces the same current as would be produced by all the electrons inside the sphere which drift slowly under the action of the electric field.

Thus, having considered electric conductivity for the isotropic case, we can conclude that σ is mainly determined by the area of Fermi surface S_F and the free-path length of current carriers.

With a closed anisotropic Fermi surface the current density vector $\vec{j}$ in the general case does not coincide in direction with the electric field $\vec{E}$. Electric conductivity then becomes a second-rank tensor σ_{ij} $(i, j = 1, 2, 3)$, but the diagonal components of tensor σ_{ij} for the main axes of Fermi surface p_x, p_y, p_z are again determined by the area S_F of the surface and the free-path lengths l_x, l_y and l_z along the corresponding axes.

Let us discuss some particular examples of the effect of the shape of the Fermi surface on electric conductivity of metals.

1. Fermi surface is a cylinder located along the $\vec{p}_x$ axis. A cylindrical Fermi surface is an infinite (open) surface in the direction $\vec{p}_x$ and passes in this direction through all cells of the reciprocal space.

Such a surface can be displaced under the action of an electric field only in the direction of the $\vec{p}_y$ axis. Let the electric field $\vec{E}$ be directed, for instance, along the axis $-\vec{p}_y$. The displacement of the surface of the cylinder along the $+\vec{p}_y$ axis by a distance $\Delta p_y = |e| E\tau_y$ gives rise to an electric current which is formed by electrons in the doubled volume *1* (Fig. 161).

The component of velocity of electrons along the $\vec{p}_x$ axis on the surface of the cylinder is absent, since the velocity $v_g = \frac{\partial \varepsilon}{\partial \vec{p}}$ of a

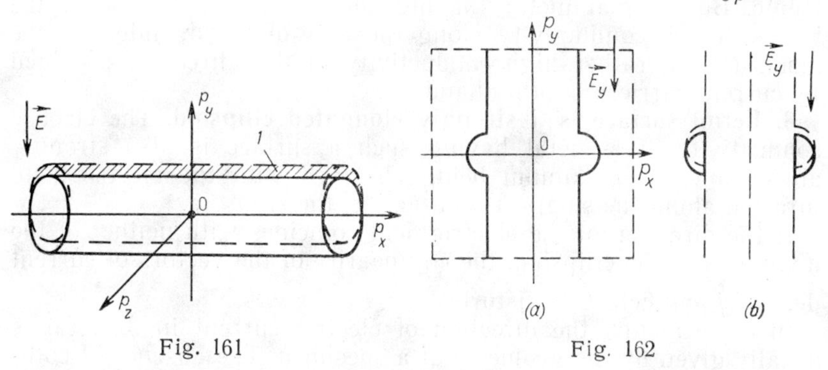

Fig. 161

Fig. 162

quasi-particle should be orthogonal to the Fermi surface and the surface itself be parallel to the direction $\vec{p}_x$. With a cylindrical Fermi surface the motion of electrons along the axis of the cylinder is therefore forbidden. In this direction of the electric field the energy band behaves as a filled one.

With an arbitrary direction of electric field $\vec{E}$ relative to the cylindrical Fermi surface, electric current can only be produced by the component E_y of the field along the $\vec{p}_y$ axis, which is equal to $E \cos \varphi$, where φ is the angle between $\vec{E}$ and $\vec{p}_y$. The dependence of electric conductivity of such a metal on angle φ is as follows:

$$\sigma(\varphi) = \sigma_{max} \cos \varphi \tag{3.17}$$

where σ_{max} is the maximum electric conductivity that corresponds to an electric field E directed perpendicular to the axis of the cylindrical Fermi surface.

2. Fermi surface is a weakly warped cylinder located along the $\vec{p}_y$ axis (Fig. 162*a*). In that case the cylindrical portions of the surface which are parallel to the $\vec{p}_y$ axis do not contribute to the electric conductivity in the direction of this axis (i.e. to the conductivity at $E = E_y$). The electric conductivity in the direction $\vec{p}_y$ is only determined by the electrons located on the crimp. Its magnitude coincides with the electric conductivity of a metal whose Fermi surface has the form of a "ring" (Fig. 162*b*) of the same shape as the crimp.

The conductivity in the direction perpendicular to the axis of the warped cylinder is, as in the former case, rather high. Hence it follows that the electric conductivity of a metal having the Fermi surface in the form of a warped cylinder is strongly anisotropic. But in real metals having such a surface in one of the bands, a low conductivity along the axis of the cylinder can be masked by a rather high conductivity in this direction provided by current carriers in other bands.

3. Fermi surface is a strongly elongated ellipsoid. The electric conductivity of a metal having such a surface is also strongly anisotropic, its maximum being observed with the electric field directed along the smallest semi-axis of the ellipsoid.

If the direction of the electric field coincides with neither of the main axis of the ellipsoid, the collinearity of the vectors of current density $\vec{j}$ and field $\vec{E}$ is disturbed.

In experiments, the direction of electric current in a metal is usually given by the geometry of a specimen. The absence of collinearity between $\vec{j}$ and $\vec{E}$ implies that with a current flowing in an anisotropic specimen, an electric field in a transverse direction, similar to the electric field in Hall effect, is formed in addition to the electric field along the specimen, i.e. to the voltage drop on it.

In a metal having the Fermi surface in the form of a strongly elongated ellipsoid the nature of dependence of conductivity on field orientation is close to that in a metal with the Fermi surface in the form of a slightly warped cylinder.

3-3. TEMPERATURE DEPENDENCE OF ELECTRIC CONDUCTIVITY

Let us discuss the dependence of electric conductivity on temperature for a metal with an isotropic spectrum of current carriers. In Lifshits' formula for electric conductivity (3.11), which is applicable in this case, only the free-path length l of carriers is temperature-dependent. The temperature dependence of this quantity in an isotropic metal therefore completely determines the temperature dependence for electric conductivity.

In a model of a metal where electrons on the Fermi surface are regarded as an ideal gas of quasi-particles, the free-path length l is inversely proportional to the number of acts of scattering N_{sc}.

In the phenomenon of electric conductivity, an act of scattering is thought of as a process causing an electron to depart from the number of Fermi particles contained in volumes *1* and *3* (Fig. 160*b*). Departure from these volumes can occur owing to a change of the momentum of a quasi-particle either in magnitude [changing the energy of the particle $\varepsilon = f(p^2)$] or in direction.

Scattering of electrons is caused by their interaction with oscillations of the lattice (phonons), with other electrons, ionized or neutral impurities, lattice defects, etc. For Fermi electrons, the most probable are processes in which the change of energy is small compared with the Fermi energy. At scattering, an electron practically always remains on the Fermi surface (or in direct vicinity of it). In other words, scattering of Fermi electrons is in most cases a process close to a constant-energy process when departure from volumes *1* and *3* is only equivalent to the rotation of the momentum vector in $\vec{p}$-space or to transition into volume *2* (Fig. 160*b*).

In adequately pure and perfect monocrystalline specimens in the region of not very low temperatures compared with Debye temperature Θ_D the main process of scattering which determines electric conductivity is the scattering of electrons on phonons. Let us recall that the energy of phonon ε_{ph} at temperature T practically does not exceed kT. On the other hand, the fact that an electron gas in a metal is degenerated at any temperature corresponds to the inequality $kT \ll \varepsilon_F$ or $\varepsilon_{ph} \ll \varepsilon_F$. It then follows that interaction of an electron and phonon in a metal is accompanied by only a slight variation of the energy of the electron. The momentum of the electron in this process practically remains the same in magnitude and can only be varied through rotation in phase space.

Note, however, that not every act of electron-phonon interaction can result in scattering, since at a small change of momentum the electron remains within the volumes *1* and *3* and continues to contribute to electric current. In other words, a single interaction of an electron and a phonon, which has a small momentum, is not identical to the scattering in the process of electric conductivity. Scattering can be caused by either a single act (at a large change of momentum) or many successive acts of interaction if the total variation of the momentum is such that the electron passes from volumes *1* and *3* to volume *2* (Fig. 160*b*).

Owing to numerous acts of electron scattering, a dynamic equilibrium is established between the displacement of the Fermi surface in the electric field, whose action results in a continuous replenishment of volumes *1* and *3*, and the transfer of electrons back into volume *2* so that volumes *1* and *3* are exhausted. With no scattering, the Fermi surface would be displaced periodically in the electric field by a magnitude of the order of the Fermi momentum, which would produce an alternating current of a high amplitude. It should be recalled that the cause of periodic motion of an electron in a metal under the action of a constant electric field and the nature of this motion have been discussed in Sec. **2-4**, Chapter Two.

We now shall find the number of individual acts of interaction of an electron at various temperatures.

Let us start from the region of high temperatures $T > \Theta_p$.

As has been shown in Chapter One, the whole spectrum of thermal vibrations of the lattice is excited in this region. The number of phonons of each frequency is proportional to T and their distribution over frequencies ω (or energies $\varepsilon_{ph} = \hbar\omega$) is such that phonons with an energy $\varepsilon_{ph} \sim k\Theta_D$ prevail in the total acoustic spectrum. The energy of optical phonons is of the same order of magnitude.

Thus, at high temperatures $T > \Theta_D$, the phonon gas can be described adequately by means of Einstein model with the characteristic frequency ω_E which is close to the limit frequency of vibrations of the lattice. In other words, the concept of phonon gas is applicable for this region of temperatures, provided that the energy of an individual phonon ε_{ph} is assumed to be $k\Theta_D$ and the number of phonons increases proportional to T.

The total thermal energy of the lattice E_{lat}^{III} is then also proportional to T, which is equivalent to a constant value of specific heat $C_v = \frac{\partial E_{lat}^{\mathrm{III}}}{\partial T}$ at $T > \Theta_D$ in accordance with the Dulong-Petit law.

Let us now estimate the momentum of phonons. As has been mentioned, the spectral density of acoustic phonons has maxima at oscillation frequencies close to the limiting ones. The corresponding momenta of phonons attain the order of magnitude of the boundary momentum. Thus, phonons with a momentum p_{ph}, equal approximately to $\frac{k\Theta_D}{v_{son}^{\flat}}$, are predominant in the spectrum of acoustic oscillations at $T > \Theta_D$.

The magnitude of p_{ph} can be estimated by means of expression (1.30) for Θ_D. Then we have: $p_{ph} \sim \frac{k\Theta_D}{v_{son}^{\flat}} \approx \frac{2\hbar}{a}$. It may be shown that the momenta of the phonons that prevail in the optical band of oscillations are also of the order of magnitude of $\hbar/a$.

The momentum of an electron p_{el} on the Fermi surface is close in magnitude to $\pi\hbar/a$. Consequently, for the region of high temperatures $T > \Theta_D$, the momentum of a phonon p_{ph} is practically of the same order of magnitude as that of an electron, p_{el}. This means that at every act of electron-phonon interaction the momentum of the electron can vary by a magnitude of the order of the initial momentum, i.e. scattering of the electron occurs practically at each interaction with a phonon.

The number of scattering centres N_{sc}, which in this case coincides with the number of phonons, increases proportional to temperature. It then follows that at $T > \Theta_D$ the free-path length l of electrons, determined by N_{sc}, is inversely proportional to temperature.

Finally, using Lifshits' formula (3.11), we find the temperature dependence for electric conductivity in the region of high temperatures to be as follows:

$$\sigma \sim \frac{1}{T} \tag{3.18}$$

At $T > \Theta_D$, the resistivity $\rho = 1/\sigma$ of a metal increases as T.

Let us now consider the region of low temperatures $T \ll \Theta_D$. Here, only acoustic phonons with an energy $\varepsilon_{ph} \lesssim kT$ are excited in the oscillation spectrum. The momentum of phonons p_{ph} in its order of magnitude is equal to kT/v^0_{son}. Since the momentum of an electron on the Fermi surface $p_{el} \approx \pi\hbar/a$ is close to $k\Theta_D/v^0_{son}$, the ratio p_{ph}/p_{el} is of the same order as T/Θ_D. In other words, at low temperatures $T \ll \Theta_D$ the inequality $p_{ph} \ll p_{el}$ holds true. It then follows that a single act of electron-phonon interaction gives no scattering. The process of scattering can only be caused by a series of successive interactions.

Let us estimate the number of interactions required to scatter an electron at $T \ll \Theta_D$. Let the momentum $\vec{p}_{el}$ of the electron be rotated through a small angle α and become equal to $\vec{p}'_{el}$ through an interaction with a phonon. The corresponding triangle of momenta is shown in Fig. 163. Since $p_{ph} \ll p_{el}$ and $\varepsilon_{ph} \ll \varepsilon_{el}$, the magnitude of the momentum $\vec{p}_{el}$ remains practically the same $|\vec{p}'_{el}| \approx |\vec{p}_{el}|$, but the electron now contributes less to electric conductivity because the projection $\vec{p}'_{el}$ onto the initial direction $\vec{p}_{el}$ has been reduced by Δp_{el} (see Fig. 163):

$$\Delta p_{el} = p_{el} - p'_{el} \cos\alpha \approx p_{el}(1 - \cos\alpha) \tag{3.19}$$

The angle α is expressed through the momenta $\vec{p}_{ph}$ and $\vec{p}_{el}$:

$$\sin\alpha = p_{ph}/p_{el} \ll 1$$

whence it follows that

$$\sin\alpha \approx \alpha \approx p_{ph}/p_{el} \quad \text{and} \quad 1 - \cos\alpha \approx \alpha^2/2 = p^2_{ph}/2p^2_{el}$$

Substituting T/Θ_D for p_{ph}/p_{el}, we obtain for Δp_{el}:

$$\Delta p_{el} \approx \frac{1}{2} p_{el} \left(\frac{T}{\Theta_D}\right)^2 \qquad (3.20)$$

Owing to scattering, the momentum $\vec{p}_{el}$ of the electron should vary by a magnitude of the order of the momentum itself. With the constant-energy scattering, this is equivalent to the change of its projection in the initial direction also by a magnitude of the order of p_{el}. The number of acts of interaction giving rise to this process, equal approximately to the ratio $p_{el}/\Delta p_{el}$, increases sharply as the temperature is reduced proportional to $(\Theta_D/T)^2$. Essentially the inverse quantity $(T/\Theta_D)^2$ determines the effectiveness of a single act of electron-phonon interaction with scattering at $T \ll \Theta_D$.

Let us now discuss the temperature relationship for the free-path length l at low temperatures, assuming, as before, that the main mechanism of relaxation is the scattering of electrons on phonons.

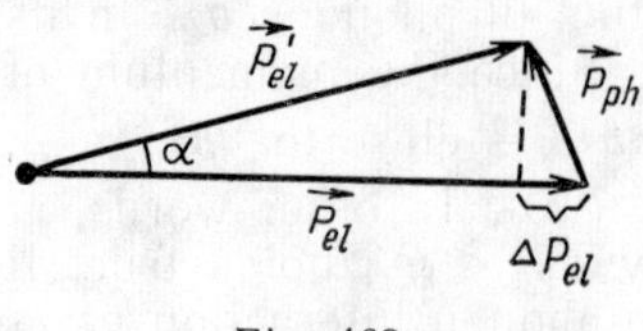

Fig. 163

The free-path length is determined, on the one hand, by the number of interaction centres N_{in}, which coincides with the number of phonons, and on the other hand, by the effectiveness of each centre, which is proportional to $(T/\Theta_D)^2$. As is known, the number of phonons in a three-dimensional lattice depends on temperature at $T \ll \Theta_D$ as T^3. Hence N_{in} is also proportional to T^3. With electrons being scattered on phonons, the equivalent number of scattering centres is

$$N_{sc} \sim \left(\frac{T}{\Theta_D}\right)^2 N_{in} \qquad (3.21)$$

This relationship is a consequence of the fact that only a large number of successive interactions, of the order of $(\Theta_D/T)^2$, can produce scattering. This is equivalent to a reduction of the number of scattering centres N_{sc} by a factor of $(\Theta_D/T)^2$ compared with the number of centres of interaction N_{in}. Since $N_{in} \sim T^3$, it then follows from (3.21) that $N_{sc} \sim T^5$ and $l \sim 1/N_{sc} \sim T^{-5}$. Thus, for scattering of electrons on phonons, the temperature dependence for electric conductivity at $T \ll \Theta_D$ is governed by the relationship

$$\sigma \sim T^{-5} \qquad (3.22)$$

The equivalent expression for resistivity $\rho \sim T^5$ is called Bloch's law. The exponent in this relationship is related to the number of dimensions of the lattice. For instance, with a two-dimensional lattice the number of phonons is proportional to T^2, and therefore, ρ should obey the law $\rho \sim T^4$.

If the electric resistance of a metal were only determined by the scattering of electrons on phonons, then it would be practically zero at very low temperatures ($T \rightarrow 0°$ K). Such a conclusion would be valid for a metal with an ideal crystal lattice having no foreign atoms. Actually there always are various disturbances of the periodicity of a lattice (dislocations, vacant places, impurity atoms, etc.). It results in that electrons in a real metal lattice are subject to other kinds of scattering apart from being scattered on phonons.

Note that with the scattering on impurities and lattice defects, the free-path length is determined by the mean distance between impurities and by the velocity of electrons, being only slightly dependent on temperature. For that reason the resistance of a real metal at very low temperatures, when the phonon mechanism of scattering (the so-called residual resistance ρ_0) can be neglected, is determined only by the concentration c and the nature of the additional centres of scattering.

With a low concentration c of impurities and defects in a metal, the resistance ρ_{met} can be represented as a sum of two terms

$$\rho_{met} = \rho_0(c) + \rho_{id}(T) \tag{3.23}$$

where $\rho_{id}(T)$ is the temperature-dependent resistance of the ideal metal and $\rho_0(c)$ is the residual resistance depending only on concentration c. This constitutes Matthiessen's empirical rule [69] which is based on that the resistance owing to scattering on additional centres is additive and independent of (or only weakly dependent on) temperature.

This concept corresponds to the model of a metal having an ideal crystal lattice within which relatively scarce additional centres of scattering are rigidly fixed. The magnitude of residual resistance of a metal may therefore serve as a criterion of the purity of the metal and the perfection of its crystal lattice.

The purity of monocrystals is estimated in practice by the ratio of the resistance at room temperature to that at the boiling point of liquid helium ($T = 4.2°$ K), $\rho_{300° K}/\rho_{4.2° K}$. For crystals of high purity this ratio can attain a value of an order of 10^3 and higher.

The temperature dependence of resistance of a metal for which Matthiessen's rule holds is shown in Fig. 164.

The resistance of an ideal lattice $\rho_{id}(T)$ at any temperatures has been calculated by Bloch [70] and Grüneisen [71] for an idealized model based on the following assumptions:

(1) the electron gas in the lattice is regarded as an ideal one,

(2) description of thermal vibrations is made by using Debye's model of phonon spectrum, and

(3) the radius r_D of Debye screening of the Coulomb potential of ions is taken equal to zero (assuming the screening of the electrostatic potential to be complete).

Under these assumptions the following formula has been obtained for the resistance $\rho_{id}(T)$ of an ideal lattice:

$$\rho_{id}(T) = \frac{KT^5}{M\Theta_D^6} \int_0^{\Theta_D/T_i} \frac{z^5\, dz}{(e^z - 1)(1 - e^{-z})} \tag{3.24}$$

where M is the atomic weight of the metal and K is a constant related to the specific volume. For the regions of low ($T < \Theta_D$) and high ($T > \Theta_D$) temperatures, formula (3.24) gives

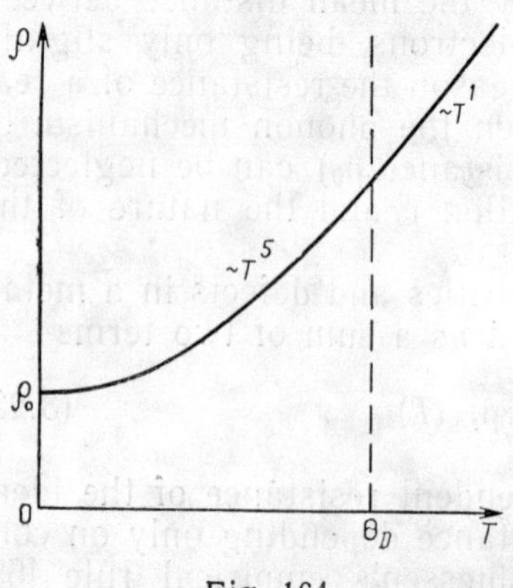

Fig. 164

$$\rho_{id}(T) \approx 124.4 \frac{K}{M\Theta_D^6} T^5$$

$$\text{at } T < 0.1\Theta_D \tag{3.25}$$

$$\rho_{id}(T) \approx \frac{K}{4M\Theta_D^2} T$$

$$\text{at } T > 0.5\Theta_D$$

It follows from formulae (3.25) that the ratio of the resistance of an ideal metal at a low temperature T_1 (where $T_1 < 0.1\Theta_D$) to that at a high temperature $T_2 (T_2 > 0.5\Theta_D)$ is

$$\frac{\rho_{id}(T_1)}{\rho_{id}(T_2)} = \frac{497.6}{\Theta_D^4} \cdot \frac{T_1^5}{T_2} \tag{3.26}$$

This relationship is determined by only one parameter, i.e. Debye temperature Θ_D. Then for a metal for which the temperature dependence of resistance is close to the idealized one, the quantity $G = \frac{\rho(T_1)}{\rho(T_2)} \cdot \frac{T_2}{T_1^5}$ must be independent of temperature. The departure of G from its constant value characterizes imperfection of a metal from the standpoint of Bloch-Grüneisen's model.

Bloch-Grüneisen formula holds well for simple metals at $T > \Theta_D$. In this region of temperatures the ratio ρ/T is practically constant. Using the measured value of resistivity of a metal at a certain temperature $T' > \Theta_D$ we can calculate the parameter Θ_D in formulae (3.24) and (3.25). This parameter, relating to the region of high temperatures, is a constant quantity which is termed the limiting value of Debye temperature Θ_D^{lim}. The same parameter Θ_D found by Bloch-Grüneisen formula, differs noticeably from Θ_D^{lim} at lower temperatures.

For one-zone metals for which the Fermi surface is almost spherical (i.e. alkali metals), the ratio Θ_D/Θ_D^{lim} remains practically constant up to the region of low temperatures. For sodium, for instance, this ratio Θ_D/Θ_D^{lim} remains constant to an accuracy within 5 per cent down to the temperature $T = 0.1\Theta_D$. It then follows that anisotropy of the electron spectrum has a substantial effect on variation of the parameter Θ_D. For metals with a high anisotropy of the Fermi surface Bloch-Grüneisen formula (3.24) is obviously inapplicable in the region of low temperatures $T \ll \Theta_D$. This is linked with large differences of momenta of electrons for different directions in the lattice, the result being that the effectiveness of the electron-phonon interaction at low temperatures depends substantially on direction.

Note that the Debye temperature Θ_D found from the electric conductivity for most metals differs from the value of this parameter determined through specific heat. This difference is linked with that longitudinally polarized phonons are most essential for the scattering at the flow of an electric current, whereas specific heat of the lattice is due to equal contribution from oscillations of all types of polarization.

Among the assumptions underlying Bloch-Grüneisen's model the one least justified is that on full screening of the electrostatic potential of an ion, which is equivalent to assuming that $r_D = 0$. Houston [72] succeeded to show that using the potential of an individual ion in the form of (2.2) with a finite r_D, it is possible to select the radius of screening such that the parameter Θ_D in the formula for temperature dependence of resistivity be a constant Θ_D^{lim} practically within the whole range of temperatures.

It is essential that the value of r_D required for this is only one-third to one-fifth of the radius of ion r_i. Houston's result is of interest for verifying the validity of OPW-approximation of the wave function of an electron in metal. This result shows indirectly that valence electrons moving in a metal actually penetrate ion-cores. Scattering of electrons on a Coulomb potential occurs only in the direct vicinity of the centre of an ion at distances of (0.2-0.3) r_i.

To conclude this section, it will be noted that the scattering of electrons on phonons at low temperatures is actually somewhat more effective than in the simplified model considered, the result being that the resistance of a real metal obeys the law $\rho \sim T^m$, where the exponent $m < 5$. Generally speaking, the reduction of resistances with temperature occurs in common simple metals not so rapidly as would be expected from Bloch-Grüneisen's formula (3.24).

3-4. ELECTRIC CONDUCTIVITY OF METALS IN A MAGNETIC FIELD. GENERAL CONSIDERATIONS

A magnetic field H varies the nature of motion of current carriers in a metal thus changing its electric conductivity. Our discussion of galvano-magnetic phenomena in metals will be restricted here when the magnetic field is directed perpendicular to electric current. With the vectors $\vec{j}$ and $\vec{H}$ directed so, the resistance of the metal is termed the transverse magnetoresistance.

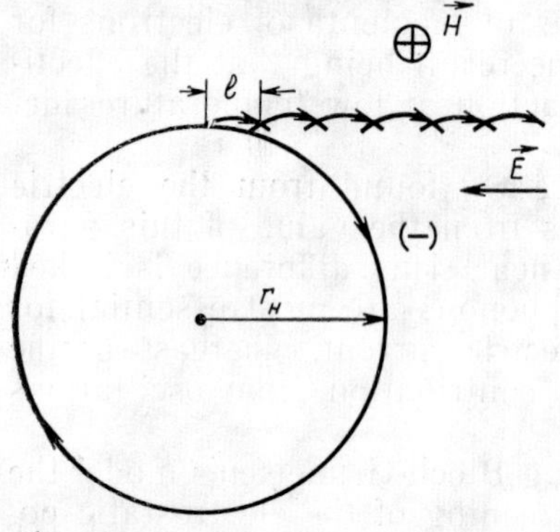

Fig. 165

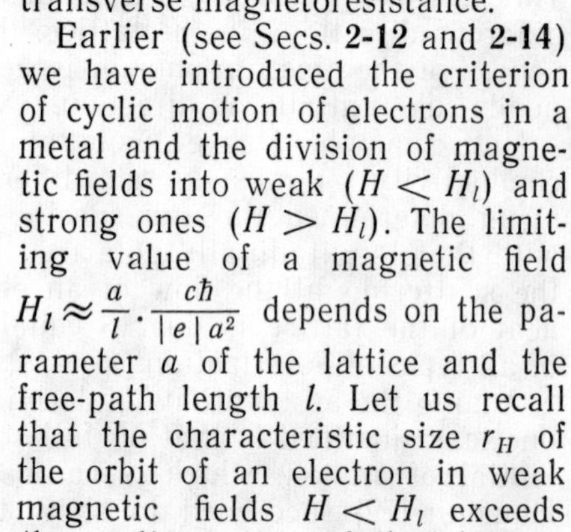

Earlier (see Secs. **2-12** and **2-14**) we have introduced the criterion of cyclic motion of electrons in a metal and the division of magnetic fields into weak ($H < H_l$) and strong ones ($H > H_l$). The limiting value of a magnetic field $H_l \approx \frac{a}{l} \cdot \frac{c\hbar}{|e| a^2}$ depends on the parameter a of the lattice and the free-path length l. Let us recall that the characteristic size r_H of the orbit of an electron in weak magnetic fields $H < H_l$ exceeds the free-path length l so that the cyclic motion of the electron is not realized, the electron moving between two successive acts of scattering along a rather short arc of the path. For that reason its motion under the action of an electric field $\vec{E}$ occurs practically along a straight line, as in the absence of magnetic field (Fig. 165).

In strong magnetic fields $H > H_l$, on the contrary, the free-path length l exceeds the size r_H of the orbit, and therefore, the electron can make a number of full acts of motion during the relaxation time τ.

Thus, the nature of motion of the electron varies substantially when the free-path length becomes commensurable with the characteristic size of the orbit. In pure metals at low temperatures this can even occur in rather weak fields. It then follows that the perturbation of the path of an electron caused by a magnetic field depends not only on the field strength H but also on the free-path length l. In this connection, the concept of effective magnetic field

H_{eff} has been introduced, the value H_{eff} at the given temperature is determined by the following expression

$$H_{eff} = H \frac{\rho(\Theta_D)}{\rho(T)} = H \frac{\sigma(T)}{\sigma(\Theta_D)} \tag{3.27}$$

The ratio $\sigma(T)/\sigma(\Theta_D)$ shows how many times the free-path length [see Lifshits' formula (3.11)] is varied with the temperature changed from Θ_D to T. Debye temperature has been taken as the initial one for sake of convenience.

The introduction of the effective magnetic field H_{eff} is based on that the effect produced by a field $H = 1$ oersted on the electric conductivity of a metal at temperature T is equivalent to that of a field $\left(\frac{l(\Theta_D)}{l(T)}\right)^{-1}$ oersted at Debye temperature.

In other words, the effect of a magnetic field on electric conductivity obeys a certain law of similarity; this effect is the same if the magnitude of the effective field H_{eff} remains constant (i.e. if the product $H \cdot l$ is constant). Using low temperatures makes it possible to attain very high values of the effective magnetic field (up to 10^9 oersted).

Let us discuss in more detail the electric conductivity of a metal in weak and strong magnetic fields.

3-5. WEAK MAGNETIC FIELDS

The analysis of the effect of a magnetic field on the transverse magneto-resistance will be begun from discussing the simplest case of a one-zone metal with a spherical Fermi surface.

In the absence of electric field $\vec{E}$, the Fermi surface is located symmetrically relative to the centre of the Brillouin zone $\vec{p} = 0$, so that each electron with the momentum $+\vec{p}$ will always have a corresponding electron with an equal opposite momentum $-\vec{p}$. According to this feature the whole ensemble of Fermi electrons may be divided into electron pairs with opposite momenta.

Consider the motion of these electrons in a constant magnetic field $\vec{H}$. The path of each electron is curved owing to the action of the Lorentz force $\vec{F}_H = -\frac{|e|}{c}[\vec{v}_F \cdot \vec{H}]$. But during the relaxation time, the electrons in a pair deviate from the straight path by the same distance in opposite directions so that their deviations in the magnetic field are pairwise compensated.

Consequently, with only a magnetic field applied to the system of electrons in a metal, no electric current can be produced.

Consider the behaviour of electrons when the electric ($E = -E_x$) and magnetic ($H = H_z$) fields cross one another. The action of the electric field causes the Fermi surface to be displaced by $\Delta p_x = \frac{|e| E_x l}{v_F}$ in the direction $+p_x$ (Fig. 166). The result is that non-compensated electrons moving with velocities of the order of Fermi velocity in the direction $+\vec{p}_x$ appear in volume *1*. Simultaneously, an equal number of Fermi electrons that have had the velocity component in the direction $-\vec{p}_x$ vanish from volume *2*.

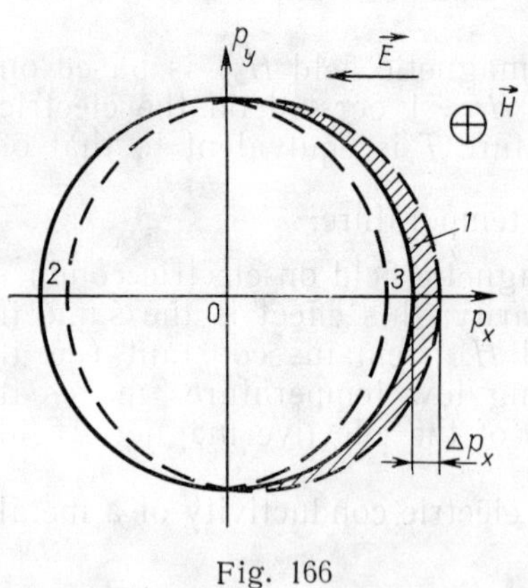

Fig. 166

With $H = 0$, the removal of particles from volume *2* had the same contribution to the electric conductivity as the filling of volume *3* with Fermi particles (see Sec. **3-2** of this chapter). But this is no more true of the magnetic field, since we have to take into account the displacement of Fermi particles in the lateral direction under the action of the Lorentz force. In this connection, with a magnetic field applied, it is more convenient to consider the pairs of electrons in which one electron is on the left-hand and the other, on the right-hand side of the Fermi surface displaced by Δp_x. A pair is composed of two such electrons that had oppositely directed Fermi momenta at $E = 0$, $H = 0$.

If the Fermi momenta on the left-hand and right-hand sides of the Fermi surface were equal in magnitude, the deviations of the electrons in a pair along the $\vec{p}_y$ axis were equal and mutually opposite, as in the case of the central symmetry of Fermi surface. But with the Fermi surface displaced by Δp_x its central symmetry is disturbed; it causes disturbance of the mutual compensation of displacements along the p_y axis in electron pairs. Let this phenomenon be discussed in more detail.

Let the projections of Fermi momenta of the electrons in a pair onto the $\vec{p}_x$ axis be equal to $+p_{Fx}$ and $-p_{Fx}$. With the Fermi surface being displaced by Δp_x under the action of an electric field $\vec{E}$, the projection of the momentum for the electron located on the right-hand side of the surface becomes equal to $p_{Fx} + \Delta p_x$, and that for the electron on the left-hand side, $-p_{Fx} + \Delta p_x$. The Fermi velocity of each of the electrons located on the Fermi surface is changed by $\Delta v_x = \frac{\Delta p_x}{m^*} = \frac{|e| E_x \tau}{m^*}$. The projections of the Fermi velocities of the electrons of the pair considered onto the $\vec{p}_x$ axis are then equal to $v_{Fx} + \Delta v_x$ and $-v_{Fx} + \Delta v_x$.

Using the equation of motion (2.58), we calculate the increment of momentum Δp_y of each electron during the relaxation time τ in the direction of the $\vec{p}_y$ axis under the action of Lorentz force $\vec{F}_H$. For the electrons on the right-hand and left-hand sides of the Fermi surface these increments are:

$$(\Delta p_y)_r = + \frac{|e|}{c} (v_{Fx} + \Delta v_x) H\tau$$

$$(\Delta p_y)_l = + \frac{|e|}{c} (-v_{Fx} + \Delta v_x) H\tau \tag{3.28}$$

The total increment $(\Delta p_y)_{pair}$ of the transverse momentum of the electron pair during time τ is:

$$(\Delta p_y)_{pair} = 2 \frac{|e|}{c} \Delta v_x H\tau \tag{3.29}$$

This quantity is equal for all electron pairs on the Fermi surface and depends only on the drift addition Δv_x to the Fermi velocity.

Thus, owing to the displacement of the Fermi surface by the electric field by Δp_x, a non-compensated variation of the momentum of each pair of electrons $(\Delta p_y)_{pair}$ in the direction $-\vec{p}_y$ is formed in the magnetic field. This variation of the momentum is equivalent to an additional displacement of the Fermi surface by $\frac{1}{2}(\Delta p_y)_{pair}$ in the direction $-\vec{p}_y$ and a transverse flow of electrons in the same direction.

Making the same calculations as in the derivation of the Lifshits formula (3.11), and transforming it to the form of (3.16), we may show that the density of transverse current j_y for the isotropic quadratic law of dispersion of electrons is

$$j_y = |e| n \frac{1}{2} \cdot \frac{(\Delta p_y)_{pair}}{m^*} = |e| n \Delta v_x \frac{|e| H}{m^* c} \tau \tag{3.30}$$

where n is the concentration of electrons. The quantity $|e| n \Delta v_x$ determines the component j_x of the electric current along the $\vec{p}_x$ axis. Relationship (3.30) can be re-written as follows

$$j_y = j_x \omega \tau \tag{3.31}$$

Hence it follows that an electric current $\vec{j}$ with the components $\vec{j}_x$ and $\vec{j}_y$ along the axes $\vec{p}_x$ and $\vec{p}_y$ is formed in an infinite piece of metal placed into crossing electric $\vec{E} = -\vec{E}_x$ and magnetic $\vec{H} = \vec{H}_z$ fields. The total current $\vec{j}$ makes an angle φ with the direction of the electric field $\vec{E}$, its tangent being $\tan\varphi = \frac{j_y}{j_x} = \omega\tau$. The angle φ is termed Hall's angle.

In a finite-size specimen of metal, the transverse current of electrons causes the formation of an excess charge of electrons at the lower portion of the specimen. This in turn causes the appearance of a transverse electric field E_y directed from above downward (Fig. 167). The electric field E_y is called Hall's field. It prevents electrons from being deviated toward the lower boundary of the specimen. Hall's field increases until the transverse component of the current of electrons becomes zero. A dynamic equilibrium is then established between the transverse current of electrons j_y (3.30) produced by the Lorentz force and the current in the opposite direction E_y appearing in the Hall field. This equilibrium is expressed by the following equality:

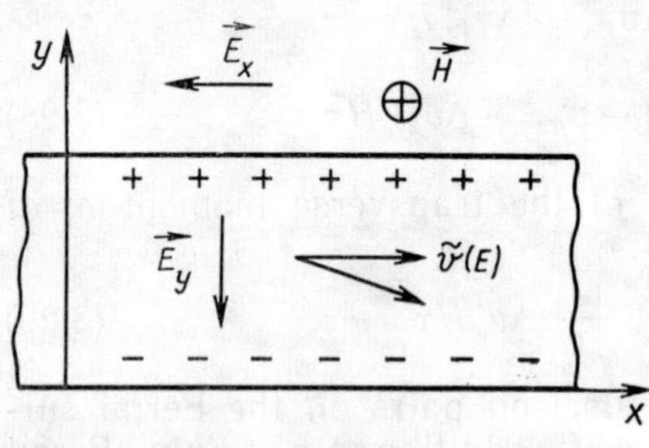

Fig. 167

$$\frac{|e|\,n}{m^*}\,|e|\,E_y\tau = \frac{|e|\,n}{m^*}\,\Delta v_x \frac{|e|}{c} H_z\tau \tag{3.32}$$

from which the magnitude of the Hall field E_y can be found. It follows from (3.32) that

$$E_y = \frac{|e|}{c} H_z \Delta v_x \tag{3.33}$$

Expression (3.33) shows that Hall's field E_y is determined by the drift addition Δv_x to the Fermi velocity of electrons.

It will be noted in this connection that Hall's field cannot straighten the curvature of paths of the electrons on the right- and left-hand sides of the Fermi surface that has been caused by the application of the magnetic field. These electrons move with velocities close to v_F with the Lorentz force of the order of $\frac{|e|}{c} H_z v_F$ acting upon them. In order to straighten their paths, an equal transverse force of opposite direction would be required, exceeding the electric strength $|e|E_y$ of Hall's field $v_F/\Delta v_x$ times (approximately 10^5 times!). Actually the Hall field is only related to that the transverse flow of electrons, which is determined by the difference between the flows produced by electrons in pairs, is equal to zero. The corresponding expression (3.32) is of the form as if the Hall field were straightening only the paths of electrons at their motion with the additional drift velocity Δv_x. Otherwise, the equality (3.32) implies that the electric force $|e|E_y$ formed by the Hall field conterbalances only a certain effective Lorentz force $F^* = \frac{|e|}{c} H_z \Delta v_x$.

The electric current j_x along the $\vec{p}_x$ axis is proportional to Δv_x. Thus, it follows from (3.33) that the Hall field is proportional to the product of the current density j_x by magnetic field H_z:

$$E_y = R j_x H_z \tag{3.34}$$

The coefficient of proportionality R has been termed the Hall coefficient. For the simple case of a one-zone metal considered this coefficient is

$$R = \frac{1}{|e| nc} \tag{3.35}$$

Note again the fact of high importance that the magnitude of Hall's field E_y is independent of Fermi velocity v_F and is fully determined solely by the drift addition Δv_F. For that reason, in cases where this will not cause contradictions, we can use the model in which electrons move with only the drift velocity Δv_x and possess no Fermi velocity.

With crossed electric $\vec{E} = -\vec{E}_x$ and magnetic $\vec{H} = \vec{H}_z$ fields applied to a metal specimen of finite dimensions, the Hall field E_y is established within a time interval of an order of $(10^{-9}\text{-}10^{-10})$ sec and results in vanishing of the transverse current of electrons. Owing to it, the mean drift velocity of electrons in direction $-\vec{E}_x$ in the magnetic field is maintained equal to the drift velocity at $H = 0$. It then follows that the electric conductivity along the x axis must not change, in the first approximation, after the Hall field has been established.

Actually, however, the additional drift velocities of electrons in the field E_x are not the same, but are distributed statistically about a certain mean value; we may denote it as $\tilde{v}(E_x)$. The main part of electrons move with a velocity sufficiently close to $\tilde{v}(E_x)$, but owing to the statistical distribution of velocities there are electrons whose velocity is either greater or lower than $\tilde{v}(E_x)$. These electrons will be deviated by the Hall field E_y and magnetic field H_x in different directions, since the Hall field can only straighten the paths corresponding to the mean velocity $\tilde{v}(E_x)$. (Here we use the model of electrons moving solely with the drift velocity which has been discussed earlier.)

The force formed by the Hall field in a metal is lower than the effective Lorentz force for the electrons moving with velocities greater than $\tilde{v}(E_x)$ and higher for those whose velocity is less than $\tilde{v}(E_x)$. Then, considering their contribution to the electric current along the $\vec{p}_x$ axis, we may assume that the electrons whose drift velocities exceed $\tilde{v}(E_x)$ are deviated downward, and those with velocities less than $\tilde{v}(E_x)$, upward from the straight path (Fig. 168).

Thus, taking into account statistical distribution of velocities, we come to the fact that for a part of electrons the projections of the drift velocity onto the direction of the electric field $\vec{E}_x$ are smaller than at $H = 0$ owing to their deviation upward or downward under the action of the magnetic field. This is equivalent to a reduction of the free-path length of these particles, and therefore, to a reduced electric conductivity along the $\vec{p}_x$ axis. We are to estimate this reduction of electric conductivity. Let $\bar{\alpha}$ be the mean angle by which the electrons deviate from the straight-line path. The mean free-path length in the direction of the x axis in the magnetic field $l_x(H)$ can be written as

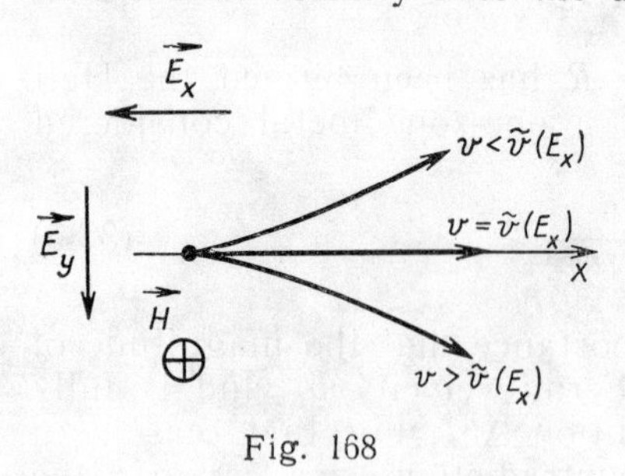

Fig. 168

$$l_x(H) = l_x(0)\cos\bar{\alpha} \tag{3.36}$$

where $l_x(0)$ is the free-path length at $H = 0$.

The deviation by angle $\bar{\alpha}$ is caused by the difference between the effective Lorentz force and the force of the Hall field. Since each of these forces is proportional to the magnetic field H, then the angle $\bar{\alpha}$ in weak magnetic fields must also be proportional to H and small in magnitude. Let $\bar{\alpha} = \sqrt{2\beta}\,H$, where β is a parameter independent of H. Expression (3.36) can then be written as

$$l_x(H) \approx l(0)(1 - \beta H^2) \tag{3.37}$$

which takes into account that for small $\bar{\alpha}$, $\cos\bar{\alpha} \approx 1 - \frac{\bar{\alpha}^2}{2} = 1 - \beta H^2$. Hence for the conductivity $\sigma_x(H)$ proportional to $l_x(H)$ we get:

$$\sigma_x(H) \approx \sigma_x(0)(1 - \beta H^2) \tag{3.38}$$

The transverse magneto-resistance $\rho_x(H)$, equal to $1/\sigma_x(H)$ in the isotropic case considered, is found by the expression:

$$\rho_x(H) \approx \rho_x(0)(1 + \beta H^2) \tag{3.39}$$

which follows from (3.38) at $\beta H^2 \ll 1$. The relative variation of resistance in the magnetic field is:

$$\frac{\rho_x(H)-\rho_x(0)}{\rho_x(0)}=\frac{\Delta\rho_x(H)}{\rho_x(0)}=\beta H^2 \qquad (3.40)$$

We shall now discuss a more general case of a one-zone metal with a closed anisotropic Fermi surface. It may be easily shown that the relationship obtained for the transverse magneto-resistance (3.40) will not qualitatively change. Indeed, in weak magnetic fields an electron passes between two successive acts of scattering only a small portion of the path determined by the shape of the corresponding section of the Fermi surface by a plane perpendicular to the magnetic field, so that the singularities of this path cannot appear in the first approximation.

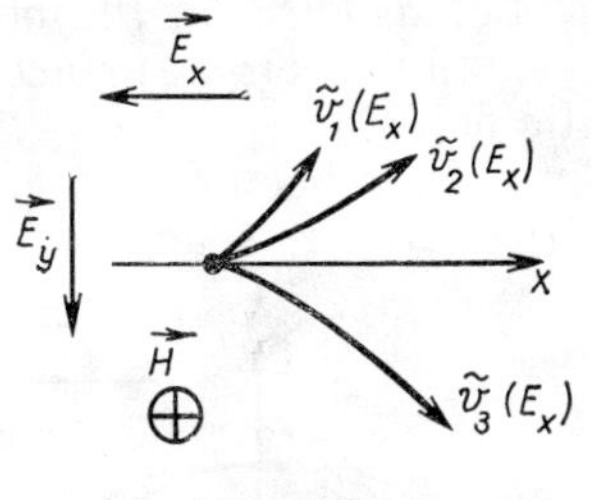

Fig. 169

But the anisotropy of the Fermi surface causes the anisotropy of electron paths, so that velocities of electrons are more scattered about $\tilde{v}(E_x)$ than in the isotropic case, since the electrons located in various points of the Fermi surface now differ more strongly from each other in their properties. For that reason the coefficient β in formula (3.39) for metals with an anisotropic Fermi surface must be greater than that for metals with an isotropic constant-energy surface. The dependence of the form (3.40) is observed in weak magnetic fields and with an open Fermi surface. The latter is then equivalent to a strong anisotropy of the properties of electrons on the Fermi surface, which only increases the coefficient β.

More complicated is the case when the Fermi surface in a one-zone metal is composed of several electron-type surfaces. Various Fermi surfaces are displaced in different ways in an electric field $\vec{E}_x$, and therefore, electrons on them acquire different mean drift velocities $\tilde{v}_i(E_x)$, where i is the number of surface ($i=1, 2, 3, \ldots$). The Hall field E_y formed in the magnetic field H_z ensures that the resulting transverse electric current equals zero. Then the effective Lorentz force $F^*_{Hi}=\frac{|e|}{c}\tilde{v}_i(E_x)H_z$ calculated for each mean value of the drift velocity $\tilde{v}_i(E_x)$ is no more balanced by the Hall field force $|e|E_y$.

In that case the departure of electrons in each group from a straight-line path (note that the path is considered from the standpoint of its contribution to electric conductivity σ_x) is determined by the finite difference $F^*_{Hi}-|e|E_y$ and is no more a weak effect connected solely with scattering of drift velocities from their mean values in the groups. This situation for the case of three groups of electrons is shown in Fig. 169. This circumstance must induce the transverse magnetoresistance to increase more strongly.

With several constant-energy surfaces, the coefficient β in formula (3.40) becomes substantially larger than in the previous case of one surface.

3-6. WEAK MAGNETIC FIELDS. HOLE-TYPE FERMI SURFACES

Let us discuss the behaviour of current carriers on a hole-type constant-energy surface (Fig. 157) in crossed electric and magnetic fields.

As has been found earlier, the contribution of holes to the electric current is additive to that of electrons. Let us determine the magnitude and sign of the contribution of holes to the Hall effect. We shall proceed similarly to what has been done in the previous section when analyzing the Hall effect on electrons.

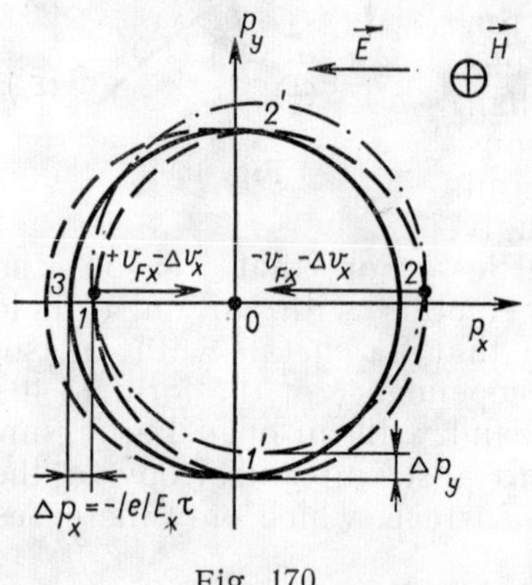

Fig. 170

Let the electric field $\vec{E}$ be directed along $-\vec{p}_x$ and the magnetic one $\vec{H}$, along $-\vec{p}_z$. The electric field causes the hole-type constant-energy surface to displace by $\Delta p_x = |e| E\tau$ in the direction $+\vec{p}_x$ (Fig. 170). This disturbs the central symmetry of distribution of electrons over momenta, so that quasi-particles having a non-compensated component of momenta in the direction $-\vec{p}_x$ appear in volume *1*. Simultaneously, particles that have compensated the flow of particles located in volume *3* will now vanish from volume *2*.

It will be recalled that the velocity of electrons on a hole-type surface is opposite in direction to their momentum, so that the particles in volumes *1* and *3* produce an electric current corresponding to the transfer of a negative charge in the inverse direction relative to the field $\vec{E}$.

In the presence of a magnetic field the particles in volumes *2* and *3* deviate under the action of the Lorentz force in different directions, which violates the equivalence between the empty volume *2* and filled volume *3*. For that reason we again will consider electron pairs in which one of the electrons is on the right-hand and the other, on the left-hand side of the Fermi surface. With $E = 0$, the momenta of the electrons in a pair were equal in magnitude and opposite in direction. Let their projections onto the $\vec{p}_x$ axis be denoted respectively as $+p_{Fx}$ and $-p_{Fx}$ for particles on the right and left.

With the Fermi surface being displaced by a distance Δp_x to the right, the projection of the momentum of an electron on the left-hand side reduces in magnitude and becomes equal to $-p_{Fx} + \Delta p_x$. Simultaneously, the projection of the momentum of an electron on the right increases and becomes equal to $+p_{Fx} + \Delta p_x$. The Fermi velocity of an electron on the left is directed along the $+\vec{p}_x$ axis, its projection onto this axis being equal to $v_{Fx} - \Delta v_x$, where Δv_x is the magnitude of variation of the Fermi velocity during the relaxation time τ under the action of the electric field. Accordingly, for an electron on the right, the projection of Fermi velocity onto the $\vec{p}_x$ axis is $-v_{Fx} - \Delta v_x$. By analogy with expressions (3.28) we can write the variation of the momentum of each electron in the direction of the $\vec{p}_y$ axis under the action of the Lorentz force. We shall write down only the total variation of the lateral momentum of an electron pair:

$$(\Delta p_y)_{pair} = 2 \frac{|e|}{c} \Delta v_x H \tau \tag{3.41}$$

The variation of the momentum is positive and equivalent to displacing the hole-type Fermi surface by $\frac{1}{2} (\Delta p_y)_{pair}$ in the direction $+\vec{p}_y$. Owing to this displacement, electrons having a non-compensated component of Fermi velocity directed upward will appear in volume *1'* and electrons that have had a non-compensated component of Fermi velocity in the inverse direction will vanish from volume *2'*. This is equivalent to the formation of a flow of negatively charged particles upward along the $+\vec{p}_y$ axis. In a finite-size specimen of metal, negative charges will be accumulated at the top of the specimen and a Hall field E_y directed upwards will be formed.

Thus, as distinct from their contribution to the electric conductivity, the contribution of holes to the Hall effect is opposite in sign to that of electrons. The Hall field E_y for holes is again found by equating to zero the lateral flow of particles, its magnitude being determined by the drift addition Δv_x and independent of the Fermi velocity of holes. It is easy to show that expressions similar to formulae (3.33), (3.34) and (3.35) hold true for holes.

When analysing the Hall effect, we have regarded holes as particles with $m^* < 0$, $e < 0$, which follows naturally from the construction of a hole-type Fermi surface. But the conclusions made earlier remain valid if holes are regarded as particles with $m^* > 0$, $e > 0$ filling a constant-energy surface. There is no need to discuss it in more detail since the proof can easily be made on the basis of the earlier analysis.

That the Hall effect on holes is opposite in sign to that on electrons is in good agreement with the experimental data for the metals possessing both electron-type and hole-type constant-energy surfaces. The effects from holes and electrons in such metals compensate one another completely or partially. If the concentration of electrons n and that of holes p are equal and the properties of these particles are the same (i.e. if their free-path lengths and relaxation times are equal), then the contribution of electrons to the Hall effect exactly compensates that of holes, so that the Hall field is zero.

When an impurity of a different valence is introduced in such a metal, the Hall field will depend on the concentration of this impurity which determines the difference between the concentrations of electrons n and holes p. Replacing the impurity of a lower valence by one of a higher valence in the same metal results in the Hall e. m. f. changing sign.

If in a two-zone metal with $n = p$ the Hall field is zero, then the effective Lorentz force causes the particles to deviate from the straight-line path of their motion with the drift component of velocity Δv_x. The coefficient β in formula (3.40) then must attain its maximum.

If the concentrations of electrons and holes are not equal or the Hall e. m. f. is not fully compensated at $n = p$ since the particles of different types are not identical in their dynamic properties, then a Hall field is formed corresponding to the transverse electric current of zero magnitude. The Hall field reduces somewhat the departure of particles from straight paths as they move with the drift velocity Δv_x, but for each group of particles the action of the Lorentz force is not fully compensated. This means that for metals containing both electrons and holes the coefficient β in formula (3.39) for the magnetoresistance is always large.

3-7. STRONG MAGNETIC FIELDS

The shape of paths of current carriers, which has no essential importance in weak magnetic fields, becomes critical in strong fields. This is linked with the fact that an electron moving in a strong field has time during the passage along its free-path length to describe a complex closed curve on the Fermi surface or to cover a large portion of an open curve. In both cases the nature of its motion is uniquely determined by the shape of constant-energy surfaces and their orientation in the magnetic field.

Before discussing the conductivity of a metal in a strong magnetic field, let us recall the nature of motion of free particles with the electric $\vec{E}$ and magnetic $\vec{H}$ fields being crossed.

With $E = 0$ and $H \neq 0$, a free particle having the initial velocity $\vec{v}_0$ in a plane perpendicular to the magnetic field describes

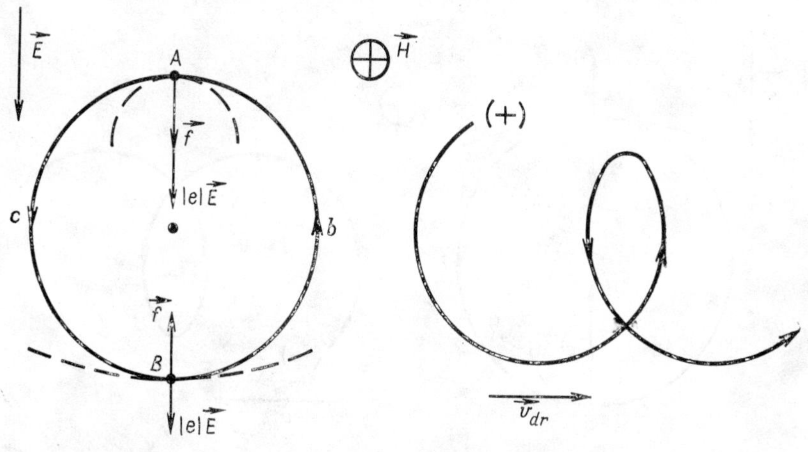

Fig. 171

a circle of radius $r_H = \frac{m_0 v_0 c}{|e| H}$. The direction of motion of a negative particle in the magnetic field is shown in Fig. 116. A positively charged particle precesses in the oppositce direction. In both cases the particle is acted upon by the same centripetal force $\vec{f}_{cp} = \frac{e}{c} |\vec{v}_0 \vec{H}|$ in all points of its path.

Let an electric field $\vec{E}$ directed as shown in Figs. 171 and 172 be applied in addition to the magnetic field. The action of the electric field gives an additional force $e\vec{E}$ of constant direction, which results in a continuous variation of the magnitude of the velocity of the particle and, as a consequence, in a change of the shape of its path. For instance, the forces $\vec{f} = \frac{|e|}{c} [\vec{v}\vec{H}]$ and $|e|\vec{E}$ acting on a positively charged particle in point A in Fig. 171 coincide in direction, whereas in point B they act in opposite direction.

In addition, when moving from A to B, a positively charged particle is accelerated by the electric field, so that its velocity in point B is higher than that in point A. As a consequence, the radius of curvature of the path in point B must be larger than in point A (dotted lines in Fig. 171 show the curvature of the path that must be in points A and B). When moving from point A to point B, the curvature of the path must monotonously increase, and when moving from B to A, decrease.

These variations result in that a positively charged particle acted upon by crossing fields $\vec{E}$ and $\vec{H}$ describes a trochoid, its trochoidal motion being accompanied with its drift in a direction

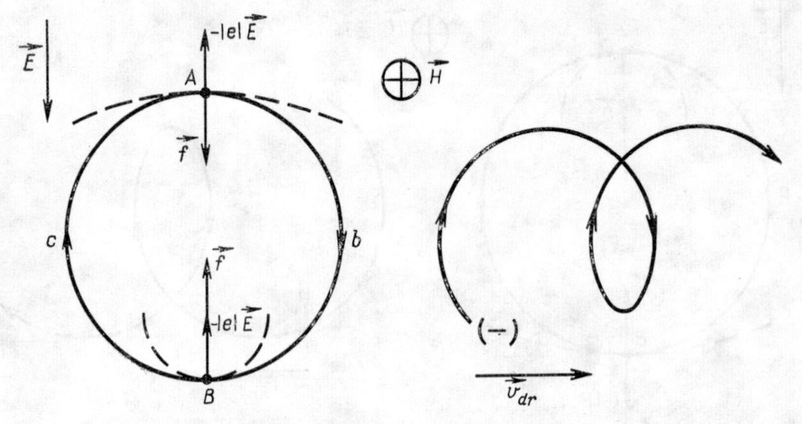

Fig. 172

perpendicular to $\vec{E}$ and $\vec{H}$: the particle is displaced to the right with each turn of the trochoid. With a negatively charged particle, the direction of its precession in the magnetic field must be changed to the opposite one, the same must be done with the direction of its interaction with the electric field.

The motion of a negatively charged particle in crossing fields $\vec{E}$ and $\vec{H}$ of the original orientation is shown in Fig. 172. Comparing Figs. 171 and 172, we can see that the paths of a positive and negative particles pass into one another upon a mirror reflection in a plane perpendicular to $\vec{E}$. The drift of the particles of both signs occurs in the same direction perpendicular to vectors $\vec{E}$ and $\vec{H}$.

The drift velocity $\vec{v}_{dr}$ depends on neither magnitude nor sign of the charge of the particle and is fully determined by the magnitude and direction of the fields $\vec{E}$ and $\vec{H}$. It may be shown that

$\vec{v}_{dr}$ is related to vectors $\vec{E}$ and $\vec{H}$ as follows:

$$\vec{v}_{dr} = \frac{c}{H^2}[\vec{E}\vec{H}] \tag{3.42}$$

The motion of charged particles in an electric $\vec{E}$ and magnetic $\vec{H}$ fields of an arbitrary mutual orientation can easily be understood if we depart from the above case of crossing fields $\vec{E}$ and $\vec{H}$. For this, we decompose the electric field $\vec{E}$ into two components: $\vec{E}_{\parallel}$ parallel to the magnetic field, and $\vec{E}_{\perp}$ perpendicular to it. The motion of a particle will then be composed of a uniformly accelerated motion along a magnetic line of force under the action of force $e\vec{E}_{\parallel}$ and the motion along the trochoid in a plane perpendicular to the magnetic field under the action of forces $\frac{e}{c}|\vec{v}_{\perp}\vec{H}|$ and $e\vec{E}_{\perp}$.

These data on the motion of particles of unlike signs in an electric and a magnetic field can be used for discussing the motion of current carriers in real metals, with the following to be had in view:

(a) the paths of current carriers in the metal at application of a magnetic field $\vec{H}$ are determined by the shape of the constant-energy surface and the orientation of the field $\vec{H}$ and may differ strongly from a circle;

(b) under the action of external fields $\vec{E}_x$ and $\vec{H}_z$, an additional Hall field $\vec{E}_y$ is formed in a finite-size piece of the metal, the current carriers moving in the resulting electric field $\vec{E}_{res} = \vec{E}_x + \vec{E}_y$;

(c) free motion of particles in the metal can only occur along the free-path length, the path being changed sharply at scattering of the particle.

With a drift in crossing fields $\vec{E}$ and $\vec{H}$, no displacement of a free particle in the direction of vector $\vec{E}$ occurs. It then follows that in an infinite metal placed into crossing fields $\vec{E}$ and $\vec{H}$ with an infinitely large free-path length of current carriers, there would be no conductivity in the direction of the electric field. The combined action of the fields would cause in that case only a transverse current of carriers to appear in the direction of the drift. The electric conductivity of a real metal in crossing fields can be connected with two causes: the formation of a Hall field in a finite-size specimen of the metal and the scattering of carriers,

The effect of scattering on electric conductivity can be related to that the cyclic motion of the carriers along the trochoid is disturbed and there appears a component of displacement of the particle in the direction of the electric field $\vec{E}$.

This process can be visualized by the motion of particles in crossing fields $\vec{E}$ and $\vec{H}$ at $l \sim r_H$ (i.e. when the magnitude of the magnetic field in the metal relates to an intermediate region $H \sim H_l$ between weak and strong fields). Figure 173*a* shows the motion of a positively charged particle in a metal at $l = \infty$ and $l \sim r_H$. Crossed circles in the figure denote the points where scattering occurs,

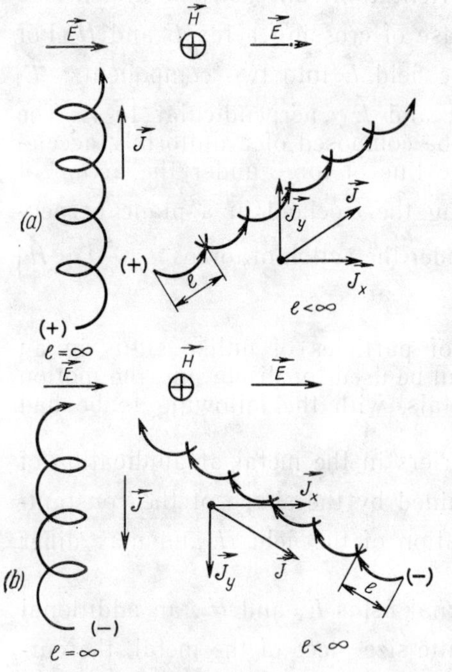

Fig. 173

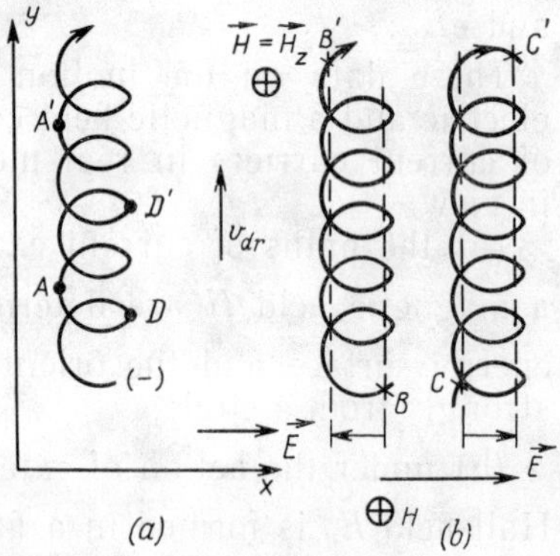

Fig. 174

which is accompanied with the loss of the drift addition $v(E)$ to the Fermi velocity or the stoppage of the particle in the model of motion selected earlier, from which the Fermi velocity has been excluded. The similar motion of a negatively charged particle at $l = \infty$ and $l \sim r_H$ is shown in Fig. 173*b*.

Let us now discuss the mechanism of electric conductivity in a strong magnetic field at $l \gg r_H$ as illustrated in Fig. 174 which shows the motion of a negatively charged particle (electron). Let the electron begin its motion in point A (for instance, after being scattered) and be scattered again in point A' located in the same plane perpendicular to the electric field $\vec{E}$, as point A (Fig. 174*a*). In that case the electron would acquire no additional energy in the electric field during the relaxation time τ, and therefore, no addition $\Delta v_x(E)$ to the Fermi velocity v_F.

A similar situation arises at the motion of an electron between two acts of scattering in points D and D' in Fig. 174*a*. But when the points where acts of scattering occur do not lie in the same plane perpendicular to $\vec{E}$, a certain effective displacement of the electron along the field $\vec{E}$ or in the reverse direction occurs during the relaxation time. Two such cases are illustrated in Fig. 174*b*.

When passing from point B to point B', an electron is displaced against the electric field $\vec{E}$, its Fermi velocity increasing by $\Delta v_x(E)$ dependent on the field strength E and the distance of its displacement. On the other hand, the passage from point C to point C' results in a displacement of the electron in the direction of the field $\vec{E}$ and a decrease of its Fermi velocity. Note that the variation of the velocity of electron $\Delta v_x(E)$ in both cases is in the same direction opposite to vector $\vec{E}$. In other words, the Fermi velocity of an electron in a certain point of the crystal, which is displaced relative to the other point in the direction opposite to the electric field $\vec{E}$, is always larger by a definite magnitude $\Delta v_x(E)$, provided no scattering has occurred between these points.

As will be seen from Fig. 174*b*, the mean distance $\overline{\Delta x}$ through which an electron can be displaced along the x axis parallel to $\vec{E}$ between two acts of scattering coincides in the order of magnitude with the size of the Larmor orbit $r_H \sim \frac{m^* v_F c}{|e| H}$. The whole passage between the points in which scattering occurs (for instance, between B and B' in Fig. 174*b*) is performed during the relaxation time τ, the path of motion being composed of several turns of a trochoid. In the course of the passage, the electron oscillates along the x axis and changes periodically its Fermi velocity. Through the averaging of these oscillations of the Fermi velocity there remains the difference between the final and initial velocities equal to $\Delta v_x(E)$.

The total variation of velocity $\Delta v_x(E)$ is formed on the last turn of the trochoid during the time of the order of the period of Larmor precession $T_H = \frac{2\pi m^* c}{|e| H}$ which constitutes only a small fraction of the relaxation time τ. It will be noted for comparison that for the case illustrated in Fig. 174*a* the averaging of oscillations of the Fermi velocity during the relaxation time at the passage from point A to point A' gives a zero variation of the velocity: $\Delta v_x(E) = 0$.

The increment of velocity $\Delta v_x(E)$ acquired during the time of the order of T_H under the action of the electric field $\vec{E}$, is, evidently $\Delta v_x(E) \sim \frac{|e|E}{m^*} T_H$, which is proportional to $1/H$ as T_H. But this increment $\Delta v_x(E)$ responsible for the contribution of the given electron to electric conductivity, must be related to the whole period of time τ during which the electron is moving without scattering. For that reason, the time-averaged real contribution to electric conductivity along the x axis is related to the mean increment of the Fermi velocity $\bar{v}_x(E)$ of the electron, which is of the order of $\frac{T_H}{\tau} \Delta v_x(E)$ and in turn proportional to E/H^2.

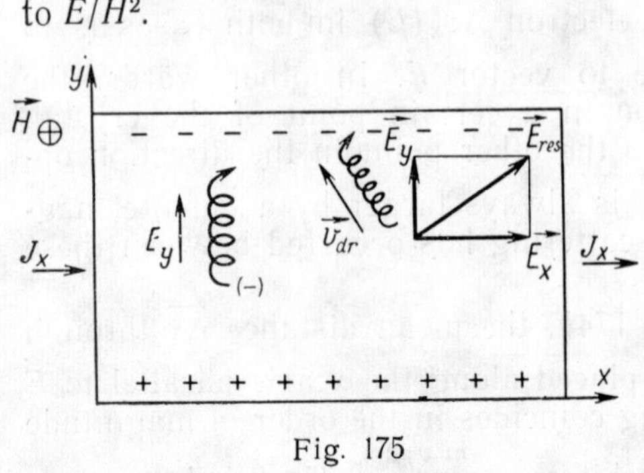

Fig. 175

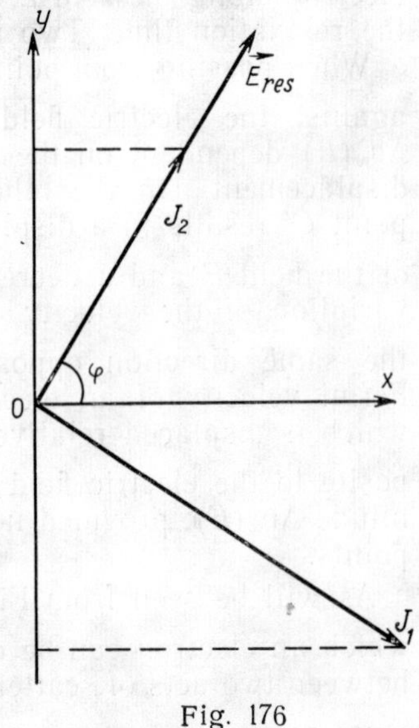

Fig. 176

Thus, the conduction current J_x along the x axis, which is determined by the concentration of electrons and the mean increment of Fe·mi velocity $\bar{v}_x(E)$ during the relaxation time is also proportional to the ratio E/H^2. This current is produced solely through the processes of electron scattering. For convenience, it will be termed collision current. The corresponding electric conductivity $\sigma_x(H)$ depends on magnetic field as $1/H^2$, and the corresponding transverse magnetoresistance $\rho_x(H)$ increases proportional to H^2.

Let us now analyse the electric conductivity of a finite-size metallic specimen with different structures of Fermi surfaces.

a. A one-zone metal with a closed Fermi surface. Consider the motion of a group of current carriers (electrons) in a piece of metal of finite dimensions. Let the electric field $\vec{E}_x$ be directed along the x axis, and the magnetic field $\vec{H}_z$, along the z axis (Fig. 175).

At the first instant upon application of the fields, electrons begin to drift in the direction of the y axis thus forming an excess negative charge at the upper boundary of the specimen (the charge is formed in a thin superficial layer whose thickness is a few interatomic distances for good metals). The excess charge at the surface forms a Hall field $\vec{E}_y$ directed upward. After the Hall field has been established, the drift of electrons occurs in the resulting electric field $\vec{E}_{res}$ equal to the sum $E_x + \vec{E}_y$. The drift velocity $\vec{v}_{dr}$ is proportional to E_{res}/H [see (3.42)] and directed perpendicular to E_{res} (Fig. 175). This motion of electrons has a corresponding drift current $J_1 \sim \frac{E_{res}}{H}$.

This drift is not related to electron scattering and is determined solely by the magnitude and direction of the fields $\vec{E}_{res}$ and $\vec{H}$. In addition, scattering of electrons produces another drift of the field $\vec{E}_{res}$ which is connected with the appearance of the collision current $J_2 \sim E_{res}/H^2$. The directions of the currents $\vec{J}_1$ and $\vec{J}_2$ are shown in Fig. 176, where φ is the angle between $\vec{E}_{res}$ and the x axis.

The Hall field E_y increases until the current component along the y axis becomes equal to zero. The magnitude of the field E_y is determined from the condition that the projections $\vec{J}_1$ and $\vec{J}_2$ onto the y axis are equal, which may be written as:

$$J_1 \cos\varphi = J_2 \sin\varphi \tag{3.43}$$

Since $J_1 \sim \frac{E_{res}}{H}$, $J_2 \sim \frac{E_{res}}{H^2}$, $E_{res} \sin\varphi = E_y$, and $E_{res} \cos\varphi = E_x$, it follows from (3.43) that

$$E_y \sim E_x \cdot H \tag{3.44}$$

$$\cot\varphi \sim 1/H \tag{3.45}$$

The conduction current J_x may be thought of as a sum of two components (see Fig. 176):

$$J_x = J'_x + J''_x = J_1 \sin\varphi + J_2 \cos\varphi \tag{3.46}$$

The components J'_x and J''_x of the conduction current are:

$$J'_x = J_1 \sin\varphi \sim \frac{E_{res}}{H} \sin\varphi = \frac{E_y}{H} \tag{3.47}$$

$$J''_x = J_2 \cos\varphi \sim \frac{E_{res}}{H^2} \cos\varphi = \frac{E_x}{H^2}$$

The current component J_x'' decreases rapidly with an increase of the magnetic field, the angle φ increasing and tending to 90 degrees.

In very strong magnetic fields the component J_x'' can be neglected, the main contribution to the current J_x being made by the component J_x':

$$J_x \approx J_x' \sim \frac{E_y}{H} \tag{3.48}$$

Noting (3.44), it then follows that in strong magnetic fields

$$J_x \sim \frac{E_x H}{H} \sim E_x \tag{3.49}$$

and $\sigma_x(H)$ becomes independent of the magnetic field.

This means that in specimens of real metals with a single group of current carriers a continuous increase of the Hall field $E_y \sim$ $\sim E_x H$ fully compensates the reduction of the collision current J_2. In strong magnetic fields the conduction current is mainly determined by the drift of electrons in the Hall field $\vec{E}_y$ which practically coincides with the resulting field $\vec{E}_{res}$. Thus drift is directed along the x axis. The transverse magnetoresistance becomes saturated in strong magnetic fields as a result of that drift.

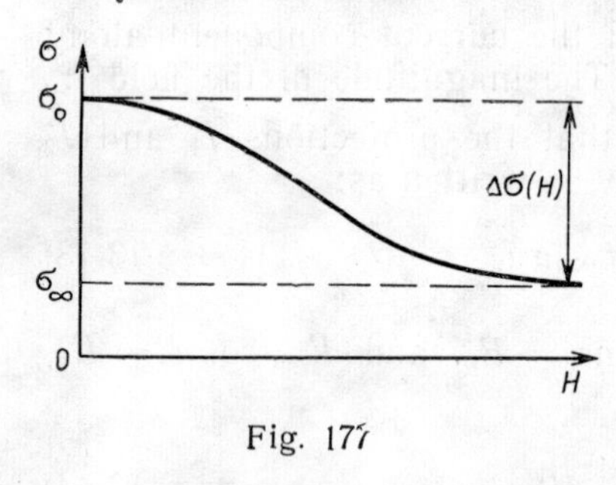

Fig. 177

Thus, the following dependence of electric conductivity on magnetic field is observed in a one-zone metal. In weak fields the conductivity first decreases with an increase of the field by the law $\sigma = \sigma_0(1 - \beta H^2)$, then there is a region of transition to strong fields where the dependence of σ on H deviates from the quadratic law, and finally, the conductivity in strong fields tends to a constant σ_∞ which is lower than the conductivity σ_0 in the absence of magnetic field (Fig. 177).

This may be explained by the fact that an electron in a strong field passes during the relaxation time τ a large distance (of the order of l) along a complex trochoidal path, but is displaced during that time against the electric current by a substantially smaller distance Δx, the result being that it undergoes a substantially greater number of scattering acts per unit length of the specimen than at $H = 0$.

b. A two-zone metal with closed constant-energy surfaces. We shall analyse the dependence of the transverse resistance on magnetic field in a metal having current carriers of two types, i.e. electrons and holes. Consider separately a case when the concentrations of electrons n and holes p are not equal and another case when they are equal.

It will be recalled that in multi-zone metals of any valence the constant-energy surfaces are formed through overlapping of zones, the concentrations of electrons and holes in a pure metal being always equal to one another. The inequality of the concentrations n and p in such metals is only possible with the presence of the impurities whose valence is either greater or lower than that of the metal.

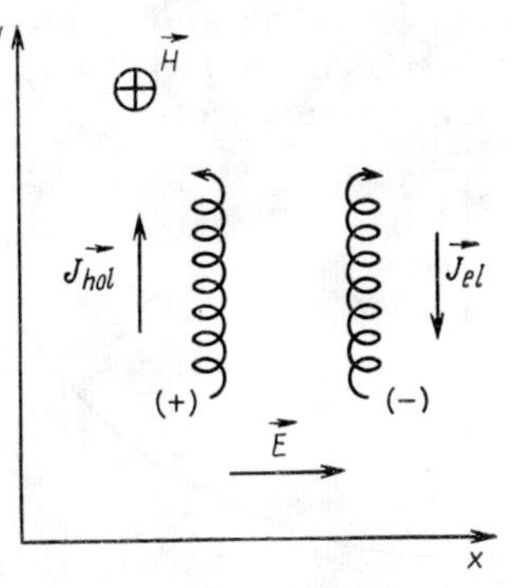

Fig. 178

For simplicity, we assume that the metal has only two overlapping zones and correspondingly only two closed constant-energy surfaces of the electron and hole type.

The earlier analysis of a one-zone metal with an electron-type or hole-type constant-energy surfaces has made it possible to build a model of motion of carriers in a magnetic field in which Fermi velocities are neglected and all operations are carried out only with the additional drift velocities appearing under the action of external forces. The model, which is very convenient owing to its good visualization, is based on division of the ensemble of Fermi electrons into pairs whose momenta at equilibrium are equal in magnitude and mutually opposite in direction.

Having thus substantiated the model of this motion, we need not always to make a complicated analysis of the motion of electrons with the velocities of the order of Fermi velocities. Note that in this model a hole may be more conveniently regarded as a particle with $m^* > 0$, $e > 0$, since it is then an "electron-like" particle with a positive effective mass, which makes it possible to visualize its drift motion.

Thus, we consider the simultaneous drift of electrons and holes in a metal specimen placed into crossing fields $\vec{E}$ and $\vec{H}$. A negatively charged electron and a positively charged hole then drift in the same side in the direction of vector $[\vec{E} \cdot \vec{H}]$ (Fig. 178). Consequently, the electric currents $\vec{J}_{el}$ and $\vec{J}_{hol}$ that are produced by the drifts of electrons and holes respectively are directed oppositely and the resulting transverse current J is equal to their difference $J_{hol} - J_{el}$.

When the drift currents J_{el} and J_{hol} are equal in magnitude, the transverse current J becomes zero (which corresponds to the equality of concentrations $n=p$ and relaxation times of electrons and holes). Then, no Hall field E_y is formed in a finite-size specimen of metal. It then follows that the conduction current J_x in strong magnetic fields is fully determined by the mechanism of collisions and is proportional to E_x/H^2, as has been shown earlier. The corresponding conductivity σ_x depends on magnetic field as $1/H^2$.

A feature characteristic of the particular case considered is that, both for weak and strong magnetic fields, the transverse magnetoresistance $\rho_x(H)$ increases proportional to H^2. But the mechanisms of appearance of electric conduction in a weak and a strong field

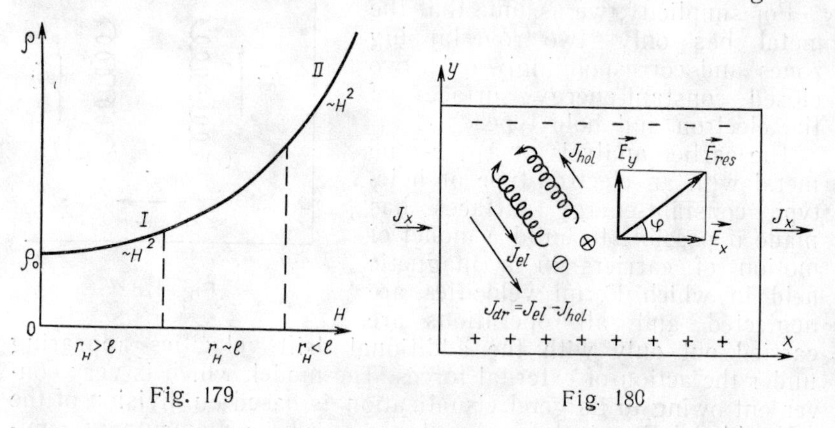

Fig. 179 Fig. 180

are of different nature (Fig. 179), though result in the same dependence on magnetic field. It has therefore been agreed to distinguish between the sections of the first (I) and second (II) quadraticity on the curves of dependence of magnetoresistance on H, related respectively to the regions of weak and strong fields. Figure 179 shows a typical curve of the transverse magnetoresistance for a metal with equal concentrations of electrons and holes.

When the concentrations of carriers of different types are not equal (i.e. $n \neq p$), the drift currents of electrons and holes cannot compensate one another and the resulting transverse current $J = J_{hol} - J_{el}$ in a finite-size specimen of metal establishes a Hall field E_y whose direction for the given orientation of the external electric and magnetic fields depends on which type of carriers in the metal is predominant in its concentration. For instance, the direction of the Hall field $\vec{E}_y$ in Fig. 180 corresponds to $n > p$.

After the Hall field $\vec{E}_y$ has been established, particles drift in the resulting electric field $\vec{E}_{res} = \vec{E}_x + \vec{E}_y$. As for the case of a one-zone metal discussed in "a", the sum of the projections onto the y axis of the drift current J_{dr}, directed perpendicular to $\vec{E}_{res}$ and proportional to E_{res}/H, and also of the collision current J_{col}, which is parallel to $\vec{E}_{res}$ and proportional to E_{res}/H^2, must be equal to zero. But, as distinct from a metal with carriers of only one type, the drift current J_{dr} is equal here to the difference between the corresponding drift currents of electrons and holes and is determined by $(n-p)$.

In very strong magnetic fields the drift current J_{dr} becomes substantially greater than the collision current J_{col} and the resulting electric field $\vec{E}_{res}$ practically coincides with the Hall field $\vec{E}_y$ whose magnitude, proportional to E_xH, then exceeds substantially the external electric field E_x. Under such conditions the conduction current J_x is mainly decided by the drift current J_{dr} in the Hall fields. This gives a constant value of electric conductivity in the limit of strong fields, similar to what occurs in a one-zone metal.

A circumstance of high importance should then be emphasized. However strong the magnetic field would be, we cannot fully ignore the collision current J_{col}. For any H, its projection onto the y axis always compensates the projection of the drift current J_{dr} onto the same axis and ensures that the lateral current J_y is zero (see Fig. 176). In the limit of strong fields, the angle φ between J_{col} and the y axis (or between J_{dr} and the x axis) reduces unlimitedly, but the equality $J_{col}\cos\varphi = J_{dr}\sin\varphi$ remains always true.

The limiting value of electric conductivity σ_∞ is the lower and is attainable in stronger fields, the less is the difference of the concentrations of electrons and holes $|n-p|$. This may be explained as follows. In the limit of strong fields the conduction current J_x tends to the value of drift current proportional to $|n-p|\frac{E_{res}}{H}$. The resulting field E_{res} in this region practically coincides with the Hall field $E_y \sim E_x \cdot H$. Consequently, the conduction current is proportional to $|n-p|\frac{E_xH}{H} = |n-p|E_x$ and the conductivity σ_∞ is decided solely by the difference $|n-p|$ and is independent of external fields. The smaller this difference, the lower the drift current, so that the inequality $J_{dr} \gg J_{col}$, owing to which the conductivity becomes saturated, holds true for stronger fields. In the limit of $n = p$, no saturation occurs with any high strength of the magnetic field.

Figure 181 shows the curves of dependence of the magnetoresistance on magnetic field for various values of the difference $\Delta n = |n-p|$. Given in the same figure is a parabolic dependence of ρ on H, which corresponds to the case of equal concentrations of electrons and holes $n = p$.

c. A one-zone metal with an open Fermi surface. In order to study the effect of open paths on the dependence of magnetoresistance on magnetic field, we shall initially consider a strongly anisotropic (i.e. strongly extended in one direction) Fermi surface of the electron type. An open surface may be regarded as the limiting case of anisotropy when the Fermi surface reaches the boundaries of the Brillouin zone in the direction of its extension.

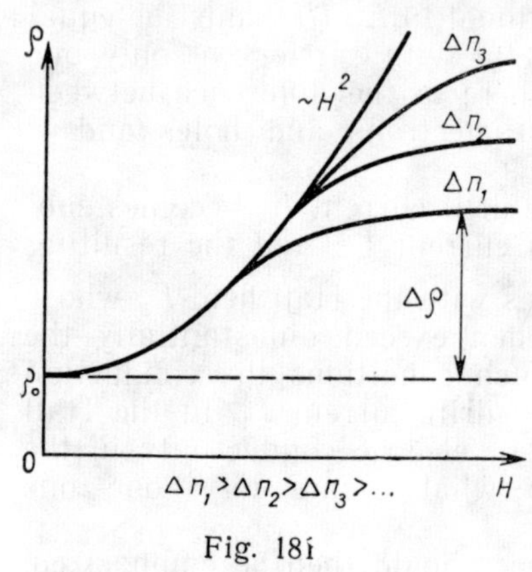

Fig. 181

Figure 182a shows a central section of an anisotropic Fermi surface by a plane perpendicular to the magnetic field and the directions of the main axes of the surface. The path of an electron in real space, corresponding to the section in Fig. 182a is illustrated in Fig. 182b. Both figures relate to a case when

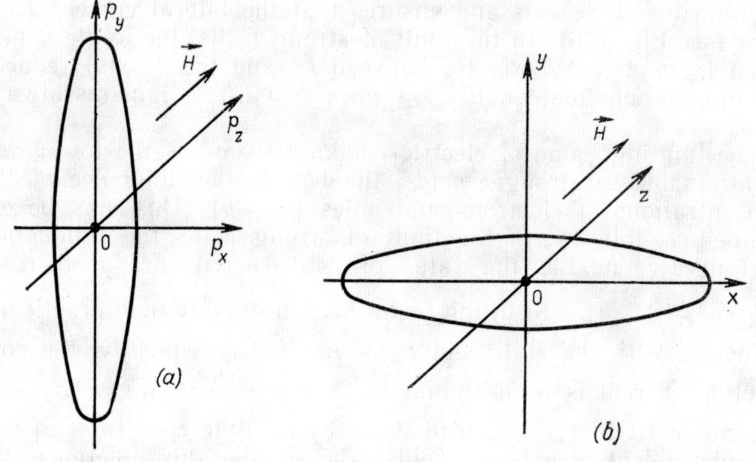

Fig. 182

the magnetic field $\vec{H}$ is perpendicular to the direction of extension of Fermi surface. If an electric field $\vec{E}$ is applied orthogonal to $\vec{H}$, the pattern of motion of electrons in the metal will depend substantially on the orientation of the field $\vec{E}$ relative to the main axes of the surface.

An electric field orthogonal to $\vec{H}$ can evidently be orientated arbitrarily in the plane $p_x p_y$, and in particular, be directed along the extension of the surface ($\vec{E} = \vec{E}_y$) or perpendicular to it ($\vec{E} = \vec{E}_x$). Let us discuss these two orientations of the field in more detail.

Let the electric field be initially directed along the $\vec{p}_x$ axis ($\vec{E} = \vec{E}_x$) perpendicular to the extension of the Fermi surface. In

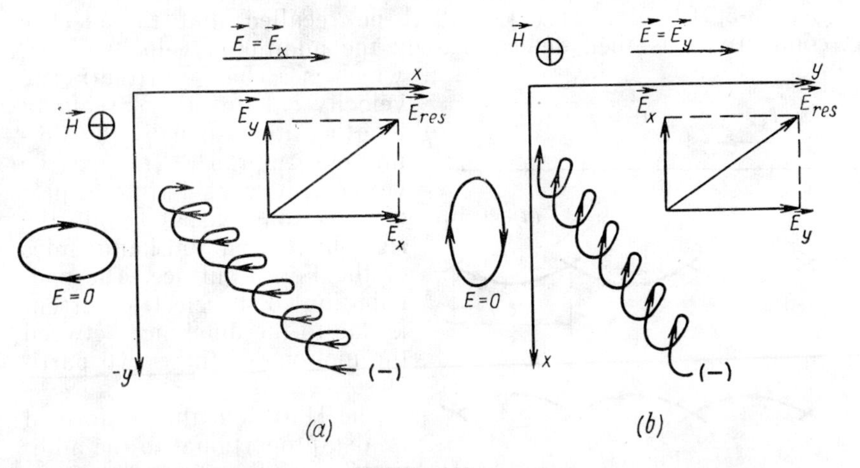

Fig. 183

a finite-size specimen of metal, a Hall field will be established and the drift of electrons will occur in the resulting field $\vec{E}_{res}$. The path of the drift motion of an electron then is as shown in Fig. 183*a*. Shown in the same figure is the shape of the path at $E = 0$.

With an increase of the anisotropy of the Fermi surface the turns of the path become more and more extended and finally the length of one turn exceeds the free-path length. Scattering then occurs before an electron travels a full turn of the path. Under such circumstances the condition of strong magnetic fields (i.e. the condition of cyclic motion of carriers) is disturbed. This means that if the anisotropy of the Fermi surface is increased for the same magnitude of the magnetic field, then at a definite value of this anisotropy the conditions of motion of carriers in a strong field change to those in a weak field.

The paths of electrons in a weak magnetic field between the points of successive acts of scattering represent a small portion of a turn of a trochoid and are curved only slightly (compare Figs. 165 and 184), the mean direction of a path being coincident with the direction of the electric field $\vec{E}$. It is also evident that paths of the type *a* and *b* shown in Fig. 184 have equal probability to exist. They correspond to the motion of Fermi electrons located respectively on the left-hand (type *a*) and right-hand (type *b*) sides of the Fermi surface.

The nature of motion of electrons in a weak magnetic field has been discussed in qualitative terms earlier in this book (see Sec. **3-5**, Chapter Three). Let it be recalled that the electric conductivity is then determined by the additional velocity $\tilde{v}(E)$, which is a change of the Fermi velocity along the free-path length, rather than by the Fermi velocity proper. This may be verified by considering a pair of electrons located on the right-hand and left-hand sides of the Fermi surface. The contribution to the electric current is due to the difference between the motions of these two particles.

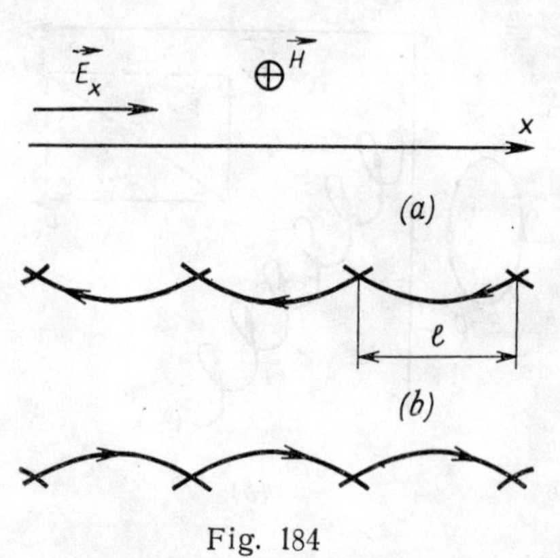

Fig. 184

The Hall field that is formed is also proportional to the additional velocity $\tilde{v}(E)$. The force of the Hall field counterbalances the effective Lorentz force and straightens the respective curvature of the paths corresponding to the additional velocity $\tilde{v}(E)$. For this pattern of motion of carriers the conduction current is proportional to $\tilde{v}(E)$, and the conductivity σ is dependent here on magnetic field only owing to the scattering of the additional velocities about their mean value $\tilde{v}(E)$.

Thus, in the whole interval of magnetic fields in which the conditions of a weak field are maintained for the given value of anisotropy of the Fermi surface, the following dependence of the transverse magnetoresistance $\rho(H)$ on magnetic field is observed: with an increase of the magnetic field, ρ increases initially by a quadratic law and then comes to saturation.

The saturation takes place because the mechanism of growth of the resistance in the magnetic field, linked with the spread of the additional velocities of electrons, is effective only within a limited interval of fields at which the mean angle $\bar{\alpha}$ of deviation of the paths from a straight line is small (see Sec. **3-5**, Chapter Three). In stronger fields, angle $\bar{\alpha}$ becomes a constant; it is linked physically with that the majority of the electrons have additional velocities close to the mean value $\vec{\tilde{v}}(E)$. The conductivity decreases initially owing to a small portion of the electrons whose velocities differ noticeably from $\vec{\tilde{v}}(E)$. With an increase of the magnetic field, electrons having these velocities are deviated more and more strongly by the effective Lorentz force and practically cease to contribute to the electric conductivity, which then continues to be formed by the motion of the electrons of the main group.

Thus, with an increase of the magnetic field, the contribution to the electric conductivity is due to the lower number of electrons which have a small spread of their additional velocities. If the conditions of a weak field are maintained, this process causes saturation. The total change of the magnetoresistance $\Delta\rho$ at the transition from a low to a high strength H of the field for an anisotropic Fermi surface is rather large, of the order of ρ_0.

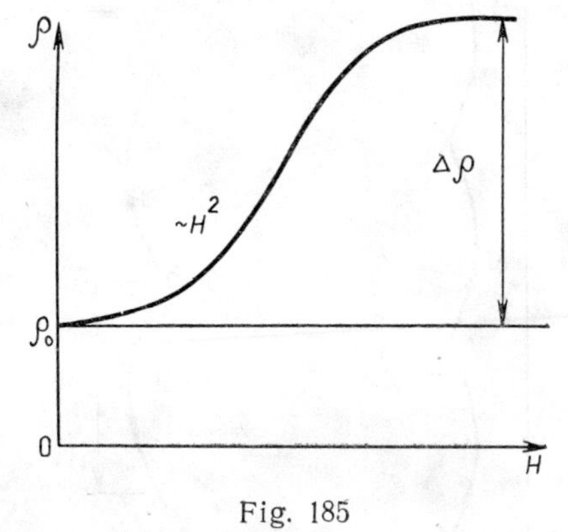

Fig. 185

The dependence of ρ on H for the case considered is illustrated in Fig. 185. [The total change of the resistance $\Delta\rho$ depends on the width of the curve of spread of the velocities of electrons about $\tilde{v}(E)$.] The same dependence of ρ on H must then remain with a transition from a strongly anisotropic Fermi surface to an open one, provided that the electric field remains perpendicular to the mean direction of openness. The difference consists only in that, for an open Fermi surface, the motion of carriers under conditions of a weak magnetic field is retained with any strength H of the field. The shape of the paths of electrons then remains such as shown in Fig. 184.

Let us now consider a case when the orientation of the electric field coincides with the direction of extension of an anisotropic Fermi surface $(\vec{E} = \vec{E}_y)$. Then, after establishing a Hall field, electrons begin to drift in the resulting electric field $\vec{E}_{res}$ as shown in Fig. 183*b*.

With a sufficiently high anisotropy, an electron is scattered before finishing a single turn along its path. But in that case the direction of motion, which is rigidly determined by the shape of the anisotropic Fermi surface, is practically perpendicular to the external electric field. The shape of paths of electrons in real space is illustrated in Fig. 186.

The mechanism of formation of the conduction current is then of the same nature as the mechanism of the collision current considered earlier. As has been shown, the electric conductivity σ is then dependent on magnetic field as $1/H^2$.

A continuous decrease of electric conductivity may be explained by that, as the magnetic field H is being increased, this shape of the paths of electrons (Fig. 186) gives a continuous rise of the number of scattering acts per unit length of the specimen of metal along the y axis.

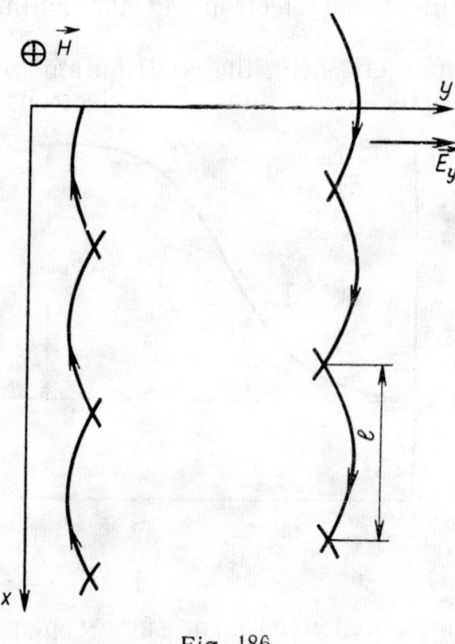

Fig. 186

The transition from a strongly anisotropic Fermi surface to an open one, as in the earlier case of the orientation of the electric field perpendicular to the direction of extension, only means that the described conditions of the motion of carriers with an open surface are retained for any strength of the magnetic field, the transverse magnetoresistance increasing unlimitedly proportional to H^2.

To conclude this section, we give the main types of dependence of the transverse magnetoresistance on a strong magnetic field in a metal with an open Fermi surface. Let α be the angle between the conduction current $\vec{J}$ (i.e. the external electric field $\vec{E}$) and the mean direction of openness of the Fermi surface.

In the general case, as has been shown by Lifshits and Kaganov [11], three main types of dependence of ρ on H and α are possible in strong magnetic fields:

(1) for the given direction of the magnetic field, there are no open sections of the Fermi surface and the volumes of hole-type and electron-type portions of the Fermi surface are equal to one another ($n = p$); in that case, for any α:

$$\rho(H) \sim H^2 \tag{3.50}$$

(2) for the given direction of the magnetic field, there are no open sections, but the volumes of the electron-type and hole-type

portions of the Fermi surface are not equal ($n \neq p$); then for any α:

$$\rho(H) = B = \text{const} \tag{3.51}$$

(3) for the given direction of the magnetic field, there is a layer of open sections with a single mean direction of openness; then

$$\rho(H, \alpha) = AH^2 \cos\alpha + B \tag{3.52}$$

where A and B are constants.

The relationships for strong magnetic fields discussed in this chapter have been obtained on the basis of quasi-classical approach to the motion of current carriers. They are valid for the monotonous (non-oscillating) portion of the magnetoresistance in the region of magnetic fields $H < H_a$ in which quantum effects are still insignificant. For large groups of current carriers in metals (in which the size of the Fermi surface is of the same order of magnitude as the size of the Brillouin zone), these relationships hold true for all really attainable magnetic fields. They cease to be valid for small groups of carriers in polyvalent metals and semi-metals in the ultra-quantum region of magnetic fields. The electric conductivity of a metal then requires a special analysis, which is beyond the scope of this book.

CHAPTER FOUR

EXPERIMENTAL METHODS FOR STUDYING THE ENERGY SPECTRUM OF ELECTRONS IN METALS

4-1. GENERAL REMARKS

Experimental methods for studying the band structure of metals are mostly based on investigation of various physical effects in metals occurring in the presence of an external magnetic field. A feature common of all these methods is the use of strong magnetic fields in which cyclic motion of current carriers is observed. This is linked with that the nature of cyclic motion is determined by the topology and shape of the Fermi surface, and therefore, allows a certain information on that surface to be obtained.

According to the type of effect used and the magnitude of the magnetic field methods for studying the electron energy spectrum can be divided into two groups. The first of them includes:

(a) galvano-magnetic effects in strong magnetic fields;
(b) the Azbel-Kaner cyclotron resonance in metals;
(c) the Gantmacher size radio-frequency effect; and
(d) the Pippard magneto-acoustic resonanse.

Of the methods of the second group, the following may be mentioned:

(a) quantum-mechanical magneto-acoustic resonance, otherwise called gigantic oscillations of ultrasonic absorption in metals; and
(b) oscillational quantum effects.

The division of physical effects and related methods of studying the spectra into two groups is linked with what concepts, whether quasi-classical or quantum, are used to interprete the given effect. The methods of the first group may be conditionally termed quasi-classical by this feature. The related phenomena are observed in such magnetic fields where the energy of orbital splitting $\hbar\omega$ is small compared with the energy of external actions.

Notwithstanding the cyclic nature of motion in that region of magnetic fields, the energy spectrum of electrons behaves as a quasi-continuous one. Under the influence of external actions, electrons pass easily from one Landau level to another, i.e. vary the quantum number n. For that reason, account of the quantization of energy for these effects is not critical for the analysis of motion of electrons.

The second group of methods uses the phenomena which may conditionally be termed quantum effects. They are observed in substantially stronger magnetic fields [1]) where the distance between Landau levels, $\hbar\omega$, becomes sufficiently large compared with the energy of external actions. External factors then either cause no passage of electrons from one Landau level to another or these passages are of resonant nature.

To conclude the general characteristic of the methods for studying the energy spectrum, it should be emphasized that a complex study using various methods is actually required for determining the band structure of some or other metal. For simple metals, the starting point for such a study is the construction of the Fermi surface by Harrison's method which is used as the basis for interpreting the experimental data obtained.

4-2. GALVANO-MAGNETIC METHODS FOR STUDYING THE ENERGY SPECTRUM

The galvano-magnetic methods for studying the zonal structure of metals are based on measurements of electric conductivity and Hall effect in weak and strong magnetic fields having different orientations relative to the axes of the crystal being studied. The total complex of these data makes it possible to find the components of the tensor of magnetoresistance ρ_{ij} as a function of the magnetic field H. Interpretation of the data of galvano-magnetic measurements is essentially made on the basis of the data on electric conductivity of the metal in weak and strong magnetic fields (see Secs. **3-4-3-7** of the previous chapter).

An essential merit of the galvano-magnetic methods is that they are simple and do not require a high purity and perfection of monocrystals. These methods were historically the first to be employed for systematic studies of electrophysical parameters of metals owing to these circumstances. Thus, for instance, first indications on the complexity of energy spectrum of electrons in metals were obtained by measuring the anisotropy of the transverse magnetoresistance in strong magnetic fields at low temperatures. Exactly these data played a substantial part in developing the theory of Fermi surfaces of metals, and especially in proving the existence of open paths of electrons.

---

[1]) In one and the same metal, there is no sharp boundary between the intervals of magnetic fields in which the effects of the first and second groups are observed. These intervals actually overlap, owing to which the quasi-classical concepts (for instance, the path of motion of electrons) are used to interprete qualitatively the quantum effects.

The complex nature of the dependence of transverse magnetoresistance of some metals having open electron paths on magnetic field may be illustrated by Fig. 187 [78], which shows the anisotropy of the magnetoresistance of gold in a magnetic field $H = 23\,500$ oersteds whose vector $\vec{H}$ is located in the plane [110] of the crystal. The unusually strong dependence of resistance on the magnetic field orientation is linked with that in the given plane of crystal of gold there exist a number of close directions in which the Fermi surface is open.

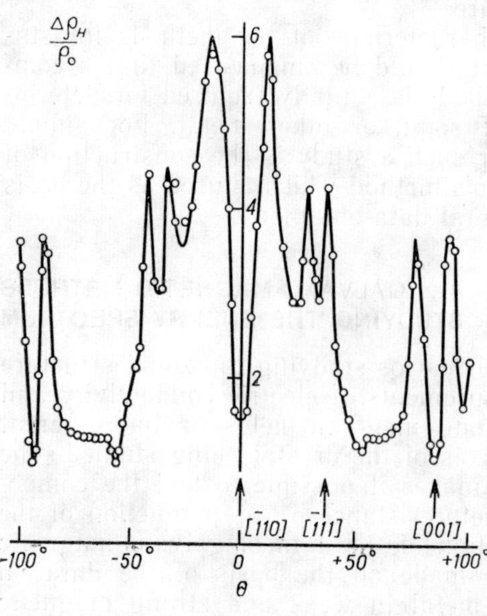

Fig. 187

It should be noted, however, that the possibilities of a detailed investigation of the Fermi surface on the basis of galvano-magnetic measurements are rather restricted. This follows from that the contributions of various zones to the electric conductivity of the metal are additive, so that a low conductivity in some direction in one zone may be masked by a high conductivity in another zone.

Interpretation of measured data on magneto-conductivity and Hall effect in a multi-zone metal whose Fermi surface is composed of a number of closed constant-energy surfaces requires, in essence, the knowledge of the particular model of the zone structure built on the basis of independent studies.

Let us discuss in detail a very important feature of the galvanomagnetic methods for studying the energy spectrum which are based on measuring the components of the galvano-magnetic tensor ρ_{ij} in weak magnetic fields $H \ll H_l$. This feature is that such measurements can be made within a very wide temperature interval from very low temperatures $T \ll \Theta_D$ to high temperatures $T \gtrsim \Theta_D$. Measurement of the components of galvano-magnetic tensor in weak fields is practically the sole means for determining the temperature relationships for various parameters of a metal (such as relaxation time τ of carriers, free-path length l, etc.) in a wide temperature region, provided that the given structure has been sufficiently reliable found by other methods at low temperatures. The point is that all the methods for studying the energy spectrum that will be discussed later are related to the cyclic motion of carriers and essentially are low-temperature methods.

In order to explain what has been said above, let us consider in more detail the structure of the components of galvano-magnetic tensor and establish the characteristics of the metal spectrum they are linked with.

We introduce the concept of mobility of current carriers. As has been shown earlier, the density j of an electric current formed by the carriers on some constant-energy surface may be expressed through the drift addition $\Delta v(E)$ to the Fermi velocity, which depends on the magnitude of the electric field. In that case the expression for the current density is of the following form: $j = |e|n\Delta v(E)$, where n is the concentration of electrons (or holes) inside the constant-energy surface. For simplicity, we reduce our discussion only to closed surfaces whose anisotropy is not very high.

Within the applicability of Ohm's law, the current density j is expressed linearly through the electric field strength E: $j = \sigma E$, and therefore, the drift addition $\Delta v(E)$ to the Fermi velocity is proportional to E. The coefficient of proportionality between $\Delta v(E)$ and E is termed the mobitily μ of current carriers:

$$\mu = \frac{\Delta v(E)}{E} \tag{4 1}$$

Thus, mobility μ is determined by the variation of the Fermi velocity in an electric field of a unit strength. The electric field strength is usually measured in volts per centimetre, and velocity, in centimetres per second, so that the unit of mobility is $cm^2/V \cdot sec$.

For an anisotropic constant-energy surface, the drift velocity $\Delta v(E)$ depends on the orientation of the electric field. The mobility of carriers is then described, as electric conductivity, by a second-rank tensor termed the tensor of mobility μ_{ij}.

It follows from (4.1) that the vector of current density can be expressed in terms of the mobility tensor as follows:

$$j_i = |e| n \mu_{ik} E_{ik} \tag{4.2}$$

By comparing formulae (4.2) and (3.1), we can establish the relationship between the tensors σ_{ik} and μ_{ik}:

$$\sigma_{ik} = |e| n \mu_{ik} \tag{4.3}$$

This expression refers to a single constant-energy surface and implies that the mobility tensor μ_{ik} is symmetrical, as also is the conductivity tensor, i.e. the equality $\mu_{ik} = \mu_{ki}$ holds true. For a number of surfaces, the total conductivity tensor is the sum of expressions of the type of (4.3) over all the existing constant-energy surfaces.

The relationship between mobility and other characteristics of current carriers can be established on a simple example of the quadratic isotropic law of dispersion. For this, we compare expression (4.3) with the Lifshits formula for electric conductivity in the form of (3.16). For isotropic mobility μ, we then have:

$$\mu = \frac{|e|\tau}{m^*} \tag{4.4}$$

In 1956, Abeles and Meiboom [79] have shown theoretically that the current density vector $\vec{j}$ in a finite-size specimen of metal in the presence of a magnetic field $\vec{H}$ of an arbitrary orientation relative to an external electric field $\vec{E}$ can be written as:

$$j_i = \sigma_{ik}(\vec{E} + \vec{E}_{Hall})_k = \sigma_{ik}\left(\vec{E} + \frac{1}{|e|n}[\vec{j} \cdot \vec{H}]\right)_k \tag{4.5}$$

where the conductivity tensor σ_{ik} is expressed in terms of the mobility tensor μ_{ik} by means of formula (4.3) and is independent of magnetic field (the subscript at the parenthesis implies that the expression includes the k-th component of the vector given in the parentheses).

Expression (4.5) is valid for the given closed constant-energy surface under the quadratic law of dispersion of current carriers. It has been shown later that, for a closed surface of not a very high anisotropy, expression (4.5) also holds true for more complicated dependences of ε on $\vec{p}$. It follows from the structure of formula (4.5) that the Hall field $\vec{E}_{Hall}$ is determined by the component of the magnetic field $\vec{H}$ perpendicular to the electric current $\vec{j}$ (or to the external electric field $\vec{E}$) and is equal to:

$$\vec{E}_{Hall} = \frac{1}{|e|n}[\vec{j} \cdot \vec{H}] \tag{4.6}$$

Equation (4.5) can be solved for the electric current density $\vec{j}$ and written as:

$$j_i = \sigma_{ik}(H)\, E_k \tag{4.7}$$

where the components of the new conductivity tensor $\sigma_{ik}(H)$, depending on magnetic field, are expressed through the components of tensor μ_{ik} and vector $\vec{H}$:

$$\sigma_{ik}(H) = |e|n\left(\mu_{ik}^{-1} + \frac{1}{c}B_{ik}\right)^{-1} \tag{4.8}$$

where the "—1" sign at the top implies that the matrix of the corresponding tensor A_{ik}^{-1} (for instance, μ_{ik}^{-1}) is an inverse one relative to the matrix A_{ik}. (Let it be recalled that μ_{ik} is independent of H.)

The antisymmetrical tensor B_{ik} can be found in a certain system of coordinates as follows:

$$B_{ik}=\begin{pmatrix} 0 & -H_3 & H_2 \\ H_3 & 0 & -H_1 \\ -H_2 & H_1 & 0 \end{pmatrix} \tag{4.9}$$

where H_1, H_2, and H_3 are the components of vector $\vec{H}$ in this system. The components of magneto-conductivity tensor $\sigma_{ik}(H)$, written in the form of (4.8), evidently obey Onsager reciprocity principle, which is expressed by the equality:

$$\sigma_{ik}(H)=\sigma_{ik}(-H) \tag{4.10}$$

The total conductivity tensor $\sigma_{ik}^{tot}(H)$ of a multi-zone metal in a magnetic field is obtained by summing up expressions of the type of (4.8) written in one and the same system of coordinates, over all constant-energy surfaces. In the end, this makes it possible to express the components of magnetoresistance tensor $\rho_{ik}^{tot}(H)=[\sigma_{ik}^{tot}(H)]^{-1}$, which can be measured directly in the experiment through the concentrations of the carriers of each type, the components of their tensors, and the magnitude of the magnetic field. For a component of the magnetoresistance tensor, Onsager principle also holds true: $\rho_{ik}^{tot}(H)=\rho_{ki}^{tot}(-H)$. For weak magnetic fields $H \ll H_l$, we can expand the complicated expression (4.8) into powers of the magnetic field and limit ourselves to the second-order terms. The parameter of expansion is the ratio H/H_l which, as may be easily shown, is equal to $\omega\tau$ or $\mu H/c$ in an isotropic case.

The components of galvano-magnetic tensor can be measured at any temperature in such a weak magnetic field H for which the inequality $\frac{\mu H}{c} \ll 1$ deliberately holds true for any component of the mobility tensor, and therefore, conditions of a weak field are retained (for good monocrystalline specimens of a metal, the strength of the magnetic field required for measurements at low temperatures is a few oersteds).

Thus, measurement of temperature dependences of the magnetoresistance tensor in very weak magnetic fields makes it possible not only to simplify expression (4.8) for $\sigma_{ik}(H)$, but also to exclude the influence of such a factor as temperature dependence of the effective magnetic field H_{eff} [see formula (3.27)].

The particular shape of the formula relating the components of magnetoresistance tensor depends on the number of constant-energy surfaces, their mutual orientation, and also on the orientation of the electric and magnetic fields. The magnetoresistance tensor is usually written in a system of coordinates related to the crystallographic axes, the electric and magnetic fields being then directed along different coordinate axes. Directions of the fields are determined by the number of unknown parameters which are to be found by measurements. These parameters include the components of the mobility tensor for each group of carriers, the angles determining the orientation of a constant-energy surface, and the concentrations of carriers.

This may be explained on the simplest example of a one-zone metal with a spherical Fermi surface of the electron type. We introduce an arbitrary system of three Cartesian axes (directed, for instance, along the three mutually perpendicular crystallographic axes). The mobility tensor μ_{ik} will evidently have the form: $\mu_{ik} = \mu\delta_{ik}$. The unknown parameters which characterize the current carriers are the isotropic mobility μ and the concentration n of electrons.

Let the magnetic field $\vec{H}$ be directed along the axis 3: $\vec{H} = \vec{H}_3$. Noting that $\mu_{ik}^{-1} = \frac{1}{\mu}\delta_{ik}$, the tensor of magnetoresistance $\rho_{ik}(H) = \sigma_{ik}^{-1}(H)$ can be written according to (4.8) as:

$$\rho_{ik}(H) = \frac{1}{|e|n}\begin{pmatrix} 1/\mu & -H_3/c & 0 \\ H_3/c & 1/\mu & 0 \\ 0 & 0 & 1/\mu \end{pmatrix} \tag{4.11}$$

In order to find the two unknown parameters μ and n, it suffices, for instance, to direct the electric current $\vec{j}$ along the axis 1 and measure the two values of the electric field, E_1 and E_2, in the direction of axes 1 and 2 (E_2 is Hall's field).

Let it be recalled that the direction of current is given in the experiment by the geometry of the crystal. In the case considered, the specimen of metal must be cut in the form of a rectangular bar with the long side along the axis 1, as shown in Fig. 188. Shown in the same figure are electrodes for measuring the potential difference along the specimen (u_1) and in a lateral direction (u_2). The ratios of u_1 and u_2 to the distances between the corresponding electrodes give the magnitudes of electric fields E_1 and E_2.

Using expression (4.11) for the tensor $\rho_{ik}(H)$, we can write:

$$E_1 = \rho_{11}j_1 = \frac{1}{|e|n\mu}j_1$$

$$E_2 = \rho_{21}j_1 = \frac{1}{|e|nc}H_3j_1 \tag{4.12}$$

from which μ and n can be found by the measured values of E_1, E_2, and j_1.

Note that expression (4.5) for the current density does not take into account the scatter of drift velocities about the mean value $\tilde{v}(E)$. As has been shown in Secs. **3-5** and **3-7**, Chapter Three, the transverse magnetoresistance in a metal with a single closed constant-energy surface increases by a quadratic law in weak magnetic fields only owing to the scatter of drift velocities and becomes saturated in strong fields. If we neglect the scatter of drift velocities, then, as follows from the shape of tensor $\rho_{ik}(H)$ (4.11), the transverse magnetoresistance becomes independent of magnetic field H. This occurs owing to that the force formed by the Hall field E_2 compensates the effective Lorentz force F_H^* at any magnitude of the magnetic field and completely straightens the paths of electrons moving with the same drift velocity $\tilde{v}(E_1) = \Delta v(E_1)$.

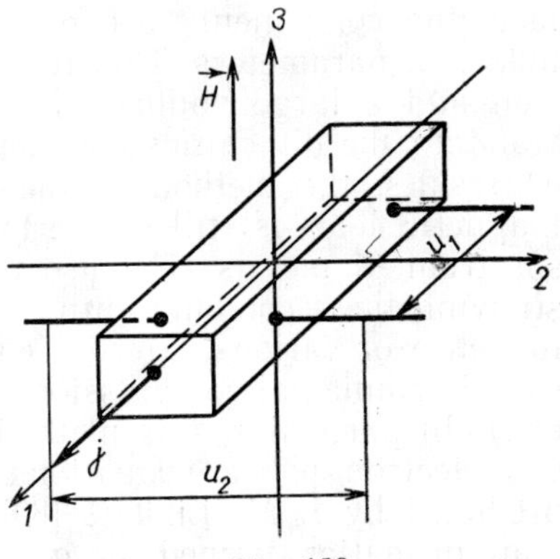

Fig. 188

It can be easily shown that the transverse magnetoresistance in an infinite specimen of metal in the absence of Hall field ($E_2 = 0$) will grow by a quadratic law with any magnitude of the magnetic field because the paths of electrons are curved through the action of the force F_H^*. For this, we invert the magnetoresistance tensor $\rho_{ik}(H)$ (4.11) and write the corresponding magneto-conductivity tensor $\sigma_{ik}(H)$ as follows:

$$\sigma_{ik}(H) = |e|n \begin{bmatrix} \dfrac{\mu}{1+\dfrac{\mu^2 H_3^2}{c^2}}, & \dfrac{\dfrac{\mu^2 H_3}{c}}{1+\dfrac{\mu^2 H_3^2}{c^2}}, & 0 \\ -\dfrac{\mu^2 \dfrac{H_3}{c^2}}{1+\mu^2 \dfrac{H_3^2}{c^2}}, & \dfrac{\mu}{1+\dfrac{\mu^2 H_3^2}{c^2}}, & 0 \\ 0 & 0 & \mu \end{bmatrix} \tag{4.13}$$

With the current directed along the axis 1 and the Hall field equal to zero ($E_2 = 0$) as before, we get:

$$j_1 = \frac{|e|n\mu}{1+\dfrac{\mu^2 H_3^2}{c^2}} E_1 \tag{4.14}$$

Hence it follows that the transverse magnetoresistance $\rho'_{11}(H) = E_1/j_1$ of a specimen which is infinite along the axis 2 is a quadratic function of magnetic field:

$$\rho'_{11}(H) = \frac{1}{|e|n\mu}\left(1 + \frac{\mu^2 H_3^2}{c^2}\right) = \rho_0\left(1 + \frac{\mu^2 H_3^2}{c^2}\right) \qquad (4.15)$$

This simple example shows, to some extent, how the components of the magnetoresistance tensor of a metal can be calculated and measured. For multi-zone metals with a complicated Fermi surface, the components are expressed through a large number of unknown parameters. This results in a rather complicated expression and a large volume of calculations required. But, notwithstanding the difficulties connected with processing of the measured results, the method of the galvano-magnetic tensor in weak magnetic fields is still in use for investigations of electron energy spectrum of metals. The use of this method is of interest when studying the recombination of the initial band spectrum under the influence of various external effects (such as strong three-dimensional or uniaxial compression, introduction of alloying impurities, etc.). In particular, this method was employed for studying various electron phase transitions under pressure, which had been predicted by I. M. Lifshits [80], and also transitions into a new state of matter, termed the gapless state, in alloys of semi-metals antimony and bismuth.

4-3. THE AZBEL-KANER CYCLOTRON RESONANCE

The phenomenon of the cyclotron resonance in metals was predicted theoretically by Azbel and Kaner in 1956 [75] and has been named after them. Experimentally, it was observed on tin by Fawcett [81] in the same year.

The phenomenon is essentially in that electrons of a metal, provided certain conditions are observed, undergo a resonant acceleration under the action of an alternating electromagnetic field, similar to what occurs with electrons in a cyclotron.

Let the vector of a variable electric field $\vec{\mathscr{E}}$ in a linearly polarized electromagnetic wave be parallel to the surface of a metal specimen and let a constant magnetic field $\vec{H}$ be also parallel with that surface and perpendicular to vector $\vec{\mathscr{E}}$. This mutual location of external fields and of the surface of a specimen such as shown in Fig. 189, has been called the Azbel-Kaner geometry.

Let $\vec{\mathscr{E}}_0(0)$ and $\vec{\mathscr{H}}_0(0)$ be the amplitudes of alternating electric and magnetic fields on the surface of a metal. The strengths of the fields $\vec{\mathscr{E}}(0)$ and $\vec{\mathscr{H}}(0)$ on the surface in dependence on time can be written as:

$$\vec{\mathscr{E}}_0 = \vec{\mathscr{E}}_0(0)\, e^{i\Omega t} \quad \text{and} \quad \vec{\mathscr{H}}(0) = \vec{\mathscr{H}}_0(0)\, e^{i\Omega t}$$

where Ω is the frequency of the electromagnetic field.

As is known from electrodynamics, an alternating electromagnetic field attenuates by an exponential law at the depth of a metal. The amplitudes of the electric $\vec{\mathscr{E}}_0(y)$ and magnetic $\vec{\mathscr{H}}_0(y)$ fields at a depth y can be expressed through $\vec{\mathscr{E}}_0(0)$ and $\vec{\mathscr{H}}_0(0)$ as follows:

$$\begin{aligned} \vec{\mathscr{E}}_0(y) &= \vec{\mathscr{E}}_0(0)\, e^{-y/\delta} \\ \vec{\mathscr{H}}_0(y) &= \vec{\mathscr{H}}_0(0)\, e^{-y/\delta} \end{aligned} \tag{4.16}$$

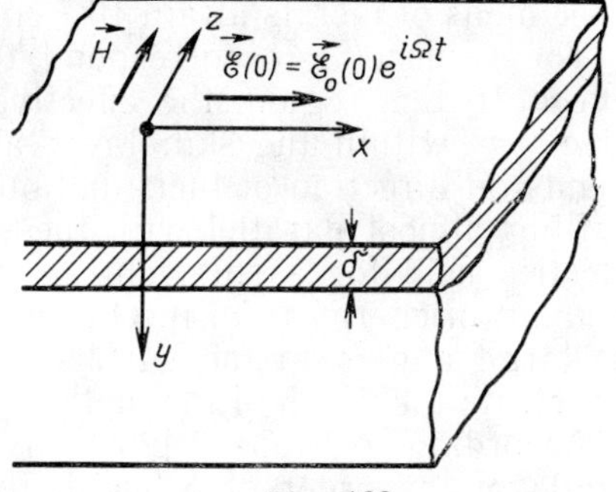

Fig. 189

The parameter δ defines the effective depth of penetration of the electromagnetic field into a metal and is termed the depth of the skin layer.

Since the depth of skin layer δ is one of the main parameters determining the conditions of formation of cyclotron acceleration of electrons, we shall discuss in more detail the dependence of δ on characteristics of the metal and frequency Ω of an alternating field.

The physical cause that an electromagnetic field penetrates a finite depth in a metal consists in its interaction with electrons of the metal owing to which the energy of the electromagnetic field is absorbed. The nature of this interaction depends essentially on the ratio between the effective depth of penetration δ and the free-path length l of electrons.

With $l \ll \delta$, all Fermi electrons located within the layer of a thickness of an order of δ interact with an equal effectiveness with the electromagnetic field. The values $l \ll \delta$ correspond to the region of normal skin effect. It has been shown in electrodynamics that the depth of skin layer can then be determined by the classical expression:

$$\delta = \frac{c}{\sqrt{2\pi\mu\sigma\Omega}} \tag{4.17}$$

where μ and σ are respectively the magnetic susceptibility and conductivity of the metal.

The normal skin effect is observed in metals at temperatures close to room temperature up to the frequencies of oscillations relating to the SHF range. Indeed, for a good conductor ($\sigma \sim \sim 5 \times 10^5$ ohm^{-1} cm^{-1} $\approx 5 \times 10^{17}$ sec^{-1}) at room temperature and frequency $\frac{\Omega}{2\pi} \sim 3 \times 10^{10}$ Hz, the depth of skin layer δ, according to formula (4.17), is of the order of magnitude of 10^{-4} cm. This magnitude of δ exceeds by several orders the free-path length l, which is 10^{-7}-10^{-8} cm under the same conditions.

With a reduction of temperature the free-path length increases and may become greater than the depth of skin layer. Thus, at temperatures of liquid helium the typical values of the free-path length for electrons on a Fermi surface in good monocrystalline specimens of metals attain 10^{-2} cm and more.

The values $l > \delta$ correspond to the region of anomalous skin effect. In this region, the effectiveness of interaction between the electrons within the skin layer and the electromagnetic field depends on direction of their motion. For instance, the electrons travelling almost parallel with the surface of a metal remain in the electric field for a sufficiently long time to absorb a substantial amount of energy from it. On the other hand, the electrons moving at large angles to the surface rapidly leave the skin layer and penetrate the depth of the metal.

According to Pippard [28], only the electrons moving within an angle of the order of δ/l with the surface of metal can interact effectively with the electromagnetic field. The concentration of effective electrons n_{eff} is $\gamma \frac{\delta}{l} n$, where n is the total concentration of electrons in the metal and γ is a proportionality factor of an order of unity. The expression for the depth of skin layer δ in the region of anomalous skin effect includes the conductivity σ_{eff} which is mainly contributed to only by the electrons effectively interacting with the electromagnetic field. For an isotropic metal for which formula (3.16) holds true, this effective conductivity is:

$$\sigma_{eff} = \frac{|e| n_{eff} \tau}{m^*} = \gamma \frac{\delta}{l} \cdot \frac{|e|^2 n\tau}{m^*} = \gamma \frac{\delta}{l} \sigma \qquad (4.18)$$

If σ_{eff} is substituted for σ in expression (4.17) and the equation obtained is solved for δ, then for the effective depth of skin layer we shall have:

$$\delta_{eff} = \left(\frac{c^2 l}{2\pi\gamma\mu\sigma\Omega} \right)^{1/3} \qquad (4.19)$$

This formula determines the depth of penetration of an electromagnetic field into metal in the region of anomalous skin effect.

Thus, with the passage from the region of normal skin effect to that of anomalous one, the form of dependence of δ on σ and Ω changes and expression (4.17) transforms into (4.19).

Let us now discuss the phenomenon of cyclotron resonance. As has been shown by Azbel and Kaner, cyclotron acceleration of electrons in a metal under the action of an external electromagnetic field can only occur under the following conditions:

$$l > r_H, \qquad r_H \gg \delta \tag{4.20}$$

whence it follows that the cyclotron resonance is observed in the region of anomalous skin effect.

The first of inequalities (4.20) is the well-known condition of cyclic motion. It is easily understood that this condition is required for cyclotron acceleration of electrons: for an electron to pass over into the resonant conditions of acceleration and to acquire a sensible additional energy from an alternating electromagnetic field, it must complete, in any case, more than one rotation around a closed orbit before scattering occurs. The second of the inequalities must be discussed in more detail.

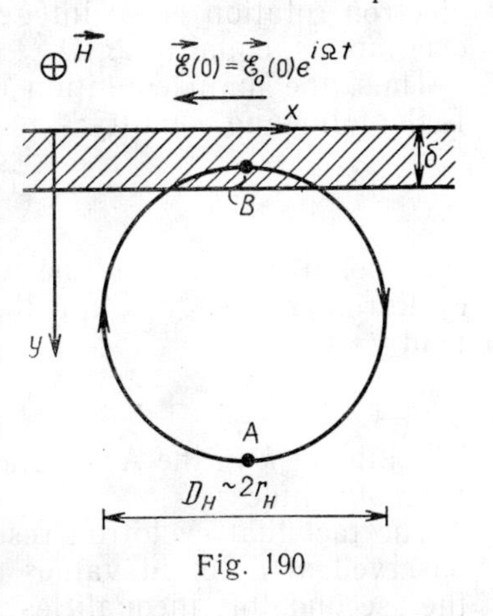

Fig. 190

In a magnetic field parallel to the surface of the metal, electrons either describe closed orbits in planes perpendicular to the surface or move along helices whose axes are parallel to the direction of the magnetic field. The paths of the first type correspond to central sections through constant-energy surfaces at $p_{\parallel} = 0$. The second type corresponds to sections at $p_{\parallel} \neq 0$.

With a closed Fermi surface of an arbitrary shape, the path of electrons (or its projection onto a plane perpendicular to the magnetic field) is a closed curve which may differ substantially from a circle. The characteristic sizes of the orbit are determined by the quantity $r_H = \frac{cp_F}{|e|H}$. The inequality $r_H \gg \delta$ implies that only a small portion of the orbit of individual electrons is located within the skin layer.

When discussing the physical essence of the cyclotron resonance, we shall assume for simplicity that the Fermi surface of the metal is sufficiently close to a sphere. An electron orbit passing through the skin layer in that case is shown in Fig. 190. It follows from the figure that at $r_H \gg \delta$ the interaction of an electron with the electric field within the skin layer occurs during a small fraction of the period T_H of its cyclotron motion. The action of the electric field is then similar to a short impact after which the electron travels along its path into the depth of the metal and continues to move there only under the action of the static magnetic field $\vec{H}$.

It is then evident that the resonant absorbtion of the energy of an electromagnetic wave occurs only if the electron each time returns into the skin layer at one and the same phase of the electric field.

This synchronism of the motion of the electron with oscillations of the electromagnetic field can only appear if the period T_H of electron rotation is an integer k times greater than the period of oscillations, equal to $2\pi/\Omega$.

Thus, the main condition of resonant acceleration of electrons is the following equality:

$$T_H = k\frac{2\pi}{\Omega} \tag{4.21}$$

This condition can also be written as a relationship between the cyclotron frequency ω and the frequency Ω of the electromagnetic field:

$$\Omega = k\omega \tag{4.22}$$

The integer k in the Azbel-Kaner resonance is termed the resonant harmonic index.

The fact that cyclotron resonance of electrons in metals can be observed at different values of $k = 1, 2, \ldots$ is a consequence of the second of inequalities (4.20): $r_H \gg \delta$. In semiconductors, where skin effect is practically non-existent because of the low conductivity and the electromagnetic field penetrating the metal can be assumed to be homogeneous, the cyclotron resonance is only possible with $k = 1$ (it is then called the semiconductor cyclotron resonance). The mechanism of acceleration of an electron with the semiconductor resonance is as follows.

A linearly polarized oscillation of an electric field may be represented as superposition of two circularly polarized oscillations in which the vectors of the electric field rotate towards one another with the cyclic frequency Ω. With the frequencies being equal, $\omega = \Omega$ ($k = 1$), the electron travelling on its orbit in the magnetic field $\vec{H}$ is constantly acted upon by the electric field rotating in the same direction as that of precession of the electron. An electric field rotating in the opposite direction practically does not interact with the electron. Under such conditions, resonant absorption of the energy of the electromagnetic wave is observed.

This mechanism of resonant interaction can only be at work with the frequencies being strictly equal to one another, $\omega = \Omega$, since only then the energy from an alternating electric field can be transmitted to the electron during a sufficiently long interval of time (during the whole relaxation time τ).

With any value of the harmonic index $k > 1$, the condition of resonant interaction between the electron and field is disturbed.

This mechanism of energy transfer in the Azbel-Kaber resonance, consisting in that the electron acquires synchronous impacts from the electric field during its short passages through the skin layer at $r_H \gg \delta$, differs substantially from the mechanism of energy transfer in the semiconductor resonance. If the inequality $r_H \gg \delta$ becomes less and less strict and finally δ becomes of the same order of magnitude as r_H, the conditions of Azbel-Kaner cyclotron resonance pass over into those of the semiconductor resonance. It is then evident that resonant transfer of energy on harmonics higher than the first ($k > 1$) will gradually diminish and vanish at $r_H \lesssim \delta$.

Simultaneous fulfilment of inequalities (4.20) for reasonable values of the frequency Ω of electromagnetic field (in practice, $\Omega/2\pi$ does not exceed 10^{11} Hz) is only possible with a sufficiently large free-path length l of electrons which is only observed at the temperatures of liquid helium in rather pure and perfect monocrystalline specimens of metal. This means that the cyclotron resonance is a purely low-temperature effect and disappears with an increase of temperature.

Let us analyse the conditions of synchronism of (4.22). For an arbitrary law of dispersion, the cyclotron frequency $\omega = \frac{|e| H}{m^* c}$ depends on the component of the momentum $p_\parallel$ of the electron parallel to the constant magnetic field $\vec{H}$. This dependence is linked with that the cyclotron mass m^* of the electron, equal to $\frac{1}{2\pi} \cdot \frac{\partial S}{\partial \varepsilon}$ [see expression (2.73)], is determined in the general case by the position of the secant plane $p_\parallel = \text{const}$ of the Fermi surface in which the orbit of the electron is located in $\vec{p}$-space: $m^* = f(p_\parallel)$.

If the relationship $m^* = f(p_\parallel)$ is monotonous, then the electrons belonging to different sections will undergo resonant acceleration at different frequencies Ω of the electromagnetic field. Quite large groups of electrons having frequencies close to Larmor precession are located on the Fermi surface near the sections for which the condition $\frac{\partial m^*}{\partial p_\parallel} = 0$ holds true.

Let it be recalled that with a spherical or ellipsoidal Fermi surface and a quadratic dispersion law, the cyclotron mass of electrons is independent of $p_\parallel$. In such cases, all electrons on the Fermi surface precess with the same cyclotron frequency and at the same time participate in resonant acceleration.

With a more complicated dispersion law, the condition of the extremal value of cyclotron mass $\frac{\partial m^*}{\partial p_\parallel} = 0$ is in most cases satisfied simultaneously with the condition of the extremal value of the area S of the section of the Fermi surface by a plane perpendicular to the magnetic field $p_\parallel =$ const: $\frac{\partial S}{\partial p_\parallel} = 0$. Coincidence of these two conditions implies that resonant electrons are located in "belts" of a width of the order of $\Delta p_\parallel$ near the maximum or minimum sections of the Fermi surface, as shown in Fig. 115.

Consider a group of electrons whose cyclotron masses practically coincide. These electrons precess in the magnetic field with the same frequency. We select from them the electrons whose orbits are located near the surface of metal and pass through the skin layer, as shown in Fig. 190. These separated electrons, however, precess with arbitrary phases (relative to the phase of the electromagnetic field) and pass through the skin layer in different instants of time which are distributed with equal probability (uniformly) within the time interval equal to the period T_H. This means that among the electrons selected there are such (similar to the electron in point B of the orbit in Fig. 190) that pass through the skin layer at the instant of the maximum of the electric field when the force of interaction $-|e|\vec{\mathscr{E}}$ coincides in direction with their Fermi velocity.

For the electrons (similar to the electron in point A of the orbit in Fig. 190) lagging behind them a half-period of rotation, the instant of passage through the skin layer coincides with the maximum of an electric field of the opposite sign. For these electrons, the force of interaction with the electric field $-|e|\vec{\mathscr{E}}$ is directed opposite to their Fermi velocity. Between the first and second group, there are other electrons on the orbit passing through the skin layer when the electric field changes sign and the time Δt of their passage through the skin layer is zero on the average.

Since the time of passage Δt is small compared with the period T_H, it is more correctly to speak about the momentum of force $-|e|\mathscr{E}\Delta t$, which causes a variation of the Fermi velocity by $\Delta\vec{v}(\mathscr{E})$, rather than about the force of interaction with the electric field. $\Delta\vec{v}(\mathscr{E})$ depends on the amplitude of electric field $\mathscr{E}_0(0)$, the depth of skin layer δ, and the path of the electron and the magnitude of its Fermi energy.

It can be proved by simple transformations that, for elliptical paths of electrons, such as shown in Fig. 191, and a quadratic law of dispersion of carriers, $\Delta v(\mathscr{E})$ is proportional to $\left[\frac{(m_B^*)^{3/2}\mathscr{E}_0^2(0)\,\delta}{\omega\varepsilon_F^{1/2}}\right]^{1/2}$, where m_B^* is the effective mass of the electron in point B of its path. It then follows that the maximum increment of velocity $\Delta v(\mathscr{E})$ (and also of energy $\Delta\varepsilon$) of an electron will be at motion along a path of type 3, provided that the comparison of the magnitude of $\Delta v(\mathscr{E})$ for various paths shown in Fig. 191 is made with the same values of ω and ε_F.

The electrons of the first group, as the electron in point B of the orbit in Fig. 190, will evidently be accelerated by the electric

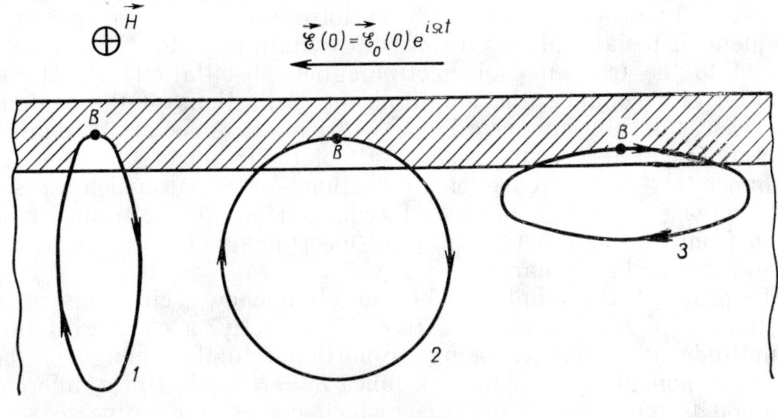

Fig. 191

field and acquire an additional velocity $\Delta v(\mathscr{E})$ with each passage through the skin layer. The electrons of the second group, as the electron in point A of the orbit in Fig. 190, are decelerated by the electric field and their velocity reduced by the same value $\Delta v(\mathscr{E})$ with each passage through the skin layer. But in both cases the change of velocity $\Delta\vec{v}(\mathscr{E})$ coincides with the direction of the momentum of the force $-|e|\mathscr{E}\Delta t$ and is directed opposite to the electric field $\vec{\mathscr{E}}$ at the instant of passage through the skin layer.

The number of passages through the skin layer of the relaxation time τ during which an electron covers along its path a distance coinciding with the free-path length l is equal to l/L, where L is the length (perimeter) of the closed orbit for electrons with $p_{\parallel} = 0$ or the length of one turn of the helix for electrons with $p_{\parallel} \neq 0$. The final velocity of electrons of the first group in the end of the accelerating cycle will be of the order of $v_F + \frac{l}{L}\Delta v(\mathscr{E})$.

Similarly, for the second group of electrons the final velocity is $v_F - \frac{l}{L}\Delta v(\mathscr{E})$. Electrons of both the first and second group contribute to the high-frequency surface current J_{sur} (the frequency of this current being coincident with the frequency of electromagnetic field Ω, since the change of their Fermi velocity $\frac{l}{L}\Delta v(\mathscr{E})$ always be the same with the reverse direction of the electric field vector $-\vec{\mathscr{E}}$).

The electrons participating in the resonance reduce the skin impedance Z of the specimen at the frequency of the electromagnetic field. In order to record the cyclotron resonance, a specimen of metal is usually placed inside an oscillating contour or a cavity tuned to the frequency of electromagnetic oscillations Ω. At the instant of resonance, the amplitude of oscillations of the contour increases sharply in accordance with an increase of its quality factor. The latter occurs as a result of reduction of the skin impedance (or an increase of conduction surfaces), which causes an increase of the coefficient of reflection of the electromagnetic field from the metal. Absorption of electromagnetic energy in the metal then reduces sharply.

In practice, the amplitude of radio-frequency oscillations on a contour can, for instance, be recorded directly, a change of this amplitude at resonance being proportional to the change of the real component of the skin impedance $Z = R + iX$. But a different method is more often employed, which makes it possible to substantially improve the sensitivity of instruments; in this method, a quantity proportional to the derivative of the real component of the skin impedance over the magnetic field $\frac{\partial R}{\partial H}$ is recorded. This method of recording the cyclotron resonance uses modulation of a constant magnetic field H at a low frequency of the audio-frequency range.

The resonant peculiarities of the skin impedance Z in a specimen being studied are observed at a slow increase of the magnetic field H. With an increase of the magnetic field, the size of cyclotron orbit r_H and the cyclotron period T_H reduce. The first peculiarities of the skin impedance, corresponding to high values of the resonant harmonic index k begin to appear when T_H is reduced down to a magnitude of the order of the relaxation time τ, i.e. with the passage to the cyclic motion of electrons.

The resonant peculiarities of impedance, resulting in a decrease of its real component, appear each time when equality (4.22) is fulfilled for successive whole values of the index k [of course, the second inequality (4.20) is then assumed to be fulfilled]. The amplitude of resonant peaks increases with a reduction of k as a result of an increase of the number of passages through the skin layer during the whole cycle of acceleration of an electron determined by the ratio $\frac{l}{L} \sim \frac{l}{2\pi r_H} \sim H$. The last resonant peak, corresponding to $k=1$, is observed with the equality of the frequencies, $\omega = \Omega$. With a further increase of the magnetic field, the main condition of cyclotron resonance (4.21) ceases to hold true.

The greater τ, the lower is the field strength H at which resonances appear, and the greater number of resonant peaks can be observed with the same frequency Ω of electromagnetic oscillations.

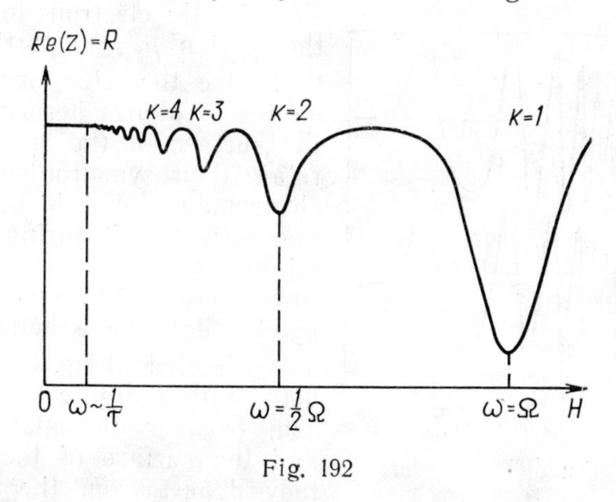

Fig. 192

An increase of the number of resonant peaks by increasing the frequency Ω at a constant relaxation time τ of the metal is limited by the possibility of forming a SHF high-quality cavity of extremely small dimensions. The limiting values of frequencies $\frac{\Omega}{2\pi}$ usually employed in experiments are of an order of 40-80 GHz.

The dependence of the real component of the skin impedance R on magnetic field at appearance of the cyclotron resonance at various harmonics of frequency ω is shown schematically in Fig. 192.

An experimental curve of the cyclotron resonance for copper, obtained at a frequency of the electromagnetic field of $\frac{\Omega}{2\pi} = 2.4 \times 10^{10}$ Hz, and a corresponding curve calculated by the Azbel-Kaner theory for the suitable values of m^* and τ are shown in Fig. 193. Note that recording of the cyclotron resonance of copper has been done by means of the modulation technique.

If for the given direction of the magnetic field $\vec{H}$ there are a number of sections of the Fermi surface corresponding to the extremal values of the cyclotron mass $m^*(p_\parallel)$, then the pattern of resonant peaks becomes more complicated, since superposition of resonances from electrons located near different sections is then observed. But these electrons, as a rule, contribute differently to the cyclotron resonance. As has been noted earlier, the increment of energy of an electron at passing through the skin layer is proportional to the effective mass to the power of 3/2. In addition, the amplitude of each resonant peak depends on relaxation time of the carriers of the given type. In practice, all other conditions being equal, the maximum amplitude is observed for the resonances in which the electrons located near the section $p_\parallel = 0$ participate. In that case the electrons describe circles in planes perpendicular to the surface of the specimen and retain their synchronism with the electromagnetic field and invariancy of the orbit during the whole relaxation time. The electrons corresponding to sections with $p_\parallel \neq 0$ describe a helix with the axis directed along the magnetic field. With a slightest non-parallelism between the magnetic field and the surface of the specimen they depart from the helix into the depth of metal and disappear from the accelerating cycle long before scattering. For that reason, resonance can be most easily observed experimentally for electrons located near the section $p_\parallel = 0$.

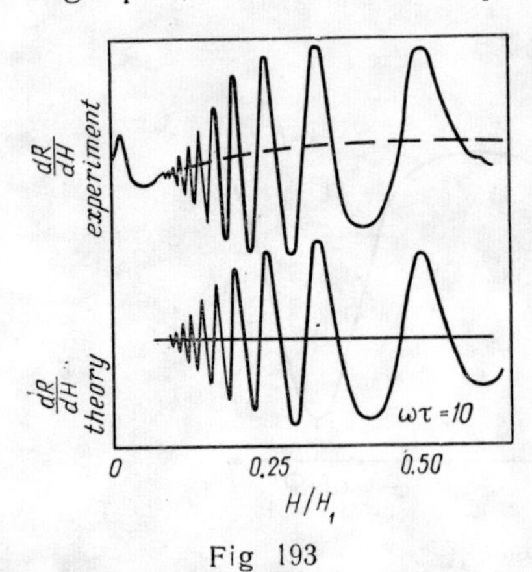

Fig 193

Let us find the periodicity with which successive resonant peaks appear at an increase of the magnetic field. For this, we find the value of the magnetic field H_k corresponding to the resonance at the k-th harmonic of cyclotron frequency from the main condition (4.22) of the resonant acceleration of electrons. Substitution of the cyclotron part into (4.22) gives

$$\Omega = k \frac{|e| H_k}{m^*_{\text{extr}} c} \tag{4 23}$$

where m^*_{extr} is the extremal cyclotron mass satisfying the condition $\frac{\partial m^*}{\partial p_\parallel} = 0$ at $m^* = m^*_{\text{extr}}$. Hence the reciprocal value of the magnetic field $\frac{1}{H_k}$ at which resonance at the k-th harmonic is observed is

$$\frac{1}{H_k} = k \frac{|e|}{m^*_{extr} c\Omega} \tag{4.24}$$

It follows from this expression that the interval of reciprocal magnetic fields $\frac{1}{H_{k+1}} - \frac{1}{H_k}$ between the k-th and $(k+1)$-th resonances is independent of the index k of the resonance and is a constant. This interval is termed the period of cyclotron resonance in a reciprocal magnetic field.

Denoting the period as $\Delta\left(\frac{1}{H}\right)$, we write the following equality:

$$\Delta\left(\frac{1}{H}\right) = \frac{|e|}{m^*_{extr} c\Omega} \tag{4.25}$$

The period of cyclotron resonance in a reciprocal magnetic field can be found from an experimental resonance curve from which the values of magnetic fields H_k (or reciprocal magnetic fields $1/H_k$) are determined. With the period $\Delta\left(\frac{1}{H}\right)$ and frequency of electromagnetic oscillations Ω being known, we can determine the extremal mass m^*_{extr} of current carriers corresponding to the given direction of magnetic field $\vec{H}$:

$$m^*_{extr} = \frac{|e|}{c\Omega} \frac{1}{\Delta\left(\frac{1}{H}\right)} \tag{4.26}$$

To find the period $\Delta\left(\frac{1}{H}\right)$, a curve of the resonance index k as a function of the reciprocal magnetic field $\frac{1}{H}$ is usually plotted. The experimental points of the curve determine the position of a straight line whose equation in the coordinates k and $\frac{1}{H}$ is $k = \frac{m^*_{extr} c\Omega}{|e|} \cdot \frac{1}{H}$. The tangent of this line, equal to $\frac{m^*_{extr} c\Omega}{|e|}$, gives the value of the cyclotron mass. For better accuracy in determining m^*_{extr}, the slope of the straight line can be found by the least square method, the relative error of m^*_{extr}, calculated, for instance, from a resonance curve containing 10 peaks, being not higher than 1 per cent. Owing to this the cyclotron resonance is at present the most accurate method for measuring the cyclotron masses of carriers in metals.

By varying the direction of the magnetic field relative to the crystallographic axes of the specimen, it is possible to determine the extremal values of cyclotron masses relating to different sections through the Fermi surface. As has been mentioned earlier, the frequency of the Azbel-Kaner cyclotron resonance is independent of the magnitude of electron-phonon interaction in metals. In other words, the cyclotron resonance makes it possible to find the renormalized cyclotron mass of electrons and holes at the Fermi level, which differs from the band cyclotron mass by a multiplier $(1+\lambda+\mu)$ (see Sec. **2-12**, Chapter Two).

Owing to a high accuracy with which the cyclotron mass can be found when measuring the Azbel-Kaner resonance, this effect can be used for establishing how various physical actions (for instance, three-dimensional or uniaxial compression of the crystal) influence the renormalization parameter λ.

To conclude this section, it should be mentioned that an experimental record of the cyclotron resonance curve makes it possible not only to determine with a high accuracy the magnitude of the cyclotron mass, but also to estimate the relaxation time τ of the corresponding group of carriers. An analytical computation of the surface impedance Z made by Azbel and Kaner [83] has shown that the real portion is proportional to the following approximate expression:

$$\mathrm{Re}\,(Z) = R \sim \left(\frac{\Omega^2 l}{c^4 \sigma}\right)^{1/3} e^{-\frac{2\pi}{\omega\tau}} \cos\frac{2\pi\Omega}{\omega} \qquad (4.27)$$

Thus, the ratio of amplitudes A_k and A_{k+1} of two successive resonant peaks corresponding to the indices k and $k+1$, approximately is:

$$\frac{A_k}{A_{k+1}} \approx e^{\frac{2\pi m^*_{\mathrm{extr}} c}{|e|\tau}\left(\frac{1}{H_{k+1}} - \frac{1}{H_k}\right)} \qquad (4.28)$$

Hence the relaxation time can be estimated by the expression:

$$\tau = \frac{2\pi m^*_{\mathrm{extr}} c}{|e| \ln (A_k/A_{k+1})}\left(\frac{H_k - H_{k+1}}{H_{k+1} H_k}\right) \qquad (4.29)$$

Thus, an analysis of the cyclotron resonance in metals makes it possible to determine important dynamic characteristics of each group of current carriers. Note, however, that observations of cyclotron resonance require a complicated experimental technique and are linked with large practical difficulties (the effect is only observed on monocrystalline specimens of a high purity and perfection, with very high requirements being set to the quality of the surface of the specimen and the accuracy of its orientation relative to the magnetic field vector, etc.).

4-4. GANTMACHER'S RADIO-FREQUENCY SIZE EFFECT

Size effects include physical phenomena in which a characteristic parameter determining the path of motion of an electron becomes commensurable with the linear dimensions of the specimen. The peculiarities of the behaviour of the parameters of the metal then make it possible to find the dimension of the orbit of the electrons corresponding to the extremal section of the Fermi surface, and therefore, to determine the dimensions of this section.

The conditions of appearance of the size effect presume that a portion of the electron paths is cut off at the boundaries of the specimen owing to the interaction of electrons with the surface. This interaction is usually described purely phenomenologically by means of what is called the coefficient of peculiarity q, which is defined as the proportion of the electrons mirror reflected from the surface (i.e. so that the angle of incidence is equal to the angle of reflection). Accordingly the proportion of electrons reflected diffusely is $1 - q$. The diffuse reflection is here thought of as such that can occur with equal probability at any angle, irrespective of the angle of incidence.

Since the de Broglie wavelength of Fermi electrons is of the same order of magnitude as the interatomic distance, while the characteristic dimensions of unevenneses of the boundary largely exceed it, it is natural to assume that the majority of electrons will be diffusely reflected from the surface of metal. Exceptions may only be the electrons moving practically parallel to the surface and having extremely small angles of incidence. It may be noted in passing, however, that some experimental facts have now become known which show that reflection of electrons incident at the surface of metal at large angles may sometimes be not of diffuse nature. These facts include, for instance, the observation of cyclotron resonances corresponding to the motion of electrons along their paths with a cut-off, which is accompanied with their reflection from the surface of metal.

The radio-frequency size effect considered in this section is not one of such special cases of interaction between electrons and surface. Its interpretation is based on the common concept that electrons impinging a boundary at large angles are diffusely reflected from it.

The radio-frequency size effect consists in appearance of singularities of the skin impedance Z of a thin metallic plate in cases when the thickness of the plate exceeds a whole number of times the diameter of the cyclotron orbit of electrons located in a plane perpendicular to the surface of a specimen. This effect was discovered by Gantmacher [76] in 1962 in a specimen of tin. He also proposed the physical interpretation of the phenomenon observed which is now widely used as a method of investigation of Fermi surfaces of metals and has been called Gantmacher's size effect.

Gantmacher's size effect is observed in metals in the region of anomalous skin effect at $\delta \ll l$. A specimen is used in the form of a plane-parallel plate whose thickness satisfies the condition:

$$\delta \ll d \ll l \tag{4.30}$$

A variable electromagnetic field of frequency Ω and a constant magnetic field H are applied to the plate. The orientation of the vectors of external fields relative to the surface of the specimen

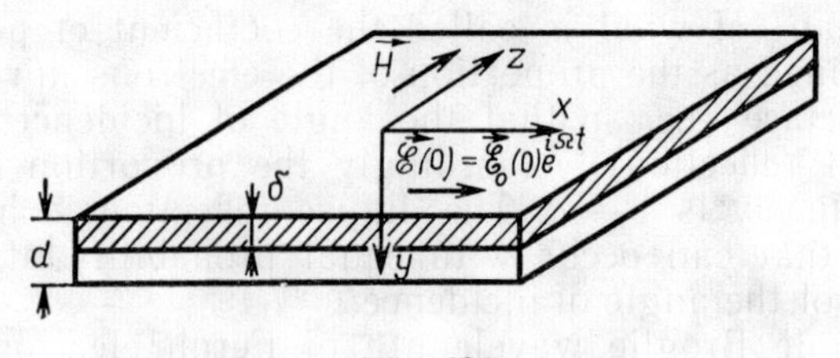

Fig. 194

is the same as in the Azbel-Kaner geometry described earlier (Fig. 194). The frequency Ω of the electromagnetic field is selected so as to satisfy the condition of quasi-stationary field during the relaxation time τ of electrons:

$$\frac{2\pi}{\Omega} = T_{\mathscr{E}} \gg \tau \tag{4.31}$$

The size effect is naturally connected with the cyclic motion of carriers in the metal and is observed at $\omega\tau > 1$. From this, and also from inequality (4.31), it follows that the frequency Ω of the electromagnetic field with the size effect must be substantially lower than the frequency ω of Larmor precession. Thus, as distinct from the Azbel-Kaner cyclotron resonance, Gantmacher's size effect is observed on radio-frequencies rather than on super-high frequencies. This feature of the phenomenon is emphasized in its name: radio-frequency size effect.

The physical essence of the phenomenon may be explained by considering first the penetration of an electromagnetic field into a metal in the region of anomalous skin effect in the presence of a constant magnetic field.

Let a magnetic field $\vec{H}$, whose magnitude obeys the inequality $H > H_l$, be directed strictly along the surface of the metal filling a half-space, and let the electric field $\vec{\mathscr{E}}$ of an electromagnetic wave be directed parallel to the surface and perpendicular to vector $\vec{H}$. We also assume that the diameter D_H of the orbit on which an electron performs Larmor precession during the relaxation time τ is substantially greater than the depth δ of skin layer.

Let us show that the electromagnetic field under these conditions can be brought in by electrons to the metal to a depth substantially exceeding the depth of skin layer.

A variable electric field penetrating the metal to the depth of skin layer interacts with electrons and accelerates them, so that a variable surface current J_{sur} appears in the layer of the thickness of an order of δ near the surface of metal.

Under conditions of anomalous skin effect ($\delta \ll l$) with the absence of magnetic field ($H = 0$), the majority of electrons pass only once through the skin layer on their free-path length, the time Δt during which they stay in the skin layer and interact with the electric fleld $\vec{\mathscr{E}}$ being small compared with the relaxation time τ (the time Δt in its order of magnitude is equal to $\frac{\delta}{l \sin \alpha} \tau$, where α is the angle made by the path of an electron with the surface of metal).

Exceptions are sliding electrons, i.e. those moving at extremely small angles to the surface. The variation of Fermi velocity Δv of each electron, and therefore, its contribution to the surface current, are determined by the angle α, and also by the magnitude of the electric field $\mathscr{E}$ during the time of stay of the electron in the skin layer, which is practically constant owing to the quasi-stationary condition (4.31) and is equal to the momentary magnitude of the field at that instant of time. It then follows that the frequency of the surface current coincides with the frequency Ω of the electromagnetic field, while the amplitude of current is proportional to the amplitude of electric field $\mathscr{E}_0$.

In the presence of a magnetic field $H > H_l$ parallel with the surface of metal, the nature of motion of electrons varies. Conditions are formed in that case under which electrons may pass many times through the skin layer, and therefore, may interact repeatedly with the electric fleld, each time gaining an energy from the latter. We consider the electrons whose cyclic orbits pass through the skin layer, as shown in Fig. 195. In the presence of a magnetic field, these electrons are resonantly accelerated and provide a resonance contribution to the surface current J_{sur}. In addition, the surface current is also contributed to by non-resonance electrons whose orbits pass through the skin layer and end at the surface of metal.

Thus, a high-frequency surface current in the presence of the magnetic field is the total effect from the motion of electrons on all possible orbits passing through the skin layer. Some of such orbits are shown in Fig. 195.

It is easy to show that no surface current can be formed under the action of only a magnetic field at $\mathscr{E} = 0$. For this, we consider, for instance, the motion of electrons only on orbits corresponding to the central section $p_{\parallel} = 0$ of the Fermi surface, assuming for simplicity these orbits to be circular. We can also make use of Fig. 195, assuming $\mathscr{E} = 0$. With these conditions posed, electrons of the central section move on their orbits with the same Fermi velocity.

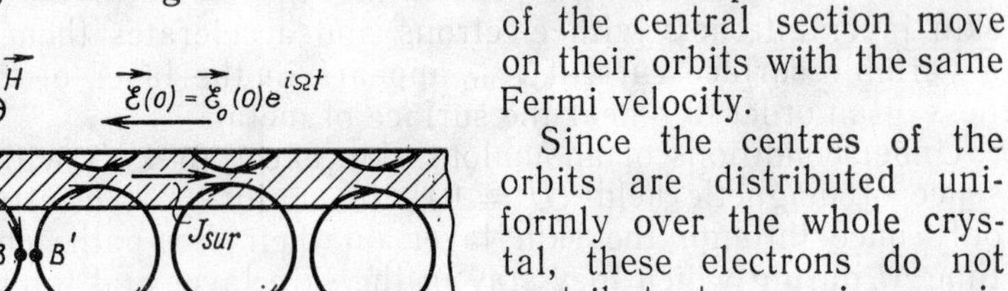

Fig. 195

Since the centres of the orbits are distributed uniformly over the whole crystal, these electrons do not contribute to a microscopic current of any direction. Indeed, the motion of an electron in each point of a given orbit is compensated by the motion of other electrons in the opposite direction on one of the neighbouring orbits. Examples may be the points A and B on one of the electron orbits (Fig. 195). With $\mathscr{E} = 0$, the motion of electrons in these points is compensated by the motion of electrons in points $A'$and B' on neighbouring orbits. Similar compensation is obviously observed on orbits corresponding to any other sections of the Fermi surface.

Thus, the balanced nature of distribution of Fermi electrons over velocities is not disturbed notwithstanding the fact that the straight paths of electrons have changed to closed or helical ones in the magnetic field.

The situation will change if an electric field $\vec{\mathscr{E}}$ parallel with the surface is present in the skin layer. The electrons on all orbits passing through the skin layer now attain an increment $\Delta\vec{v}$ to the Fermi velocity which is directed opposite to vector $\vec{\mathscr{E}}$. Though the motions with the Fermi component of velocity on neighbouring orbits on average compensate each other, the motions with the additional velocity $\Delta\vec{v}$ give an additive contribution to the surface current (see, for example, the motion of electrons in points A and A' in Fig. 195). With $D_H \gg \delta$, the time Δt of stay of electrons in the skin layer is substantially smaller than the period T_H of Larmor precession. During this time, an electron actually acquires a short "impact" from the electric field, which results in a change of velocity $\Delta v \approx \frac{|e|\mathscr{E}\Delta t}{m^*}$.

With the quasi-stationary condition (4.31), all passages of electrons through the skin layer during the time τ occur at one and the same magnitude of the electric field $\mathscr{E}$. In that case the contribution of each electron to the surface current is proportional to both Δv and the number of passages through the skin layer along the free path length. Since Δv varies in time with the frequency of the electromagnetic field, the variation of the surface current in the presence of the magnetic field occurs with the frequency Ω, rather than with the frequency of Larmor precession.

Let us now follow the motion of the electrons, which have acquired an additional Fermi velocity Δv, in the depth of metal. We first suppose that all these electrons have the same diameter D_H of their orbits. In that case, their velocity at a depth D_H becomes again parallel with the surface of metal, the result being that a current layer ("current sheet") of a thickness of an order of δ is formed at that depth. Indeed, with these electrons moving parallel with the surface of metal, no total pairwise compensation can occur, since in points of the type C the Fermi velocities of the electrons which have passed the skin layer are not equal on the average to the Fermi velocities of the electrons in points of the type C' located on deeper orbits. The appearance of a layer in which a high-frequency current flows at a depth D_H is accompanied with a splash of the electric field $\mathscr{E}'$ with the frequency Ω, this field actually reproducing the field in the skin layer.

The electric field $\mathscr{E}'$ causes acceleration of secondary electrons at the depth D_H, which in turn results in the formation of a current layer and a new splash of the electric field $\mathscr{E}''$ at the depth $2D_H$, etc.

Thus, a chain of electron orbits is formed under the conditions indicated, penetrating the depth of the metal, and a corresponding chain of splashes of the electric field is also formed. The amplitude of field splashes in a real case decreases rapidly with an increase of depth at passage from one current layer to another. This is linked, first of all, with the scatter of the diameters of orbits of the electrons belonging to different sections of the Fermi surface. Thus, for instance, if the Fermi surface is an ellipsoid, then the diameters of electron orbits vary from $D_H = 0$ in reference points (where a plane perpendicular to the magnetic field touches the Fermi surface) to $D_H = D_{H\max}$ in section $p_{\parallel} = 0$, depending on the position of the secant plane.

The scatter of the diameters of orbits results in that only a small portion of electrons whose cyclic orbits pass through the skin layer is collected at any depth in a layer of a thickness of the order of δ. A relatively numerous group of particles with approximately the same diameters of orbits is formed by the electrons located in the belt on the Fermi surface near the extremum section. Depending on the sharpness of the extremum, they give a more or less pronounced splash of the field $\mathscr{E}'$ at a depth $D_{H\text{extr}}$ coinciding with the diameter of the extremal orbit. The electrons belonging to other sections only provide a monotonous background in the form of an electromagnetic field of a low amplitude only weakly dependent on the depth in the metal.

Thus, the formation of the surface current J_{sur} is due to all the electrons whose orbits pass through the skin layer, whereas the first splash of the field at the depth $D_{H\text{extr}}$ is only formed by the electrons of the extremal section. It then follows that only a small portion of the additional energy gained by the electrons from the electric field in the skin layer is spent to form a splash of the field at the depth $D_{H\text{extr}}$. Let the ratio of the amplitude of this splash to the amplitude $\mathscr{E}_0$ of the field in the skin layer be termed the transmission coefficient ξ of the electric field. The amplitudes of subsequent splashes will evidently decrease as terms of a geometrical progression with the exponent ξ. For Fermi surfaces close to ellipsoidal ones, the transmission coefficient ξ is of the order of $(\delta/D_{H\text{extr}})^{2/3} \cdot (\omega\tau)^{5/6}$. Attenuation of the amplitudes of splashes at a depth becomes substantially greater in cases of complicated and especially of non-convex Fermi surfaces, at which the contribution of non-extremal electrons to the surface conductivity increases.

The formation of splashes of the field in a sufficiently thick specimen of metal results only in an insignificant additional absorption of the energy of the electromagnetic field because of the rapid attenuation of their amplitudes at a depth, and practically has no effect on the surface impedance. In particular, such splashes should evidently be formed with the Azbel-Kaner cyclotron resonance. But they were not discussed in the section dealing with that effect, since their appearance remains unnoticed in observations of the cyclotron resonance.

The situation changes radically when the thickness d of the layer of metal becomes commensurable with the depth to which splashes of the electromagnetic field penetrate. In that case the formation of field splashes results in a number of new physical effects.

In a plate of metal of a sufficiently small thickness d placed into a variable magnetic field, the condition

$$d = nD_{H\text{extr}} \tag{4.32}$$

will periodically hold true with various integer numbers $n = 1, 2, 3, \ldots$ (it may be recalled that the diameter D_H of an orbit is proportional to $1/H$). Condition (4.32) implies that one of the splashes of the electromagnetic field in the plate coincides with the latter's opposite side. In that case the "current sheet" corresponding to this splash forms a surface current on the opposite side of the plate, which in turn excites a variable electromagnetic field behind the plate. Thus, the formation of splashes of the field under condition (4.32) results in that the plate of metal of thickness d, exceeding substantially the depth of the skin layer $\delta (d \gg \delta)$, turns to be transparent for the electromagnetic field. Transfer of the field through the metal is effected by resonance electrons through a chain of paths.

Selective transparence repeats periodically with condition (4.32) fulfilled for various whole numbers $n = 1, 2, 3, \ldots$. This phenomenon makes it possible to correlate the diameter of the extremal orbit with the size of the specimen, and therefore, may be related to size effects. Periodical transparence of a plate is very difficult to be observed experimentally because of the extremely low amplitude of the electromagnetic field penetrating the plate. Another effect can be observed experimentally, which is also connected with the passage of splashes of the field onto the back side of a metal specimen. This phenomenon manifests itself in a periodic variation of the surface impedance of the plate. Namely this effect was detected by Gantmacher and has been called the radio-frequency size effect.

In order to characterize the Gantmacher effect, we have to elucidate the dependence of the impedance of a metallic plate on the number of units of the chain of paths that is formed in it, and also the mechanism of formation of singularities of the impedance at fulfilment of condition (4.32).

For this, we consider the formation of a chain of paths in a plate of thickness d with a gradual increase of the magnetic field H.

In relatively weak magnetic fields, where the diameter $D_{H\text{extr}}$ exceeds the thickness d of the specimen, precession of electrons relating to the extremal section of the Fermi surface is impossible (Fig. 196). In that case the additional energy which these electrons gain from the electric field in the skin layer is dispersed through their diffuse reflection from the back side of the plate and is spent to increase the energy of oscillations of the metal lattice.

In stronger magnetic fields, where the diameter $D_{H\text{extr}}$ is comparable with the thickness d of the plate [condition (4.32) being fulfilled for $n = 1$], apart from various paths of electrons ending in the surface of the metal (we still consider only the electrons relating to the extremal section of the Fermi surface), there may exist cyclic orbits, similar to what is shown in Fig. 197.

In that case, a single unit of the chain of resonant paths is formed across the thickness of the plate. The electrons moving on an orbit such as in Fig. 197, are not scattered at the surface of metal during the relaxation time τ and are repeatedly accelerated under the action of the electric field in the skin layer (the number of passages across the skin layer being of the order of magnitude of $l/D_{H\text{extr}}$.) Because of this, they provide a resonance contribution to the high-frequency surface current J_{sur}, resulting in that the amplitude of this current increases.

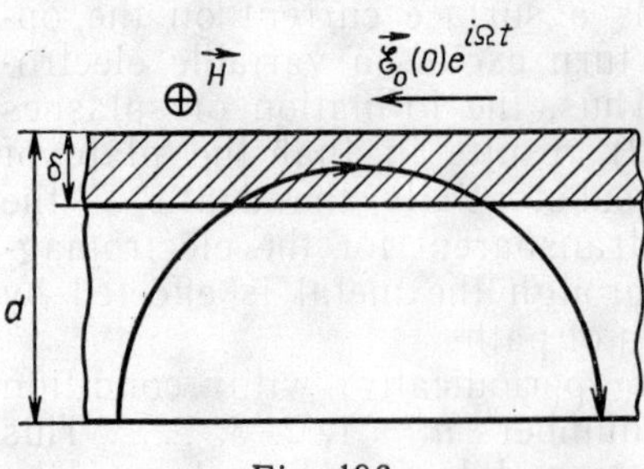

Fig. 196

This is equivalent to an increase of the surface conductivity on the plate, or more strictly, to a decrease of the real part of the surface impedance $Z = R + iX$.

At the same time, there are observed a maximum of the surface current at the back side of the plate and a maximum of the amplitude of the electromagnetic field penetrating through the plate owing to the effect of transparence. The ratio of amplitudes of the fields on the back and front sides of the plate is evidently equal to the transmission coefficient ξ of the field.

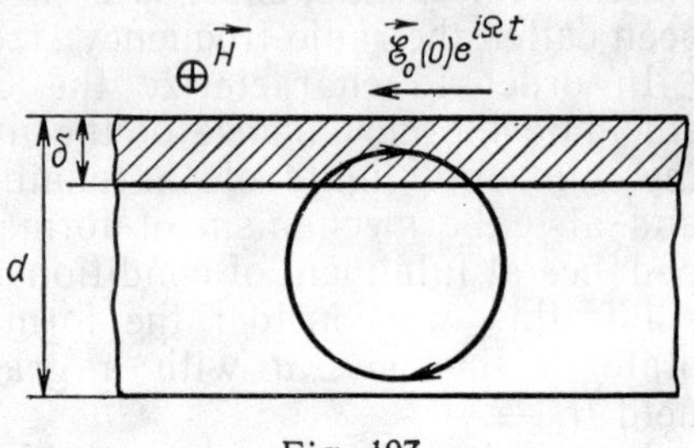

Fig. 197

In the region of magnetic fields at which $D_{H\text{extr}} < d < 2D_{H\text{extr}}$ a splash of the field at the depth $D_{H\text{extr}}$ is not coincident with the back side of the plate. In that case the second unit of the path chain is not closed. The electrons of the extremal section that are accelerated at the depth $D_{H\text{extr}}$ are dispersed through diffuse scattering from the back side of the plate (Fig. 198) and transfer to the lattice an additional energy gained by them through the splash of the electric field.

The next minimum of absorption of the electromagnetic field energy in the plate is observed at formation of a second closed chain of paths, i.e. at fulfilment of the condition $d = 2D_{H\text{extr}}$. Simultaneously, a maximum of the surface current and a minimum of the real part of the skin impedance are formed.

Our discussion shows that the absorption of the electromagnetic field energy in a metallic plate passes through a local minimum each time when the condition (4.32) is periodically fulfilled at an increase of the magnetic field. Actually, at fulfilment of this condition, the additional mechanism of scattering of energy from the back side of the plate, which is related to the diffuse reflection of electrons of the last unit of the chain of resonant paths, ceases to be effective. The amount of the additional energy carried by these electrons decreases as ξ^{2n} with an increase of the number n of units in the chain. Accordingly, the amplitude of singularity [the depth of the minimum $\mathrm{Re}(Z)$] of the skin impedance also decreases.

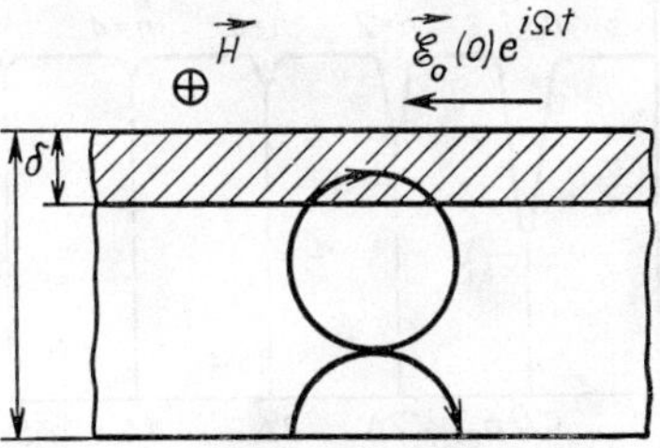

Thus, if we measure the real part of the skin impedance $\mathrm{Re}(Z)$ of the plate (as has been indicated, it is proportional to the amplitude of oscillations of the field in the cavity or oscillational contour into which the plate is placed) with a slow increase of the magnetic field H, then in addition to the monotonous variation of $\mathrm{Re}(Z)$ there will be observed periodic singularities whose amplitude decreases in the magnetic field.

Fig. 198

To calculate the period of singularities, we pass from the path of an electron in $\vec{r}$-space to its path in $\vec{p}$-space. The diameter $D_{H\text{extr}}$ of the extremal orbit in the direction of a normal to the plate can then obviously be expressed through the diameter $P_{p\text{extr}}$ of the extremal section of the Fermi surface in the direction of the electric field vector $\mathscr{E}$.

Substituting $\frac{c}{|e|H} P_{p\text{extr}}$ for $D_{H\text{extr}}$ into (4.32), we get:

$$d = n\frac{c}{|e|H_n} P_{p\text{extr}} \tag{4.33}$$

where H_n is the magnitude of the magnetic field at which the n-th singularity is observed. It follows directly from (4.33) that the period $\Delta(H)$ of repetition of singularities in the direct field is

$$\Delta(H) = H_{n+1} - H_n = \frac{c}{|e|d} P_{p\text{extr}} \tag{4.34}$$

The singularity of the maximum amplitude ($n = 1$) corresponds to the magnetic field $H_1 = \Delta(H)$. Further singularities are observed in fields multiple of $\Delta(H)$. The dependence of the real part of the skin impedance on magnetic field is shown schematically in Fig. 199. Such a dependence, obtained experimentally for a metal, makes it possible to find the period $\Delta(H)$, which according to (4.34) determines the diameter of the extremal section of the Fermi surface.

Note that experimental investigations of Fermi surfaces of metals by means of the Gantmacher size effect encounter large difficulties. In order that the effect could be practically observed, a thin monocrystalline plate is required whose outer surfaces must be optically clean and parallel to one another to an accuracy of fractions of angular minutes. Not all metals allow such plates to be prepared. Up to the present, Gantmacher's effect has been only observed for tin, bismuth, rubidium, and cadmium. The minimum thickness of the plate which is practically feasible is a few tenths of a millimetre. Inequality (4.30) will then only be observed if the free-path length l is at least 1-2 mm. Such free-path lengths are only observed in the region of liquid-helium temperatures, because of which Gantmacher's effect is related to the group of low-temperature phenomena.

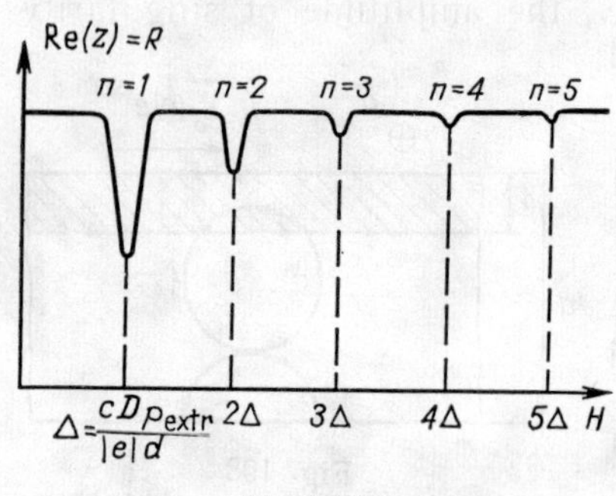

Fig. 199

To conclude this section, consider Gantmacher's effect in an oblique magnetic field making an angle φ with the surface of the specimen. In a magnetic field $\vec{H}$ strictly parallel with the surface of metal ($\varphi = 0$), electrons related to any section of the Fermi surface are repeatedly returned back into the skin layer and contribute to the surface current J_{sur}. Since a splash of the field at depth $D_{H\text{extr}}$ is formed by the electrons located in the belt near the extremal section, whose number is not large, the transmission coefficient ξ of the field is small ($\xi \ll 1$) and the amplitude of splashes decreases rapidly with an increase of the number n. In that connection, not more than three singularities of the skin impedance can be observed in a magnetic field at $\varphi = 0$.

With a field $\vec{H}$ inclined by an angle φ, the electrons having the velocity component $v_{\parallel}$ in the direction $\vec{H}$ turn to be in different conditions relative to the high-frequency electric field $\vec{\mathscr{E}}$ in the skin layer that the electrons relating to the extremal section for which $v_{\parallel} = 0$.[1]) Indeed, the electrons with $v_{\parallel} \neq 0$ drift along $\vec{H}$, leave the skin layer, and penetrate the depth of the metal. This process begins at angles of inclination of the field $\varphi \gtrsim \delta/l$. As a result of this, the contribution of the electrons with $v_{\parallel} \neq 0$ to the surface current decreases, as also does the amount of energy they gain from the electromagnetic field in the skin layer. The contribution of the electrons relating to the extremal section to the surface current, on the contrary, increases, because the motion of these electrons is not affected noticeably by the inclination of the field $\vec{H}$ at small angles φ.

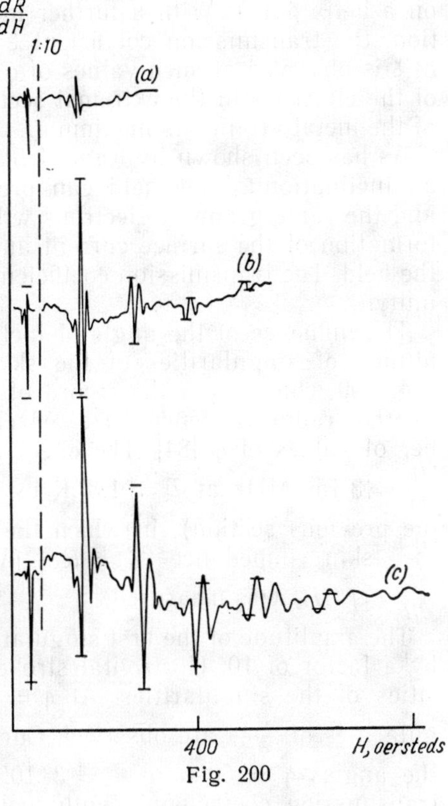

Fig. 200

Thus, in an oblique field (with $\varphi > \delta/l$), the transmission coefficient ξ increases with angle φ, resulting in an increase of splashes of the field and of the amplitude of singularities of the skin impedance. The coefficient ξ continues to increase until the electrons relating to the belt near the extremal section also begin to leave the skin layer at certain values of the angle of inclination. This process begins at angles φ of the order

[1]) The component of velocity $v_{\parallel}$ is zero for any extremum-area section of the Fermi surface, and not only for its central section corresponding to $p_{\parallel} = 0$. Indeed, $v_{\parallel} = \frac{\partial \varepsilon}{\partial p_{\parallel}} = \left(\frac{\partial \varepsilon}{\partial S}\right)_{p_{\parallel}} \cdot \left(\frac{\partial S}{\partial p_{\parallel}}\right)_{\varepsilon}$ by definition, where the subscipts $p_{\parallel}$ and ε at the derivatives indicate what quantities are constant in partial differentiation. From the condition of extremality of the section $\left(\frac{\partial S}{\partial p_{\parallel}}\right)_{\varepsilon} = 0$ there follows the equality $v_{\parallel} = 0$, since $\left(\frac{\partial \varepsilon}{\partial S}\right)_{p_{\parallel}}$ is a finite-value quantity equal to $\frac{1}{2\pi m^*}$, where m^* is the cyclotron mass corresponding to the extremal section.

of $\delta/D_{H\text{extr}}$, at which the electrons of non-extremal sections practically cease to contribute (at $\varphi \sim \delta/D_{H\text{extr}}$, an electron with $v_{\parallel} \sim v_F$ leaves the skin layer after having made a single rotation on a helix path). With a further increase of the angle of inclination, the transmission coefficient ξ reduces. The maximum value of ξ is observed at such values of φ when the relative contribution of the electrons of the extremal section to the surface conductivity of the metal attains its maximum.

As has been shown by Kaner [85], for a spherical Fermi surface, an inclination of the field can provide conditions at which one and the same group of electrons will practically participate in the formation of the surface current and the formation of a splash of the field. The transmission coefficient ξ of the field then approaches unity.

The influence of the angle of inclination of the field on the amplitude of singularities of the skin impedance is illustrated in Fig. 200 which gives the record of an experimental measurement of the radio-frequency size effect for cadmium for a number of values of φ [84]. The effect was recorded at the frequency $\frac{\Omega}{2\pi} = 3.15$ MHz at $T = 1.8°$ K by the modulation technique (see the previous section), in which the derivative of the real part of the skin impedance $Z = R + iX$ over the magnetic field, $\frac{\partial R}{\partial H} = f(H)$, was measured.

The amplitude of the first singularity in Fig. 200 is shown reduced by a factor of 10. Horizontal strokes denote the calculated intensities of the singularities. At $\varphi = 0$ (curve *a*) three weak singularities of $\frac{\partial R}{\partial H}$ are observed. Curves *b* and *c* (corresponding to the angles $\varphi = 50'$ and $\varphi = 3° 10'$) visualize the growth of the transmission coefficient ξ with inclination of the magnetic field.

4-5. PIPPARD'S MAGNETO-ACOUSTIC RESONANCE

Magneto-acoustic effects consist in a periodic variation of the absorption of ultrasound in a metal placed into a magnetic field. In order to explain physically the essence of these effects, we have first to discuss the influence of conduction electrons on the propagation of sound in a metal.

As has been established, electrons play an essential part in the formation of a metallic bond, i.e. the formation of forces acting between ions in the crystal lattice of a metal. The elastic properties of a metal are determined to a substantial extent by conduction electrons whose presence ensures the very existence of a stable system of likely charged ions.

As has been given earlier (see Chapter Two, Sec. **2-12**), account of the displacement of ions at motion of electrons results in interactions of the electrons with elastic oscillations of the lattice (phonons). This interaction is the basis underlying the electron mechanism of absorption of the energy of an ultrasonic wave, which is the most essential mechanism of absorption in normal metals with the concentration n of electrons of the order of 10^{23} cm^{-3}.

Let us make a distinction between the concept of an individual phonon and that of an ultrasonic wave excited in the lattice by an external periodic action. An ultrasonic wave is thought of as an intense flow of coherent phonons with the same frequencies and wave vectors. In order to describe a great number of coherent phonons, it is reasonable to use the concept of a classical wave field which forms a sequence of compressed and rarefied regions in the lattice, this sequence moving with the velocity of sound.

Owing to coherence of the phonons in the flow, the propagating of ultrasonic wave forms an ordered perturbation of a sufficiently high intensity in the lattice. From this standpoint, non-coherent vibrations of the lattice, whose spectrum consists of a set of various frequencies ω_i (thermal oscillations), can be more conveniently represented not as an ensemble of ultrasonic waves, but as an ensemble of phonons that participate in the processes of interaction (for instance, with electrons) as individual quantum particles of energy $\hbar\omega_i$ and momentum $\hbar\vec{k}_i$.

The formation of compressed and rarefied regions in the ionic lattice of the metal at the propagation of an ultrasonic wave in it results in a periodic variation of the density of space charge, this variation moving together with the wave. It may be explained by the valence electrons being displaced from the compressed regions (i.e. regions of higher density of ion-cores) into those rarefied, with a positive space charge being formed in the compressed regions and a negative one, in the rarefied regions, as a result of which a periodic distribution of the electric field is formed in the metal. Since the Fermi velocity of electrons v_F is several orders of magnitude greater than the velocity of sound v_{son} (mean values of v_F and v_{son} in metals are respectively $\sim 10^8$ cm/sec and $\sim 10^5$ cm/sec), the picture of this electric field for the electrons is practically fixed at each instant of time.

Being accelerated in the electric field, electrons are then scattered on phonons and thus transfer to the lattice their energy gained from the ultrasonic wave. This in essence is the electron mechanism of absorption of the energy of ultrasound.

The nature of interaction of electrons with an ultrasonic wave depends on the ratio between the wavelength λ_{son} of sound and the free-path length l.

Let us first consider the region of low-frequency oscillations at $\lambda_{son} \gg l$. The classical description of sound as of a wave process in a continuous medium is valid for this region. At $\lambda_{son} \gg 1$, the electric field connected with the sound wave in the metal is practically uniform along the free-path length. In such a field, electrons are repeatedly accelerated and repeatedly scattered on phonons. In the limit of low acoustic frequencies $\Omega < 1/\tau$ (τ being the relaxation time of electrons), the absorption coefficient Γ of sound at $\lambda_{son} \gg l$ can be expressed in terms of quasi-static characteristics of the metal and is proportional to the concentration of electrons n and the square of the frequency Ω.

At a passage to the region of high-frequency oscillations $\lambda_{son} \ll 1$, the interaction of electrons with the ultrasonic wave undergoes a qualitative change. According to the nature of their interaction, electrons may now be divided into two groups, the first, not numerous, group including the electrons moving together with the sound wave, i.e. those for which the projection of the Fermi velocity onto the direction of the wave vector $\vec{k}_{son}$ of sound is equal to the sound velocity v_{son}.

Since $v_F \gg v_{son}$, these electrons move practically in the plane of constant phase of the wave at an angle close to 90 degrees to the vector $\vec{k}_{son}$ (the direction of motion of these electrons making an angle of the order of v_{son}/v_F with the constant-phase plane). Their displacement in the direction of vector $\vec{k}_{son}$ during the relaxation time τ is substantially smaller than the wavelength λ_{son}. In other words, electrons of this group move in a practically constant electric field during the whole relaxation time and owing to it interact intensively with the ultrasonic wave.

The second group are the majority of electrons that move at an angle $\alpha < 90$ degrees to the direction of the wave vector $\vec{k}_{son}$. Electrons of this group pass a number of compressed and rarefied regions during the relaxation time. Since the electric field changes sign at each passage from one such region to another, a moving electron is alternately accelerated and braked by the electric field during a short length of time Δt of the order of λ_{son}/v_F, so that its energy is practically not changed during the relaxation time.

Thus, electrons of the second group are ineffective in absorption of energy of the ultrasonic wave and do not contribute sensibly to the attenuation of this wave.

The appearance of this group of electrons at $\lambda_{son} \ll l$ causes a variation of the frequency dependence of the attenuation coefficient Γ. Indeed, the proportion of the electrons participating in the dissipation of the sound energy at $\lambda_{son} \ll l$ is of the order of λ_{son}/l, which is proportional to $1/\Omega$. The correct order of Γ can be obtained if the classical attenuation coefficient $\Gamma \sim n\Omega^2$ is made to take into account that the number of the electrons that absorb effectively the energy of ultrasonic wave at $\lambda_{son} \ll l$ is smaller by a factor of l/λ_{son} than the number of the electrons absorbing effectively at $\lambda_{son} \gg l$. It then follows that the coefficient of attenuation of ultrasound increases linearly with frequency in the region of high-frequency oscillations.

Let us now discuss the interaction of electrons with an ultrasonic wave in the presence of the magnetic field $\vec{H}$.

In the low-frequency region $(l \ll \lambda_{son})$, the effect of the magnetic field reduces to a variation of the number of electrons on the Fermi level which determine the magnitude of attenuation coefficient Γ. As has been shown in Sec. **2-16**, Chapter Two, oscillations of the density of electron states on the Fermi level occur in a magnetic field, as a result of which oscillations of various kinetic and thermodynamical characteristics of the metal occur at $l > r_H$ and $\hbar\omega > kT$ [see (2.115)], one of these characteristics being the attenuation coefficient Γ for ultrasound. Quantum oscillations of all these characteristics are of the same nature and will be discussed together in Sec. **4-7**, Chapter Four.

The propagation of high-frequency sound $(l \gg \lambda_{son})$ in the presence of a magnetic field has some specific features. In that case, resonant acceleration of electrons in the magnetic field can be observed under certain conditions.

For clarity, we consider an ultrasonic wave of transverse polarization. With propagation of such a wave, ions of the lattice are displaced from their equilibrium positions in transverse directions, which gives a periodic distribution of the electric field in the lattice, as shown in Fig. 201.

Let the magnetic field $\vec{H}$ be directed perpendicular to the plane of wave polarization (i.e. the plane in which the vector $\vec{k}_{son}$ and the vector of the electric field $\vec{E}$ connected with the wave are located). In a strong magnetic field $H > H_l$ (i.e. at $r_H < l$) electrons precess in a plane perpendicular to $\vec{H}$. When the diameter of the orbit of an electron, D_H, in the direction of vector k_{son} exceeds an odd number of times $(2k + 1, k = 0, 1, 2, 3, \ldots)$ half the wavelength $\lambda_{son}/2$, there exist paths, such as shown in Fig. 201, on which electrons may undergo a resonant acceleration in the electric field $\vec{E}$.

A noticeable contribution to the resonant absorption of ultrasound in a magnetic field is evidently provided by the group of

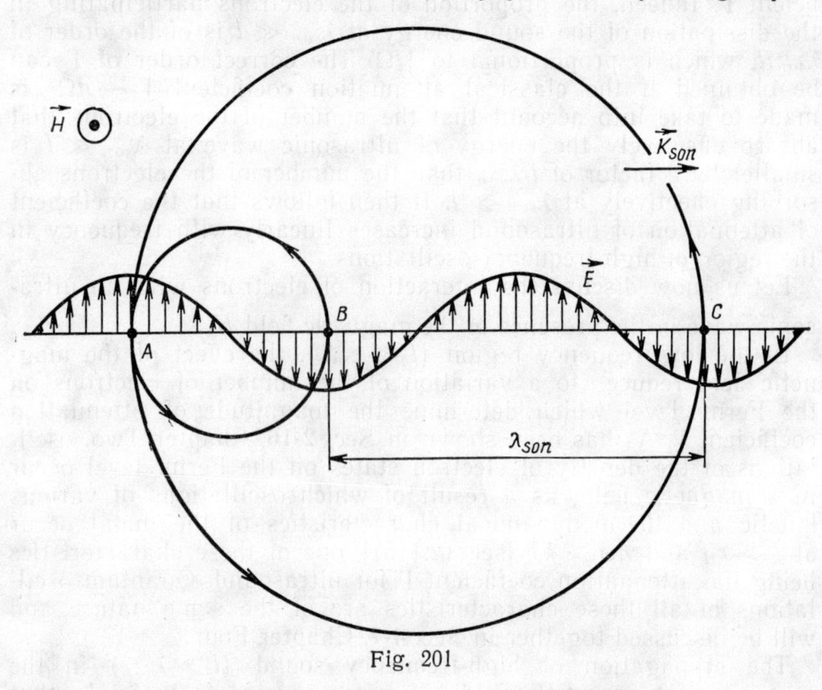

Fig. 201

electrons relating to the extremal section of the Fermi surface, for which the condition is fulfilled:

$$D_{H\mathrm{extr}} = (2k+1)\frac{\lambda_{son}}{2} \qquad (4.35)$$

The diameter $D_{H\mathrm{extr}}$ of the extremal orbit of an electron in the direction of vector k_{son} can be expressed in terms of the diameter $P_{p\mathrm{extr}}$ of the extremal section of the Fermi surface in the direction of the wave polarization vector:

$$D_{H\mathrm{extr}} = \frac{c}{|e|H} P_{p\mathrm{extr}}$$

The condition of resonant acceleration of electrons (4.35) can be re-written for $P_{p\mathrm{extr}}$ as follows:

$$P_{p\mathrm{extr}} = \frac{2k+1}{2} \frac{|e|H_k}{c} \lambda_{son} \qquad (4.36)$$

Hence it follows that the maxima of the absorption coefficient Γ for ultrasound are observed at discrete values of the magnetic field H_k

$$H_k = \frac{2}{2k+1} \frac{c}{|e| \lambda_{son}} P_{pextr} \tag{4.37}$$

The maxima of absorption are repeated periodically in the reciprocal magnetic field $\frac{1}{H}$ with a period $\Delta\left(\frac{1}{H}\right)$, which is:

$$\Delta\left(\frac{1}{H}\right) = \frac{1}{H_{k+1}} - \frac{1}{H_k} = \frac{|e| \lambda_{son}}{cP_{p\,extr}} \tag{4.38}$$

Thus, the period of maxima of absorption coefficient $\Delta\left(\frac{1}{H}\right)$ makes it possible to determine the diameter of the extremal section of the Fermi surface in a direction perpendicular to vectors $\vec{k}_{son}$ and $\vec{H}$.

Note that resonant absorption of ultrasound can also be observed when the magnetic field is located in a plane perpendicular to the vector of polarization of the ultrasonic wave, so that its direction makes an angle $\alpha \neq 0$ with the direction of vector $\vec{k}_{son}$. It may easily be seen that condition (4.35) must then be written as follows:

$$D_{Hextr} \sin\alpha = \frac{2k+1}{2} \lambda_{son}, \quad k = 0, 1, 2, \ldots \tag{4.39}$$

whence the diameter P_{pextr} of the corresponding extremal section in the direction of polarization vector is expressed in terms of α and H_k as follows:

$$P_{pextr} = \frac{2k+1}{2} \frac{|e| H_k}{c \sin\alpha} \lambda_{son} \tag{4.40}$$

The amplitude of the maxima of absorption of ultrasound increases with magnetic field H. This is linked with that an increase of the magnetic field increases the number of rotations of the electron on its orbit during the relaxation time τ which is of the order of magnitude of l/r_H and proportional to H. The effectiveness of absorption of the energy of ultrasound by each of the electrons relating to the given extremal section is then increased.

The phenomenon of periodic variation of the coefficient of absorption of ultrasound in a magnetic field, which is based on the described mechanism of acceleration of electrons in the electric field of a wave, has been called the magneto-acoustic resonance. It has been discovered experimentally by Bömmel [86] in 1955. Pippard has first indicated [87] that in this effect the size of the electron orbit becomes comparable with the wavelength of ultrasound, because of which the phenomenon has been named Pippard's geometrical resonance.

As the Azbel-Kaner cyclotron resonance and Gantmacher size effect discussed above, Pippard's resonance relates to typical quasi-classical effects in which the discrete nature of the energy levels of electrons in a magnetic field is not revealed. This effect can be observed at high values of quantum numbers n, the electron passing freely from one Landau level to another in the course of its acceleration by the electric field of an ultrasonic wave. In this region the energy spectrum behaves as a quasi-continuous one: the presence of a magnetic field is only required to form the cyclic orbit of the electron on the Fermi surface.

With an increase of the magnetic field, first resonant maxima of the absorption coefficient Γ are observed in such fields for

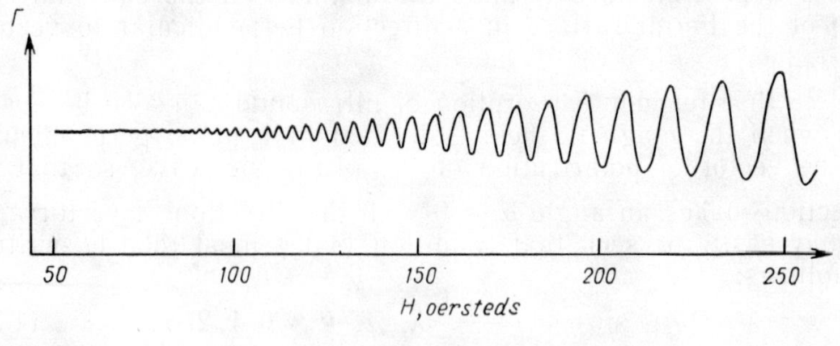

Fig. 202

which the condition $r_H \sim l$ or $\omega \sim \frac{1}{\tau}$ holds true. The amplitude of resonances then increases monotonously. The last resonant maximum of absorption is observed at $D_{H\text{extr}} = \frac{\lambda_{son}}{2}$, i.e. in the magnetic field $H_0 = \frac{2c}{|e|\lambda_{son}} P_{p\text{extr}}$. In stronger magnetic fields, the conditions of resonant acceleration of electrons cease to be valid.

Since electrons can effectively interact with sound only during a small portion of their precession period T_H when their velocity is approximately perpendicular to the wave vector, then successive fulfilment of condition (4.35) at different k results in a periodic relationship of the absorption coefficient Γ of ultrasound on reciprocal magnetic field, rather than in appearance of sharp resonant peaks.

An experimental record of magneto-acoustic resonance obtained by Ketterson and Stark [88] for magnesium at an ultrasonic frequency $\Omega/2\pi = 260$ MHz is shown in Fig. 202. The effect was observed in a specimen of magnesium of extremely high purity in wich the condition $\omega \sim 1/\tau$ was fulfilled in magnetic fields of a strength $H < 100$ oersted.

Note that simultaneous fulfilment of inequalities $l \gg \lambda_{son}$ and $l > r_H$ at reasonable values of ultrasonic frequency Ω is only possible in the region of sufficiently low temperatures, because of which Pippard's magneto-acoustic resonance is also essentially a low-temperature effect.

For experimental observations of this effect, the main difficulty consists in that a reliable acoustic contact must be ensured between the specimen of the metal being studied, on the one hand, and the exciter of ultrasound and receiver, on the other. Records are usually made of the amplitude of an ultrasonic wave passing through the metal, but it is also possible to record a wave reflected from the back face of the crystal. In the latter case ultrasound is excited in the form of short pulses and the exciter of ultrasound serves as the receiver of "sound echo" in the crystal during pauses between the pulses.

Various piezo-electric oscillators (for instance, crystals of quartz or Seignette's salt) or magnetostrictive oscillators are employed as exciters. The output signal of the receiver, which is proportional to the amplitude of the ultrasonic wave passing through the metal, makes it possible to directly determine the relative variations of the absorption coefficient Γ in a magnetic field.

As with the two effects discussed earlier, modulation of the magnetic field at a low frequency Ω_{mod} makes it possible to improve the sensitivity of measuring apparatus. The amplitude of the first harmonic of the signal at the modulation frequency Ω_{mod} is then proportional to the derivative $\frac{\partial \Gamma}{\partial H}$.

4-6. QUANTUM-MECHANICAL MAGNETO-ACOUSTIC RESONANCE

This effect consists in a resonant interaction between electrons in a metal and ultrasonic phonons of energy $\hbar\Omega$ in a strong quantizing magnetic field H at $\hbar\omega > \hbar\Omega$. The quantum-mechanical resonance manifests itself in the form of periodic sharp peaks of attenuation of ultrasound in a reciprocal magnetic field in a metal, because of which it is otherwise termed gigantic quantum oscillations of absorption of ultrasound. The effect was predicted theoretically by Gurevich, Skobov, and Firsov in 1961 [89] and has been found in many metals.

Let us find out the conditions of appearance of a resonant absorption of ultrasonic phonons for a particular case of an electron system with a quadratic dispersion law. For simplicity, we shall neglect the spin splitting of Landau levels. The energy ε of electrons in a magnetic field H directed along the z axis under such conditions is only dependent on the quantum number n and the projection of momentum onto the z axis, p_z, and can be written as $\varepsilon = \varepsilon(n, p_z) = \left(n + \frac{1}{2}\right)\hbar\omega + \frac{p_z^2}{2m_z}$ [see (2.98) in Chapter Two, Sec. **2-14**].

Having absorbed a phonon of energy $\hbar\Omega$ and momentum $\hbar\vec{q}$, an electron passes from the state with the quantum number n and the projection of momentum p_z into a new state which is characterized by n' and p_z'. This process can only occur if the initial state with the given n and p_z is filled and the final state with n' and p_z' is empty.

In the effect of absorption of ultrasound, the interaction of phonons with electrons at the Fermi level is of the principal interest. The allowed empty states onto which electrons can pass as a result of absorption of phonons are then located above the Fermi level. The passage onto these states occurs either with an increase of the quantum number n or with a constant n but with an increase of the projection p_z of momentum.

The laws of conservation of energy and momentum for this process can be expressed as follows:

$$\left(n + \frac{1}{2}\right)\hbar\omega + \frac{p_z^2}{2m_z} + \hbar\Omega = \left(n' + \frac{1}{2}\right)\hbar\omega + \frac{p_z'^2}{2m_z} + p_z + \hbar q_z = p_z' \tag{4.41}$$

where $\hbar q_z$ is the projection of the momentum of phonon onto the z axis. Hence there follows the relationship between n, n', p_z, and q_z:

$$(n' - n)\hbar\Omega = \hbar\Omega - \frac{2\hbar p_z q_z + \hbar^2 q_z^2}{2m_z} \tag{4.42}$$

Since in the case considered the energy of phonon $\hbar\Omega$ is less than the distance $\hbar\omega$ between Landau levels, the equality (4.42) can only hold true at $n' = n$. In other words, an electron that has absorbed a phonon of energy $\hbar\Omega$ cannot pass onto another Landau level. The variation of the state of the electron in this process consists only in a variation (increase) of the projection of momentum p_z. It will be noted that a passage onto the empty states with $|p_z'| > |p_z|$ is only possible when the projections of momenta of the electron p_z and phonon $\hbar q_z$ have like signs. Thus, it follows from equality (4.42) at $n' = n$ that absorption of a phonon can occur with the following condition fulfilled:

$$\hbar q_z = p_z\left(\sqrt{1+\frac{\hbar\Omega}{p_z^2/2m_z}}-1\right) \tag{4.43}$$

If the energy of motion of the electron along the magnetic field obeys in the initial state the inequality $\frac{p_z^2}{2m_z} \gg \hbar\Omega$, then condition (4.43) can be transformed into a simpler expression:

$$\hbar q_z = \frac{\hbar\Omega m_z}{p_z} \tag{4.44}$$

which can be given a clear geometrical interpretation. Indeed, in the region of not very high frequencies at which no dispersion of sound still appears, the frequency of phonon Ω is expressed through the magnitude of the wave vector q and the velocity of sound v_{son} : $\Omega = v_{son} \cdot q$. It then follows from equality (4.44) that the initial projection of momentum $p_z = p_{z0}$ at the absorption of a phonon must be as follows:

$$p_z = p_{z0} = m_z v_{son}\frac{q}{q_z} = m_z\frac{v_{son}}{\cos\theta} \tag{4.45}$$

where θ is the angle between the direction of the propagation of sound and the magnetic field $\vec{H}$ (see Fig. 203). Solving (4.45) for v_{son}, we find the condition of resonant absorption in the form:

$$v_{son} = \frac{p_{z0}}{m_z}\cos\theta = v_{z0}\cos\theta \tag{4.46}$$

Thus, the process of resonant absorption of phonons is only due to such electrons on the Fermi surface, for which the projection of velocity v_{z0} of motion along the magnetic field onto the direction of propagation of ultrasound is equal to the velocity of sound wave. These electrons move on helical paths around a magnetic line of force in such a way that the centre of their orbit remains in one and the same plane of the constant phase of the wave at any instant of time. In other words, the motion of resonant electrons along the magnetic field occurs in phase with the wave.

The electrons moving in the constant phase plane of the ultrasonic wave are accelerated in the electric field $\vec{E}$ formed by the wave during their relaxation time τ and increase the component of momentum p_z along the magnetic field by a value Δp_z proportional to $E\tau$.

It may be easily seen that resonant electrons are located in a narrow belt near a certain section of the Fermi surface by a plane perpendicular to the magnetic field, as illustrated in Fig. 204. Indeed, if condition (4.46) holds true for one of the electrons in the belt, then it must hold true for all other electrons in that belt, since they have the same velocity v_z along the magnetic field. The position on the Fermi surface of the section (or belt) for which condition (4.46) is fulfilled as uniquely determined by the angle θ between the direction of the propagation of sound and the magnetic field.

Note that condition (4.46) can be fulfilled not for any values of θ between 0 and 90 degrees, but only for the angles θ smaller than a certain limiting angle θ_{lim}. In order to show that a limiting angle must exist, let us consider how the position of the section corresponding to resonant electrons will be changed with θ increasing from zero.

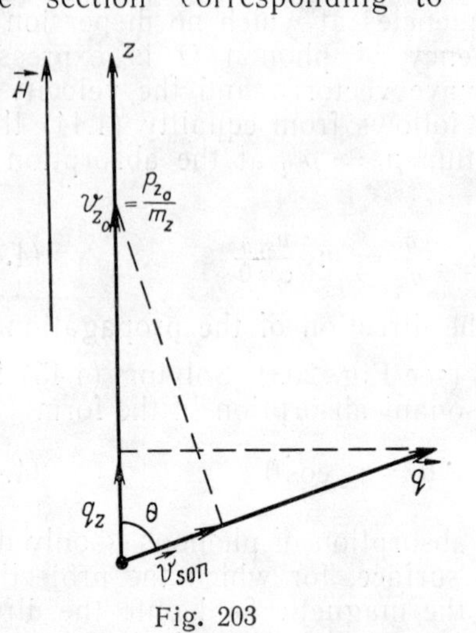

Fig. 203

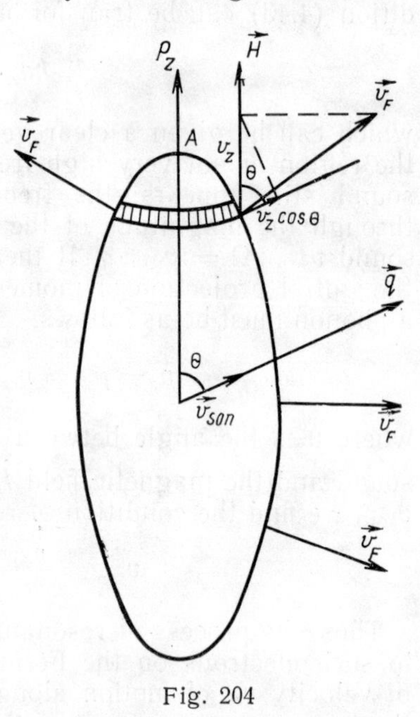

Fig. 204

At $\theta = 0$, the section is located at a height $p_z = m_z v_{son}$. Since $m_z v_{son}$ is much smaller than the Fermi momentum corresponding to the reference point A in Fig. 204, this section is very close to the central one. Let the Fermi velocity in point A be denoted as $v_F(A)$. For instance, at $\frac{v_{son}}{v_F(A)} \sim 10^{-3}$, this section differs from the central one by only 10^{-4} per cent.

As the angle θ is being increased, the section corresponding to resonant electrons moves upward and at $\theta = \theta_{lim}$ reaches the reference point A in which the plane perpendicular to the magnetic field touches the Fermi surface. Thus, the existence of the limiting angle is linked with that the value of the z-th component of velocity on the Fermi surface is limited by $v_F(A)$. The magnitude of θ_{lim} is found from the condition:

$$\cos \theta_{lim} = \frac{v_{son}}{v_F(A)} \quad \text{(Fig. 205)}$$

For angles $\theta_{lim} < \theta \leqslant 90°$, none of the electrons on the Fermi surface can move in synchronism with the ultrasonic wave, and therefore, participate in the process of resonant absorption of ultrasound.

The existence of the limiting angle θ_{lim} is a specific singularity of the phenomenon of quantum-mechanical resonance as distinct from the Pippard's magneto-acoustic resonance.

Thus, for each particular value of angle $\theta < \theta_{lim}$, the states of the electrons which can participate in resonant absorption of ultrasound relate to quite a definite section of the Fermi surface. But for an arbitrary value of the magnetic field these states (which serve as the initial states with n and p_z in the process of interaction) may be empty. In that case, naturally, no resonant absorption can occur.

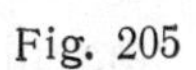

Fig. 205

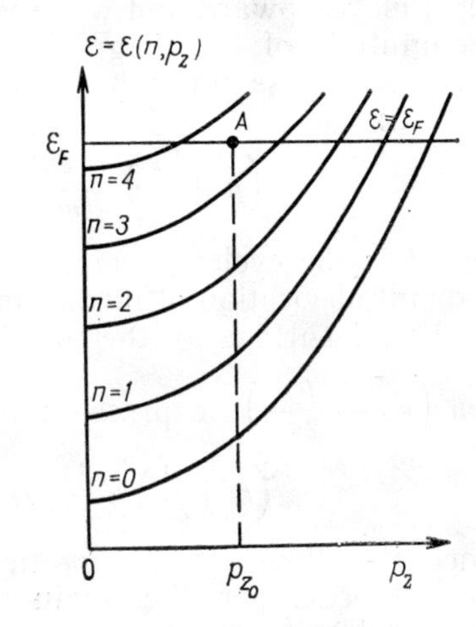

Fig. 206

It follows from the analysis of distribution of electrons in $\vec{p}$-space in the presence of a quantizing magnetic field (see Chapter Two, Sec. **2-15**) that the filled electron states on the Fermi surface are located in the belts corresponding to the intersection of the Fermi surface and Landau cylinders. These belts are discrete and their number and position on the Fermi surface depend on the magnitude of magnetic field H. Resonant absorption of ultrasound can only occur when one of the belts in which Fermi electrons are located coincides with the belt which with the given angle θ, corresponds to the position of resonant electrons on the Fermi surface. The discrete values p_z corresponding to the filled electron states on the Fermi surface can be found by the position of the points of intersection of Landau parabolae $\varepsilon = \varepsilon(n, p_z)$ with the Fermi level (Fig. 206).

Let, for a definite angle θ, the projection p_z of the momentum of electrons satisfying condition (4.46) be equal to p_{z0}. The situation shown in Fig. 206 corresponds to such a case when the Fermi electrons with the projection $p_z = p_{z0}$ are absent, since neither of the Landau parabola passes through the point A having the coordinate $p_z = p_{z0}$ on the Fermi level. With such a value of the magnetic field, no resonant absorption of ultrasound is observed.

With an increase of the magnetic field H Landau parabolae begin to move upward and will sequentially pass through point A. The magnitude of the magnetic field at which the n-th Landau parabola will pass through point A is found from the following expression:

$$\left(n+\frac{1}{2}\right)\frac{|e|\hbar H_n}{m^*c}+\frac{p_{z0}^2}{2m_z}=\varepsilon_F \tag{4.47}$$

where m^* is the cyclotron mass of electrons. At $H = H_n$, a peak of resonant absorption of ultrasound is observed. Since the section of the Fermi surface by the plane $p_z = p_{z0}$ is equal to $S(p_{z0}) = 2\pi m^*\left(\varepsilon_F - \frac{p_{z0}^2}{2m_z}\right)$, expression (4.47) can be re-written as

$$\left(n+\frac{1}{2}\right)\frac{|e|\hbar}{c}H_n=\frac{1}{2\pi}S(p_{z0}) \tag{4.48}$$

Hence it follows that gigantic oscillations of absorption of ultrasound occur periodically in reciprocal magnetic field. Their period $\Delta\left(\frac{1}{H}\right)=\frac{1}{H_{n+1}}-\frac{1}{H_n}$ determines the magnitude of the section $S(p_{z0})$ of the Fermi surface:

$$S(p_{z0})=\frac{2\pi|e|\hbar}{c}\cdot\frac{1}{\Delta\left(\frac{1}{H}\right)} \tag{4.49}$$

Formula (4.49) coincides in form with the Lifshits-Onsager formula for quantum oscillations (2.126) which determines the extremal section of the Fermi surface. But with the quantum-mechanical resonance it is possible to find any section of the Fermi surface whose position is dependent on the angle θ between the direction of the wave vector $\vec{q}$ of ultrasound and that of the magnetic field $\vec{H}$.

Thus, the quantum-mechanical resonance makes it possible to investigate the structure of the Fermi surface in substantially more details than the effects in which only one extreme section S_{extr} is determined.

It can be shown that formula (4.49), obtained here for a particular case of the quadratic anisotropic dispersion law, is valid, as the Lifshits-Onsager formula, for any arbitrary law of dispersion. For this, we have to consider quantization of the areas of electron orbits, as has been done in Chapter Two, Sec. **2-15**.

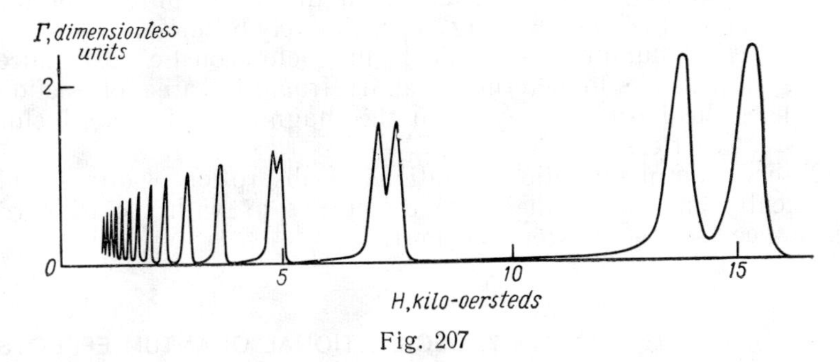

Fig. 207

Note that formula (4.49) [as also formula (2.126)] implies that the section of the Fermi surface is independent of the quantum number n at which the singularity of the absorption coefficient is observed. This is equivalent to assuming that the Fermi level is constant, which is true at $\hbar\omega \ll \varepsilon_F$.

With passing to the ultra-quantum region of magnetic fields where oscillations of the Fermi level become essential, the interval of reciprocal magnetic fields between subsequent peaks of the coefficient of absorption of ultrasound ceases to be constant and equal to $\Delta\left(\frac{1}{H}\right)$, as has been found for weaker fields. The last peak of resonant absorption occurs when the Landau parabola with $n = 1$ crosses the Fermi level in the point with $p_z = p_{z0}$.

The conditions of appearance of the quantum-mechanical resonance are evidently connected with observation of inequalities (2.115) $\omega\tau > 1$ and $\hbar\omega > kT$. With an increase of the magnetic field, peaks of ultrasonic absorption are formed in the fields for which the cyclotron frequency ω attains the greater of two quantities, either $1/\tau$ or $kT/\hbar$[1]. The quantum-mechanical magneto-

---

[1]) Note that when $\theta > 0$, Pippard's magneto-acoustic resonance may be first observed in weak fields; it then disappears and only peaks of gigantic quantum oscillations of ultrasonic absorption are further observed.

acoustic resonance is observed at temperatures of liquid helium and is essentially at low-temperature effect.

A typical experimental dependence of the coefficient of absorption of ultrasound on magnetic field with the effect of gigantic oscillations of absorption is shown in Fig. 207 which gives the curve of the quantum-mechanical resonance of electrons in bismuth at $T = 1.26°$ K obtained by Fujimori [90]. The bifurcate shape of the resonant peaks of absorption is linked with the spin splitting of Landau levels in strong fields. As can be seen, for the given orientation of the magnetic field the orbit splitting of the levels exceeds the spin splitting approximately 5 times.

Thus, the quantum-mechanical magneto-acoustic resonance makes it possible to determine, apart from the area of section $S(p_{z0})$ of the Fermi surface, also the magnitude of the g-factor of current carriers.

Observation of gigantic oscillations of absorption of ultrasound practically encounters the same difficulties as with the Pippard resonance (see the previous section).

4-7. OSCILLATIONAL QUANTUM EFFECTS

Effects of this kind are linked with quantum oscillations of the density of states on the Fermi level with a passage of electron- or hole-type Landau levels through it. The physical essence of oscillational phenomena has been discussed in detail in Chapter Two, Sec. **2-15**.

Quantum oscillations in a metal were first discovered in 1930 by Schubnikov and de Haas [91] in the Leiden cryogenic laboratory in Holland. They found the electric conductivity of bismuth to be in an oscillational dependence on magnetic field. Because of this, quantum oscillations of conductivity (or of electric resistance) of a metal have been called the Schubnikov-de Haas effect.

In 1931, de Haas and van Alphen [92], working in the same laboratory, discovered oscillations of magnetic susceptibility which have been later termed the de Haas-van Alphen effect.

During the four decades that have passed since then, oscillations of practically all known thermodynamic and kinetic characteristics of a metal depending on the number of electrons of the Fermi level have been found experimentally. For instance, apart from the known oscillations of electric conductivity and magnetic susceptibility, oscillations of heat capacity, entropy, thermal conductivity, thermo-e. m. f. thermomagnetic Q-factor, etc. were observed experimentally.

The general theory of oscillational quantum effects has been developed by Lifshits and Kosevich in 1956 [14] on the basis of their theoretical study of Fermi surfaces of metals.

The de Haas-van Alphen and Schubnikov-de Haas effects are now among the most widely used methods of studying Fermi surfaces. This is linked with that the experimental equipment required for observations of these effects may be rather simple.

Measurements of the period of oscillations $\Delta\left(\frac{1}{H}\right)$ can be easily made with different orientations of the magnetic field relative to the axes of the crystal being studied. Expressing the dependence of the period on orientation of the magnetic field by means of the Lifshits-Onsager formula (2.126) makes it possible to determine any extremal sections of the Fermi surface and, using this information, to correct the initial model of the surface constructed by the Harrison method. Practically all known Fermi surfaces of simple metals have been constructed by this method.

As has been indicated in Chapter Two, Sec. **2-16**, oscillational effects can be observed when the inequalities $\hbar\omega \gg kT$ and $\omega\tau > 1$ hold true (see 2.115). In addition, the condition of degeneration of electrons in the metal, $kT \ll \varepsilon_F$ is naturally assumed to be fulfilled. Thus, quantum oscillations also relate to low-temperature effects and are observed in sufficiently pure crystals with a high relaxation time τ of current carriers.

The amplitude of oscillations, proportional to the first power of the small parameter $\left(\frac{\hbar\omega}{\varepsilon_F}\right)^{1/2}$, is usually not large and constitutes fractions of a per cent of the quantity being measured. The amplitude of oscillations is reduced through the temperature blurring of the Fermi level and also the expansion of Landau levels owing to scattering of current carriers. In very pure specimens of metals, oscillations from small groups of carriers (having low values of cyclotron masses and high mobilities) can sometimes be observed up to temperature of the order of 50 to 60° K.

Let us consider the nature of oscillational phenomena using as an example the de Haas-van Alphen effect, whic is the one most thoroughly studied theoretically. When observing this effect in a specimen of metal, measurements are made of either the magnetic moment M of the specimen or the derivative $\frac{\partial M}{\partial H}$, equal directly to the magnetic susceptibility χ.

Experimntal records of the de Haas-van Alphen effects in single crystals of copper and antimony are shown respectively in Figs. 208 and 209. The curve of the variation of magnetic susceptibility $\Delta\chi$ in dependence of magnetic field H (Fig. 208) contains oscillations of two different frequencies. The corresponding periods $\Delta_1\left(\frac{1}{H}\right)$ and $\Delta_2\left(\frac{1}{H}\right)$ relate to one another approximately as 1 : 30. The curve has been obtained with the magnetic field directed along one of the body diagonals of the type [111] in the face-centered lattice of copper. The oscillations of the lower frequency correspond to the extremal section of the "neck", and those of the greater frequency, to the extremal section of the "belly" of the Fermi surface of copper.

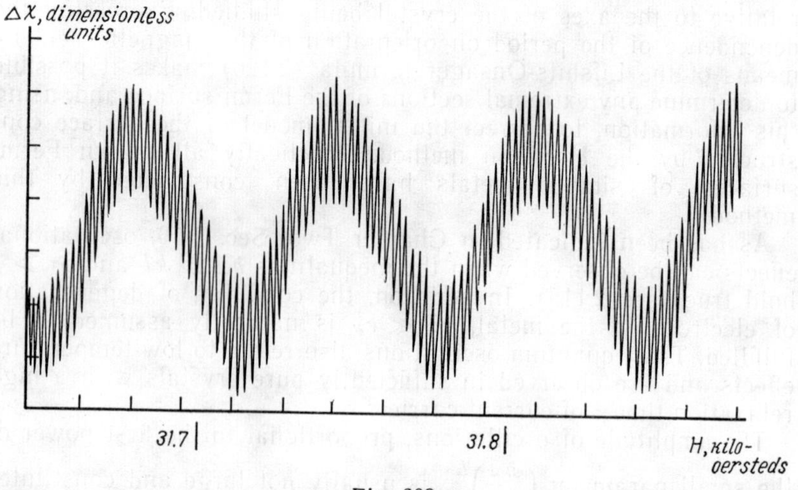

Fig. 208

The oscillational dependence $M = M(H)$ for antimony in Fig. 209 shows pulsations of two close frequencies. Such a curve is formed as a result of superposition of the oscillations corresponding to two close extremal sections of the Fermi surface. By determining the period of oscillations in the interval between the nodes of pulsations, and also the period of pulsations, we can thus separate the contributions from various extremal sections.

Note that oscillations in Figs. 208 and 209 are almost sinusoidal, their amplitude increasing in the magnetic field.

This is seen most clearly in oscillational curves shown in Fig. 210, each of which corresponds to only one section through the Fermi surface. These curves represent the de Haas-van Alphen effect in bismuth at different orientations of the magnetic field.

Let us emphasize again that oscillational dependences of various characteristics of a metal are based on oscillations of the density of states of carriers at the Fermi level. These oscillations appear as a result of that the Landau levels begin to intersect the Fermi level with an increase of the magnetic field.

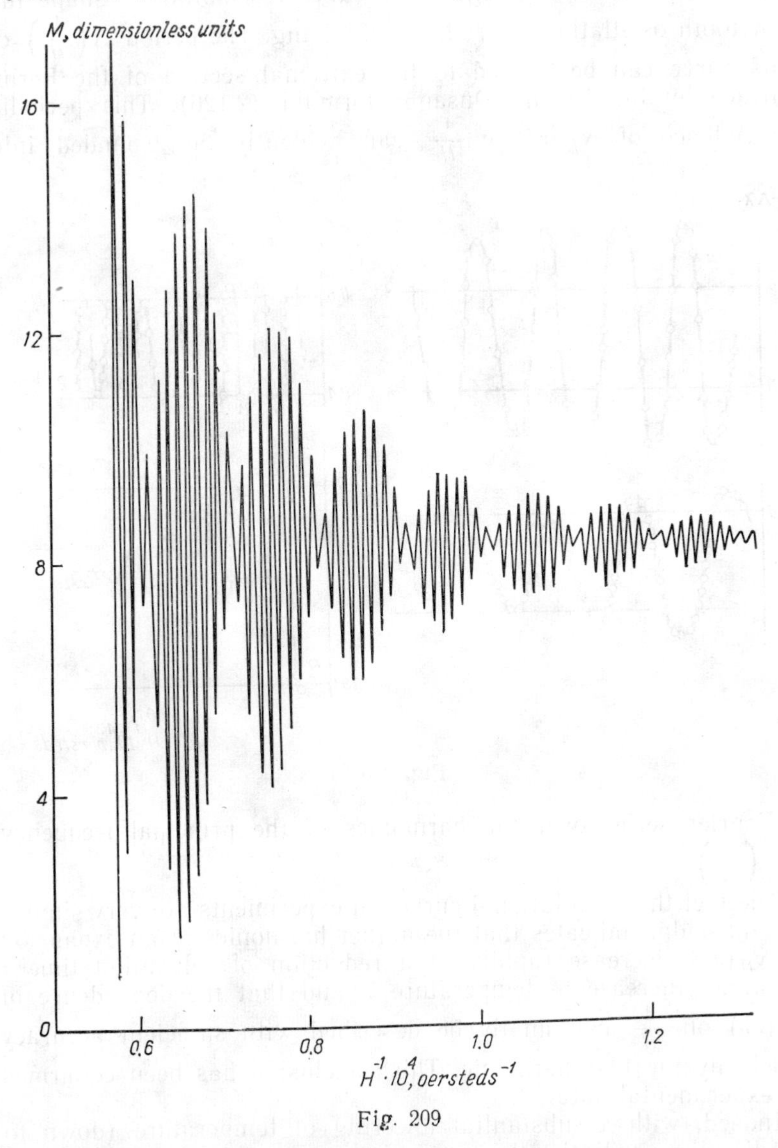

Fig. 209

Using expression (2.113) and Figs. 139 and 140, it may be easily shown that the oscillating portion of the density of states at the Fermi level $\nu_H(\varepsilon_F)$ in dependence of the reciprocal magnetic field at finite values of the relaxation time τ and temperature T must have the form of a periodic curve resembling in shape the saw-tooth oscillations in radio engineering. The period $\Delta\left(\frac{1}{H}\right)$ of this curve can be related to the extremal section of the Fermi surface by the Lifshits-Onsager formula (2.126). This periodic dependence of $\nu_H(\varepsilon_F)$ on $\frac{1}{H}$ can evidently be expanded into

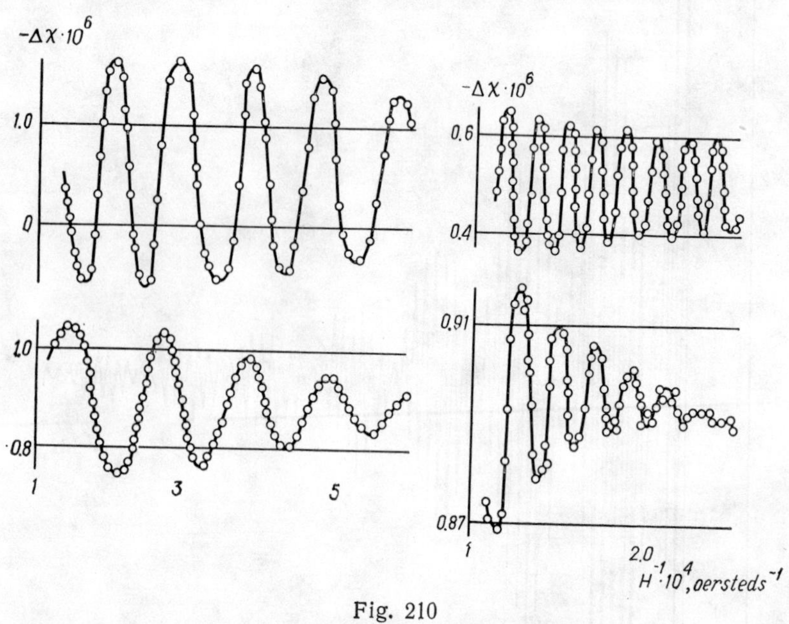

Fig. 210

a Fourier series over the harmonics of the principal frequency $\Delta^{-1}\left(\frac{1}{H}\right)$.

The fact that oscillational curves in experiments are very similar to sinusoidal indicates that the higher harmonics in an expansion of $\nu_H(\varepsilon_F)$ decrease rapidly at a reduction of relaxation time τ or at an increase of temperature T and that the dependence of $\nu_H(\varepsilon_F)$ on $\frac{1}{H}$ can finally be described with sufficient accuracy solely by the first harmonic. This conclusion has been confirmed by experimental data.

Indeed, with a substantial lowering of temperature (down to 0.1° K), the oscillational curves obtained in high-purity crystals begin to differ noticeably from sinusoidal ones. This is especially true of oscillations at low quantum numbers n, which begin to resemble saw-tooth oscillations. The dependence of the shape of oscillations on the quantum number n of oscillations is linked with that, at low values of n, i.e. near the ultra-quantum limit of magnetic fields, the ratios $\frac{kT}{\hbar\omega}$ and $\frac{\hbar/\tau}{\hbar\omega}$ become minimal, and therefore, the effect of the temperature blurring of the Fermi level and expansion of Landau levels also becomes minimal [1]).

For the same reasons, the amplitude of oscillations is an increasing function of the ratios $\frac{\hbar\omega}{kT}$ and $\frac{\hbar\omega}{\hbar/T}$.

Lifshits and Kosevich [14] have found theoretically the dependence of the oscillating portion of the magnetic moment $\tilde{M}$ of metals on magnetic field and temperature.

The principal harmonic of this quantity, having the frequency $\Delta^{-1}\left(\frac{1}{H}\right)$ in the reciprocal magnetic field, is of the form as follows:

$$\tilde{M}_1 = \text{const}\, H^{-1/2} T \frac{\exp\left(-\frac{2\pi^2 kT_D}{\hbar\omega}\right)}{\sinh\left(\frac{2\pi^2 kT}{\hbar\omega}\right)} \cos\left(\pi \frac{m^*}{m^s}\right) \sin\left(\frac{cS_{\text{extr}}}{|e|\hbar H} - \pi \pm \frac{\pi}{4}\right) \tag{4.50}$$

where m^* and m^s are the cyclotron and spin masses for the given direction.

Expression (4.50) does not take into account the variation of the Fermi level in the magnetic field. Because of this it is applicable under the condition $\hbar\omega \ll \varepsilon_F$ or, what is equivalent, for large quantum numbers, $n \gg 1$.

The quantity T_D entering the expression (4.50) has the dimension of temperature and is termed Dingle's temperature. It is related with the relaxation time τ by the formula

$$T_D = \frac{\pi\hbar}{4\tau k} \tag{4.51}$$

Thus, kT_D is nothing else as the expansion of Landau levels through the scattering of carriers. The multiplier $\exp\left(-\frac{2\pi^2 kT_D}{\hbar\omega}\right)$ determines the dependence of the amplitude of oscillations on relaxation time and is termed the Dingle factor [96]. Evidently, when $\tau = \infty$, then $T_D = 0$ and the Dingle factor is unity.

---

[1]) Let us emphasize that blurring of the Fermi levels occurs only owing to an increase of temperature, whereas expansion of Landau levels is related to scattering of carriers and is expressed through $\hbar/\tau$. At low temperatures, τ is determined by impurities and imperfections in the crystal and is practically independent of T. However, at temperatures above the region of the residual resistance, τ decreases with an increase of T. Thus, both the blurring of Fermi level and expansion of Landau levels are determined in this region by temperature.

The multiplier $\left[\sinh\left(\frac{2\pi^2 kT}{\hbar\omega}\right)\right]^{-1}$ contains the main dependence of the amplitude of oscillations on temperature T. Its structure is the same as that of the Dingle factor. Indeed, owing to the large coefficient $2\pi^2 \approx 20$ this multiplier is

$$\sinh\left(\frac{2\pi^2 kT}{\hbar\omega}\right) \approx \frac{1}{2}\exp\left(\frac{2\pi^2 kT}{\hbar\omega}\right) \quad \text{at} \quad kT \gtrsim 0.1\hbar\omega$$

Hence the dependence of the amplitude of oscillations on temperature and relaxation time can be written in the form of the general factor $\exp\left(\frac{-2\pi^2 k(T+T_D)}{\hbar\omega}\right)$. (The pre-exponential multiplier T^1 is a relatively slowly varying function of temperature and therefore plays no substantial part in the temperature dependence of the amplitude of oscillations at $kT \gtrsim 0.1\,\hbar\omega$. At $T \to 0$ the ratio $T/\sinh\frac{2\pi^2 kT}{\hbar\omega}$ tends to the limit $\hbar\omega/2\pi^2 k$.)

Thus, the effect of the processes of scattering of carriers on the nature of oscillations is formally equivalent to an increase of temperature by T_D. Namely this circumstance was the reason for introducing Dingle's temperature.

The multiplier $\cos\left(\pi\frac{m^*}{m^s}\right)$ in expression (4.50) determines the dependence of the amplitude of the first harmonic on spin splitting of Landau levels; m^*/m^s is the ratio of spin splitting to orbital splitting, equal respectively to $\frac{|e|\hbar H}{m^s c}$ and $\frac{|e|\hbar H}{m^* c}$. When m^*/m^s is increased, this multiplier turns periodically to zero at $\pi\frac{m^*}{m^s} = (2k+1)\frac{\pi}{2}$, where $k = 0, 1, 2, \ldots$. Thus, when the condition

$$m^s = \frac{2}{2k+1} m^*, \quad k = 0, 1, 2, \ldots \tag{4.52}$$

is fulfilled, the first harmonic with the frequency $\Delta^{-1}\left(\frac{1}{H}\right)$ disappears in the oscillations. The oscillations that are then observed correspond to the doubled frequency $2\Delta^{-1}\left(\frac{1}{H}\right)$, and therefore, the extremal section S_{extr} of the Fermi surface found for their period by means of the Lifshits-Onsager formula (2.126) turns to be twice its real magnitude. This phenomenon is called the spin damping of the first harmonic of oscillations[1]). It was first

[1]) As follows from the theoretical relationship $\tilde{M}(H)$, when the condition (4.52) is fulfilled, not only the first harmonic, but also all the odd harmonics disappear in the expansion of this quantity. The even harmonics disappear under the condition that $m^s = \frac{4}{2k+1} m^*$, where $k = 0, 1, 2, \ldots$.

observed on bismuth at definite orientations of the magnetic field.

Figure 211 shows the diagrams of Landau levels (at $p_z = 0$) for the values of masses m^s and m^* satisfying the condition (4.52) at $k = 0$, 1, and 2 respectively. As can be seen from the figure, the intervals between neighbouring levels become equal to $\frac{1}{2}\hbar\omega$ under this condition, as a result of which the frequency

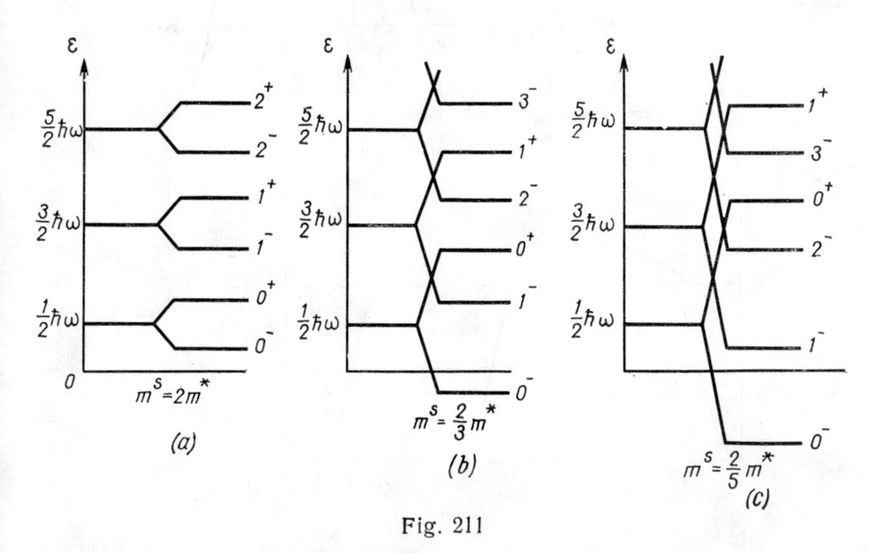

Fig. 211

of passing of the levels through the boundary of Fermi distribution increases twice. An exception is the magnitude of intervals between the lowest levels at $k \geqslant 1$, as shown in Fig. 211*b* $\left(\text{at } m^s = \frac{2}{3} m^*\right)$ and in Fig. 211*c* $\left(\text{at } m^s = \frac{2}{5} m^*\right)$.

It has been found by experiments that the maximum spin splitting of the levels, which is really observed in metals, exceeds only insignificantly the orbital splitting (the ratio m^*/m^s does not exceed 1.1-1.2). Therefore, only the spin damping at $k = 0$ or 1 can be of practical interest.

The diagrams in Fig. 211 correspond to the cases when the amplitude of the first harmonic of oscillations is strictly equal to zero.

The dependence of the amplitude of the first harmonic on the mass ratio m^*/m^s, which is given by the multiplier $\cos\left(\pi \frac{m^*}{m^s}\right)$, can be explained on the basis of the interpretation of the spin splitting of Landau levels given in Chapter Two, Sec. **2-16**. As has been shown there, the motion of the spin-splitted Landau levels through the boundary Fermi level $\varepsilon = \varepsilon_F$ can be described

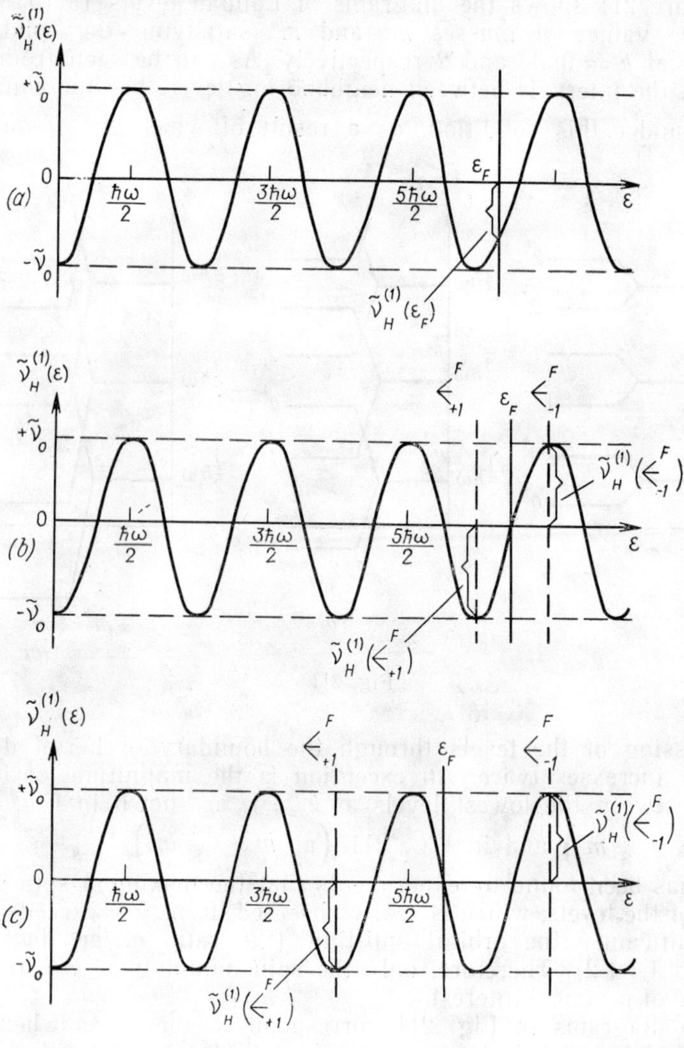

Fig. 212.

equivalently if we introduce a boundary of Fermi distribution $\Big\langle^{F}_{s=\pm 1} = \varepsilon_F \mp \frac{1}{2}\frac{|e|\hbar H}{m^s c}$ for each direction of the spin (and for each value of the spin variable $s = \pm 1$) and consider the motion of unsplitted orbital levels $\varepsilon_n = \left(n + \frac{1}{2}\right) \hbar\omega$ through these boundaries.

Using this representation, let us make the following construction for different values of m^*/m^s.

Let the spin splitting be initially absent, i.e. $m^*/m^s = 0$. We separate the first harmonic from the oscillating component of the density of states $\tilde{\nu}_H(\varepsilon)$ [the graph of dependence of the density of states $\nu_H(\varepsilon)$ on energy is given in Figs. 139 and 140] and mark with a solid vertical line the position of the Fermi level $\varepsilon = \varepsilon_F$, as shown in Fig. 212*a*. With an increase of the magnetic field in that case there can be observed a periodic motion of the maxima and minima of the density of states through the Fermi level and the corresponding periodic increase and decrease of the density of states at the Fermi level by the same magnitude, which will be denoted as $\tilde{\nu}_0$.

Now let us take into account the spin splitting. Then, in addition to the Fermi level $\varepsilon = \varepsilon_F$ for a definite value of magnetic field H we mark the two boundaries of the Fermi distribution with vertical dotted lines

$$\zeta_{-1}^{F} = \varepsilon_F + \frac{|e|\hbar H}{2m^s c} \quad \text{and} \quad \zeta_{+1}^{F} = \varepsilon_F - \frac{|e|\hbar H}{2m^s c}$$

corresponding to the two opposite orientations of the spin. The maxima and minima of the density of states will now pass through two boundaries ζ_{+1}^{F} and ζ_{-1}^{F} at a variation of the magnetic field. The total contribution to oscillations is determined by the sum of the values of the oscillating density of states at each of the boundaries.

When both boundaries pass simultaneously through the maxima and minima of the density of states, then the first harmonic of oscillations is increased.

When the maximum on one boundary is accompanied with the minimum on the other, then their contributions to the oscillations are mutually cancelled. This situation evidently corresponds to spin damping of the first harmonic. The damping at $m^s = 2m^*$ $(k = 0)$ is illustrated in Fig. 212*b*. In that case, if the maximum of density of states intersects one of the boundaries, then the neighbouring minimum intersects the other boundary. The maximum in Fig. 212*b* coincides with the right-hand boundary, and the next minimum, with the left-hand one. In the magnetic field at which this minimum reaches the right-hand boundary, the next maximum will coincide with the left-hand boundary. The situation repeats at an increase of the magnetic field,

With an increase of the magnetic field the distances between neighbouring maxima and minima, equal to $\hbar\omega/2$, increase linearly. At the same time the distance between the boundaries $\zeta_{-1}^F - \zeta_{+1}^F = \frac{|e|\hbar H}{m^s c}$, which is also equal to $\hbar\omega/2$ at $m^s = 2m^*$, also increases. At any magnitude of the magnetic field the contributions from both boundaries to the oscillating portion of the density of states are mutually cancelled. This results in that the oscillations with the frequency of the first harmonic vanish completely.

The spin damping at $m^s = \frac{2}{3} m^*$ evidently corresponds to the case when the coincidence of one of the maxima of the density of states with one of the boundaries is accompanied with the coincidence of the second minimum to the right or left from the given maximum with the other boundary (Fig. 212*c*).

Let us finally show that with an arbitrary relationship between the spin and orbital splitting the amplitude of oscillations must be proportional to the factor $\cos\left(\pi \frac{m^*}{m^s}\right)$.

It is clear from the analysis made above that the total contribution to the oscillating portion of the density of states $\tilde{\nu}_H(\varepsilon)$ at any magnitude of the magnetic field is an algebraic sum of the oscillating values of the density of states on each boundary. The first harmonic of the density of states $\tilde{\nu}_H^{(1)}(\varepsilon)$ as a function of energy ε has the period equal to $\hbar\omega$ and can be written in the form $\tilde{\nu}_H^{(1)}(\varepsilon) = -\tilde{\nu}_0 \cos\left(2\pi \frac{\varepsilon}{\hbar\omega}\right)$. Its magnitude on each of the boundaries of the Fermi distribution $\zeta_{\pm 1}^F = \varepsilon_F \pm \frac{|e|\hbar H}{2m^s c}$ is respectively

$$-\tilde{\nu}_0 \cos\left[2\pi \frac{\varepsilon_F \mp \frac{|e|\hbar H}{2m^s c}}{\hbar\omega}\right] =$$

$$= -\tilde{\nu}_0 \left[\cos\left(2\pi \frac{\varepsilon_F}{\hbar\omega}\right) \cos\left(\pi \frac{m^*}{m^s}\right) \pm \sin\left(2\pi \frac{\varepsilon_F}{\hbar\omega}\right) \sin\left(\pi \frac{m^*}{m^s}\right)\right]$$

Adding these expressions together with the signs "+" and "—", we get that the sum of contributions from the two boundaries to the density of states is equal to $-2\tilde{\nu}_0 \cos\left(\pi \frac{m^*}{m^s}\right) \cos\left(\frac{2\pi\varepsilon_F}{\hbar\omega}\right)$. The last of the multipliers in the expression obtained coincides to the accuracy of the constant phase with the oscillating factor $\sin\left(\frac{cS_{\text{extr}}}{|e|\hbar H} - \pi \pm \frac{\pi}{4}\right)$ in formula (4.50), since $\frac{2\pi\varepsilon_F}{\hbar\omega} = \frac{cS_{\text{extr}}}{|e|\hbar H}$. The second multiplier $\cos\left(\pi \frac{m^*}{m^s}\right)$ gives the sought-for dependence of the amplitude of oscillations on the ratio of spin and orbital splittings.

Expression (4.50) makes it possible to calculate the magnitude of the cyclotron mass m^* through the ratio of the amplitudes of oscillations measured at two different temperatures T_1 and T_2. Let the amplitude of oscillations be denoted as $A(T, H_n)$ where $H = H_n$ is the magnitude of the magnetic field corresponding to the passage of the Landau level with the quantum number n.

Let us assume for definiteness that $T_1 < T_2$. Then, evidently, $A(T_1, H_n) > A(T_2, H_n)$.

According to formula (4.50), the ratio of amplitudes $A(T_1, H_n)$ and $A(T_2, H_n)$ is

$$\frac{A(T_1, H_n)}{A(T_2, H_n)} = \frac{T_1}{T_2} \cdot \frac{\sinh\left(\dfrac{2\pi^2 m^* c k T_2}{|e|\hbar H_n}\right)}{\sinh\left(\dfrac{2\pi^2 m^* c k T_1}{|e|\hbar H_n}\right)} \tag{4.53}$$

This expression is only valid for the case when T_D remains constant in the interval of temperatures from T_1 to T_2. The parameter T_D is independent of temperature in the region of residual resistance, i.e. practically in the region of helium temperatures.

The expression (4.53) obtained is a transcendental equation by means of which the cyclotron mass m^* can be found for the given values of T_1, T_2, H_n, and the ratio of amplitudes of oscillations. Note that equation (4.53) determines the cyclotron mass at the Fermi level, i.e. the cyclotron mass that has been renormalized as a result of electron-phonon interaction.

In a particular case at $T_1 = \frac{1}{2} T_2$, equation (4.53) is solvable for m^*, the explicit expression for m^* being of the form:

$$m^* = \frac{|e|\hbar H_n}{4\pi^2 c k T_1} \cosh^{-1}\left[\frac{A(T_1, H_n)}{A(2T_1, H_n)}\right] \tag{4.54}$$

In practice, the temperatures T_2 and T_1 are usually taken equal respectively to 4.2° and 2.1° K. For improved accuracy, the cyclotron mass m^* is calculated at different values of H_n (i.e. at different quantum numbers n) and the values obtained are then averaged. Notwithstanding this, the accuracy with which the cyclotron mass is found from the temperature dependence of the amplitude of oscillations in the de Haas-van Alphen effect is not high. This is linked with the circumstance that expression (4.50) for the first harmonic describes only approximately the oscillating portion of the magnetic moment $\tilde{M}$.

In addition, a large error usually arises at separating the oscillating component from the experimental dependence of the moment M on magnetic field. The monotonous component is very often in a complicated dependence on H, which results in a substantial uncertainty in the determination of the amplitudes of oscillations. And finally, an error in the determination of the cyclotron mass arises owing to that the Dingle temperature T_D is not strictly constant in the temperature interval from T_1 to T_2.

Thus, the temperature dependence of the amplitude of oscillations serves, strictly speaking, only for evaluation of the magnitude of the cyclotron mass.

The comparative easiness of the observation of oscillations at an arbitrary orientation of the magnetic field relative to the crystallographic (and geometric) axes of the specimen makes it possible to find sufficiently rapidly the cyclotron masses for any sections of the Fermi surface. Such data are usually sufficient for the general description of the Fermi surface. In cases when it is required to obtain more accurate values of the cyclotron masses for some orientations of the magnetic field, this may be done by the method of Azbel-Kaner cyclotron resonance.

The ratio of amplitudes of oscillations for two successive values of magnetic fields H_n and H_{n+1} determines the Dingle temperature T_D. Indeed, according to formula (4.50), the ratio of amplitudes $A(T, H_n)$ and $A(T, H_{n+1})$ is

$$\frac{A(T, H_n)}{A(T, H_{n+1})} = \left(\frac{H_{n+1}}{H_n}\right)^{1/2} \frac{\sinh\left(\dfrac{2\pi^2 m^* ckT}{|e|\hbar H_{n+1}}\right)}{\sinh\left(\dfrac{2\pi^2 m^* ckT}{|e|\hbar H_n}\right)} \times$$

$$\times \exp\left[\frac{2\pi^2 m^* ckT_D}{|e|\hbar}\left(\frac{1}{H_{n+1}} - \frac{1}{H_n}\right)\right] \qquad (4.55)$$

Hence T_D is found by simple logarithmation. It is evident that T_D can only be found after the magnitude of the cyclotron mass m^* has been determined.

The relaxation time τ of carriers is calculated from the magnitude of T_D by means of formula (4.51). But the value of τ found by this method is substantially smaller (usually by one or two orders) than the relaxation time τ_0 determined in weak magnetic fields (for instance, through the value of the Hall mobility $\mu = \dfrac{|e|\tau_0}{m^*}$, which for a one-zone metal is equal to the product of the Hall coefficient $R = \dfrac{1}{|e|nc}$ by the conductivity $\sigma = |e|n\mu$, i.e. $\mu = c\sigma R$).

This difference between τ_0 and τ is connected in essence with that these quantities correspond to entirely different conditions under which scattering of current carriers occurs. Thus, the relaxation time τ_0 is an average quantity for all electrons on the Fermi surface which describes the isotropic scattering of carriers at $H = 0$. On the other hand, the relaxation time τ found from oscillations corresponds to scattering of the electrons passing out from the Landau level beyond the Fermi level. It describes the scattering of the carriers having a very small component of velocity $v_\parallel$ in the direction of the magnetic field (let us recall that at passage out of the Landau level, $p_\parallel \to 0$ for the electrons located on it). In that case, as has been shown by Brown [97], the probability of scattering of electrons on ionized impurities increases substantially. Let this be discussed in more detail.

As is known, the efficiency of scattering of electrons decreases at a reduction of temperature as a result of a rapid decrease of the density of phonons in the process of their freezing-out. At low temperatures in the region of residual resistance the main mechanism of scattering of electrons which determines the relaxation time in sufficiently pure crystals is, as earlier, the scattering on ionized impurities.

Because of the effect of screening of the Coulomb potential of the impurity by free carriers the interaction of an electron with the impurity occurs within the region whose size is determined by the magnitude of the Debye radius of screening r_D.

At a rather low concentration of impurities N_i (compared with the concentration of electrons n) the scattering centres of the effective diameter πr_D^2 are distributed in the lattice at average distances $N_i^{-1/3}$ from each other that exceed substantially r_D. In that case, as has been shown by Davydov and Pomeranchuk [98], the probability w of scattering (which determines the inverse relaxation time $1/\tau$) of an electron moving in the lattice with the velocity v is proportional to the expression $\frac{r_D^2 N_i^{1/3}}{v}$. It can be used for comparing the relaxation time τ of electrons on the Landau level at $v_\parallel \to 0$ with the relaxation time τ_0 of electrons at $H = 0$.

Since we speak of scattering of Fermi electrons, then evidently the magnitude of $1/\tau$ at $H = 0$ is proportional to $\frac{r_D^2 N_i^{1/3}}{v_F}$.

Electrons on the Landau level precess with the Fermi velocity in the plane perpendicular to the magnetic field and travel along the field with the velocity $v_\parallel$.

When the bottom of the Landau parabola comes closer to the Fermi level, then the magnitude of the velocity component $v_\parallel$ of the electrons located on that parabola decreases and becomes substantially less than v_F. In that case the probability of scattering of electrons $1/\tau$ is determined by $1/v_\parallel$. Owing to that the

Landau level has a finite width, the minimal value $v_{\parallel \min}$ of the velocity component along the magnetic field does not turn to zero.

The minimum energy of motion in the direction of the field, which is approximately equal to $\frac{1}{2} m_{\parallel} v_{\parallel \min}^2$, then coincides in the order of magnitude with the energy width of the Landau level $\hbar/\tau$. (The quantity $\frac{1}{2} m_{\parallel} v_{\parallel \min}^2$, as $\hbar/\tau$, represents the uncertainty of energy at the Landau level.)

Thus, the analysis made above makes it possible to conclude that the ratio of relaxation times τ_0/τ must be approximately equal to the ratio $v_F/v_{\parallel \min}$, which in turn coincides in the order of magnitude with the ratio $\sqrt{\varepsilon_F}/\sqrt{\hbar/\tau}\left(\text{since } \varepsilon_F \approx \frac{1}{2} m_{\parallel} v_F^2\right)$. Hence it follows that

$$\tau_0 \approx \left(\frac{\varepsilon_F}{\hbar}\right)^{1/2} \tau^{3/2} \tag{4.56}$$

The final formula (4.56), obtained on the basis of coarse approximations, makes it possible to relate τ_0 and τ only to an accuracy of a certain proportionality factor which is of the order of unity in an isotropic metal. But the derivation of this formula given above allows the difference between the relaxation times τ_0 and τ to be cleared out.

The Fermi energy ε_F of carriers can be determined by using formula (2.121) given in Chapter Two, Sec. **2-16**. This formula includes the cyclotron mass m^* and also the magnitude of the magnetic field H_n at which the passage out of the n-th Landau level is observed.

If the magnetic field in the experiment is sufficiently strong for observing the passage out of all Landau quantum levels (this is possible if the group of carriers being studied is small and the corresponding Fermi energy is low), then the quantum number n which relates to the last oscillational maximum is equal to unity (without account of the spin splitting of levels). In that case all the maxima being observed can be simply labelled.

If the magnetic field is not sufficiently strong to attain the ultraquantum region, then the quantum numbers corresponding to the oscillational maxima are found as follows. Let a maximum be observed at $H = H_n$ whose quantum number n is unknown.

Another maximum, at a distance of k numbers in the direction of weaker fields from the first, is observed in the field $H = H_{n+k}$ and has the quantum number $n + k$ (let us recall that the quantum number n decreases at an increase of the magnetic field). If we assume that $n \gg 1$, then we can neglect the variation of the Fermi energy in the region being considered. Then, writing down the formula (2.121) for the numbers n and $n+k$, we get the following equality:

$$\left(n+\frac{1}{2}\right)\frac{|e|\hbar H_n}{m^*c}=\left(n+k+\frac{1}{2}\right)\frac{|e|\hbar H_{n+k}}{m^*c}$$

Hence the number n is expressed through the known quantities k, H_n, and H_{n+k}:

$$n=k\frac{H_{n+k}}{H_n-H_{n+k}}-\frac{1}{2} \tag{4.57}$$

Thus, an experimental study of the de Haas-van Alphen effect in a metal makes it possible to determine not only the shape and dimensions of the Fermi surface, but also to find the main parameters describing the energy spectrum and the properties of current carriers, i.e. Fermi energy, cyclotron mass, and relaxation time.

In addition, if the oscillational maxima are doubled owing to spin splitting of Landau levels, then it is possible to determine the spin mass and g-factor of current carriers. These parameters can be determined also in cases when, with a certain orientation of the magnetic field, there is observed vanishing of the first harmonic of oscillations related to the effect of spin damping.

Calculations of various parameters were made by using expression (4.50) for the first harmonic of the magnetic moment. A similar expression describes, with an accuracy to proportionality factors, the first harmonic in oscillations of other thermodynamic and kinetic characteristics of the metal. This indicates to the same nature of all oscillational quantum effects. Thus, formulae (4.52)-(4.55) are also applicable for the Schubnikov-de Haas and other similar effects.

Oscillational quantum effects were considered in the book from the aspect of their generality. It does not mean, however, that oscillations of various quantities have no specific peculiarities which are essential in their observations. For instance, the Schubnikov-de Haas effect near the ultra-quantum limit of magnetic fields is very sensitive to the mutual orientation of the magnetic field $\vec{H}$ and current $\vec{j}$ through the specimen. In particular, no passage out of the Landau 0^+ level is often observed for the transverse magnetoresistance, since this level can be freed from electrons only as a result of passages with turning of the spin. In addition, the Schubnikov-de Haas effect at very high currents through the specimen is sensitive to the phenomena related with heating of electrons, and also with electron-phonon entrainment.

As another example, it may be indicated to the effect of the de Haas-van Alphen oscillations on the magnitude of the internal magnetic field which determines the quantization of the energy of electrons in a metal. Under certain conditions a noticeable difference between this field and the external magnetic field applied is observed during a substantial portion of the oscillational cycle (the Shoenberg effect [99]).

The examples given above show that a study of quantum oscillations actually can give a more detailed information on the behaviour of electrons in the metal than that obtained on the basis of their general description.

We have discussed in this chapter the main methods of studying of the electron energy spectrum of metals which have found the widest application or played an important role in the progress of the physics of metals. The quasi-classical and quantum effects described here make it possible to obtain detailed information on conduction electrons in metals. Within the frames of the elementary description of the electron structure of metals which is given in the book we have not touched on the question whether these data are sufficient for an exhaustive description of the energy spectrum. Note that the progress attained in the study of the electron spectra of metals evidently makes it possible to raise the problem of purposeful variation of the properties of a substance.

This problem relates to the development of a new direction in the solid-state physics (and in particular, in the physics of metals) devoted to studies of the properties of substances under extremal physical conditions (such as super-high pressures, extremal electric and magnetic fields, etc.) under which recombination of the electron energy spectrum is observed.

BIBLIOGRAPHY

1. Eichenwald A. A., *Elektrichestvo* (*Electricity*). Moscow-Leningrad, Gostekhteorizdat, 1932, p. 473.
2. Riecke E., *Phys. Zeitschr.*, **2** 639 (1900).
3. Drude K., *Ann. Phys.*, **1**, 566; **3**, 369 (1900).
4. Lorentz H., *Proc. Amsterdam Acad.*, **7**, 438, 585, 684 (1905).
5. Sommerfeld A., *Zeitschr. Phys.*, **47**, 1 (1928).
6. Bloch F., *Zeitschr. Phys.*, **52**, 555 (1928).
7. Bloch F., *Elektronentheorie der Metalle*. Handbuch der Radiologie, **6**, 1; 226-278 (1933).
8. Brillouin L., *Journ. Phys. Rad.*, **1**, 377 (1930).
9. Brillouin L., *Waves Propagation in Periodic Structures*. New York, 1946.
10. Lifshits I. M., *ZhETF*, **20**, 834 (1950).
11. Lifshits I. M., Azbel M. Ya., Kaganov M. I., *ZhETF*, **31**, 63 (1956); **32**, 1188 (1957).
12. Lifshits I. M., Azbel M. Ya., Kaganov M. I., *Elektronnaya teoriya metallov* (*Electronic Theory of Metals*). Moscow, "Nauka" Publishers, 1971.
13. Lifshits I. M., Kaganov M. I., *Uspekhi fizicheskikh nauk*, **69**, 419 (1959); **78**, 411 (1962); **87**, 389 (1965).
14. Lifshits I. M., Kosevich A. M., *ZhETF*, **29**, 730 (1955).
15. Lifshits I. M., Peschansky V. G., *ZhETF*, **35**, 1251 (1958); **38**, 188 (1960).
16. Abrikosov A. A., *ZhETF*, **59**, 1280 (1970).
17. Abrikosov A. A., *Vvedenie v teoriyu normal'nykh metallov* (*Introduction to the Theory of Normal Metals*). Moscow, "Nauka" Publishers, 1972.
18. Gurevich V. L., *ZhETF*, **35**, 669 (1958); **37**, 71 (1959).
19. Ziman J. M., *Adv. Phys.*, **10**, 1 (1961).
20. Ziman J. M., *Phys. Rev.*, **121**, 1320 (1961).
21. Ziman J. M., *Electron and Phonon*. Oxford, 1960.
22. Ziman J. M., *Principles of the Theory of Solids*. Cambridge, 1964.
23. Kittel Ch., *Introduction to Solid State Physics*. New York,, 1956.
24. Kittel Ch., *Quantum Theory of Solids*. New York-London, 1963.
25. Harrison W. A., *Phys. Rev.*, **104**, 1281 (1956).
26. Harrison W. A., *Pseudo-potentials in the Theory of Metals*. New York-Amsterdam, 1966.
27. Harrison W. A., *Solid State Theory*. New York-London-Toronto, 1970.
28. Pippard A. B., *Phil. Trans. Roy. Soc.*, A250, 325 (1957).
29. Pippard A. B., *The Dynamics of Conduction Electrons*. New York, 1965.
30. Mott N. F., *Proc. Phys. Soc.*, **46**, 680 (1934).
31. Mott N. F., Jones H., *The Theory of the Properties of Metals and Alloys*. Oxford, 1936.
32. Jones H., *Proc. Phys. Soc.*, **49**, 250 (1937).
33. Jones H., *Theory of Brillouin Zones and Electron States in Crystals*. Amsterdam, 1962.

34. Pines D., *Solid State Phys.*, **1**, 367 (1955).
35. Pines D., *Elementary Excitations in Solids.* New York, 1963.
36. Kikoin I. K., Senchenkov A. P., *Fizika metallov i metallovedenie*, **24**, 5; 843 (1967).
37. Barrett C. S., *Structure of Metals.* New York, 1952.
38. Khotkevich V. I., *ZhETF*, **23**, 91 (1952).
39. Born M., Oppenheimer J., *Ann. Phys.*, **84**, 457 (1927).
40. Peierls R. E., *Quantum Theory of Solids.* Oxford, 1955.
41. Lindemann F., *Phys. Zeitschr.*, **11**, 609 (1910).
42. Born M., von Kármán T., *Phys. Zeitschr.*, **13**, 297 (1912).
43. Frenkel Ya. I., *Vvedenie v teoriyu metallov.* 4th ed. "Nauka" Publishers, Leningrad, 1972.
44. Einstein A., *Ann. der. Phys.*, **22**, 180 (1907), **34**, 170 (1911).
45. Debye P., *Ann. der Phys.*, **39**, 789 (1912).
46. Walker C. B., *Phys. Rev.*, **103**, 547 (1956).
47. Hartree D. R., *Proc. Cambr. Phil. Soc.*, **24**, 111 (1928).
48. Landau L. D., *ZhETF*, **30**, 1058 (1956).
49. Herring C., *Phys. Rev.*, **57**, 1160 (1940).
50. Phillips J. C., Kleinman L., *Phys. Rev.*, **116**, 287, 880 (1959).
51. Animalu A. O. E., *Proc. Roy. Soc.*, A294, 376 (1966).
52. Heine V., Abarenkov I., *Phil. Mag.*, **9**, 451 (1964).
53. Harrison W. A., *Phys. Rev.*, **116**, 555 (1959).
54. Wigner E. P., Seitz F., *Phys. Rev.*, **43**, 804 (1933); **46**, 509 (1934).
55. Shoenberg D., *Proc. Roy. Soc.*, A281, 62 (1964).
56. Khaikin M. S., Mina R. T., *ZhETF*, **42**, 35 (1962).
57. Shoenberg D., *Phil. Trans. Roy. Soc.*, A245, 1 (1952).
58. Moore T. W., Spong F. W., *Phys. Rev.*, **125**, 846 (1962); **126**, 2261 (1962).
59. Fletcher J. F., Larson D. C., *Phys. Rev.*, **111**, 455 (1958).
60. Migdal A. B., *ZhETF*, **34**, 1438 (1958).
61. Landau L. D., *Zs. Physik*, **64**, 629 (1930).
62. Onsager L., *Phil. Mag.*, **43**, 1006 (1952).
63. Mott N. F., *Phil. Mag.*, **6**, 287 (1961).
64. Brandt N. B., Svistova E. A., *Journ. Low Temp. Phys.*, **2**, 1 (1970).
65. Hazegawa H., Howard R. E., *Journ. Phys. Chem., Solids*, **21**, 179 (1961).
66. Brandt N. B., Chudinov S. M., *Journ. Low Temp. Phys.*, **8**, No. 3/4, 339 (1972).
67. Brandt N. B., Chudinov S. M., *ZhETF*, **59**, 1494 (1970).
68. Lifshits I. M., *ZhETF*, **32**, 1509 (1957).
69. Matthiessen A., *Ann. Phys. Chem.*, **7**, 761, 892 (1864).
70. Bloch F., *Zs. Physik*, **59**, 208 (1930).
71. Grüneisen E., *Ann. Physik*, **16**, 530 (1933).
72. Houston W. V., *Zs. Physik*, **47**, 33 (1928)
73. Hall E. H., *Amer. Journ. Math.*, **2**, 287 (1879).
74. Hall E. H., *A Dual Theory of Conduction in Metals.* Cambridge, Mass., 1938.
75. Azbel M. Ya., Kaner E. A., *ZhETF*, **30**, 811 (1956).
76. Gantmacher V. F., *ZhETF*, **42**, 1416; **43**, 345 (1962).
77. Pippard A. B., *Phil. Mag.*, **2**, 1147 (1957).
78. Gaidukov Yu. P., *ZhETF*, **37**, 1281 (1959).
79. Abeles B., Meiboom S., *Phys. Rev.*, **101**, 544 (1956).
80. Lifshits I. M., *ZhETF*, **38**, 1569 (1960).
81. Fawcett E., *Phys. Rev.*, **103**, 1582 (1956).
82. Kip A. E., Langenberg D. N., Moore T. W., *Phys. Rev.*, **124**, 359 (1961).
83. Azbel M. Ya., Kaner E. A., *ZhETF*, **39**, 80 (1960).
84. Naberezhnykh V. P., Tsymbal L. T., *ZhETF*, **60**, 259 (1971).
85. Kaner E. A., *ZhETF*, **44**, 1036 (1963).
86. Bömmel O., *Phys. Rev.*, **100**, 758 (1955).

87. Pippard A. B., *Phil. Mag.*, **2**, 1147 (1957).
88. Ketterson J. B., Stark R. W., *Phys. Rev.*, **156**, 748 (1967).
89. Gurevich V. L., Skobov V. G., Firsov Yu. D., *ZhETF*, **40**, 786 (1961).
90. Fujimori Y., *Memoirs of the Faculty of Science*, Kyushy University, Ser, B, **4**, No. 2, p. 47 (1969).
91. Schubnikov L. V., de Haas W. J., *Leiden Commun.*, **207**, 210 (1930).
92. de Haas W. J., van Alphen P. M., *Leiden Commun.*, 212, (1931).
93. Shoenberg D., *Phil. Mag.*, **5**, 105 (1960).
94. Brandt N. B., Minina N. Ya., *ZhETF*, **51**, 108 (1966).
95. Brandt N. B., Razumeenko M. V., *ZhETF*, **39**, 276 (1960).
96. Dingle R. B., *Proc. Roy. Soc.* (London), A211, 517 (1952).
97. Brown R. N., *Phys. Rev.*, **2**, No. 4, 928 (1970).
98. Davydov B., Pomeranchuk I., *ZhETF*, **9**, No. 11, 1294 (1939).
99. Shoenberg D., *Phil. Trans.*, A255, 85 (1962).

TO THE READER

Mir Publishers would be grateful for your comments on the content, translation and design of this book. We would also be pleased to receive any other suggestions you may wish to make.

Our address is:
USSR, 129820, Moscow I-110, GSP
Pervy Rizhsky Pereulok, 2
MIR PUBLISHERS

INDEX

Printed in the Union of Soviet Socialist Republics